Advances in MALDI and Laser-Induced Soft Ionization Mass Spectrometry

Rainer Cramer
Editor

Advances in MALDI and Laser-Induced Soft Ionization Mass Spectrometry

Springer

Editor
Rainer Cramer
Department of Chemistry
University of Reading
Reading, Berkshire, UK

ISBN 978-3-319-04818-5 ISBN 978-3-319-04819-2 (eBook)
DOI 10.1007/978-3-319-04819-2

Library of Congress Control Number: 2015947772

Springer Cham Heidelberg New York Dordrecht London

Printed on acid-free paper

Springer International Publishing AG Switzerland is part of Springer Science+Business Media (www.springer.com)

Preface

The 'soft' ionization technique *matrix-assisted laser desorption/ionization* (MALDI) is without doubt one of the great success stories of modern mass spectrometry (MS). In particular, the further development of MALDI and in general 'soft' laser ionization, focusing on their unique characteristics and advantages in areas such as speed, spatial resolution, sample preparation, and low spectral complexity, have led to great advances in mass spectral profiling and imaging with an extremely auspicious future in (bio)medical analyses.

Nonetheless, in some of today's main MS application fields (e.g., in proteomics), electrospray ionization (ESI) rather than MALDI is mostly still the technique of choice despite MALDI's many 'unique selling points.' There are many reasons for this such as ESI's easy online coupling to liquid chromatography and formation of multiply charged ions with the associated advantages in structure analysis as well as being a superior match to high-performing mass analyzer technologies. However, MALDI and other 'soft' laser ionization techniques have been making great strides in these strong domains of ESI. Although often based on direct coupling to ESI, these have in recent years progressively been made by borrowing ideas from ESI's fundamental aspects and sometimes arguably by merging with these to become a new form of MALDI (or 'soft' laser ionization).

These two parallel mainstreams of developments have led to a multitude of promising new applications for MALDI and laser-induced 'soft' ionization. Many of these have now reached the level where they are becoming increasingly interesting to the scientific community at large, looking for alternative analytical solutions or performance that can go beyond the current standard.

This book provides a practical guide with 'road-tested' protocols, focusing on some of the most intriguing and unique techniques and applications that also promise to be analytically valuable and not just an interesting fringe phenomenon in the vast array of possible laser-based 'soft' ionization techniques. As such, these applications and techniques will also be of interest to the nonexpert scientist who would like to explore and try new methods in the (bio)analytical sciences that might offer complementary and/or superior analytical capabilities.

Finally, I would like to dedicate this book to my former supervisor, mentor, and close collaborator on many of my (MA)LDI studies. As a brilliant scientist and co-inventor of MALDI, his knowledge about 'soft' laser desorption and ionization was second to none.

In memoriam Franz Hillenkamp.

Reading, UK Rainer Cramer

Contents

Contributors

U. Bahr Institute for Pharmaceutical Chemistry, University of Frankfurt, Frankfurt, Germany

Gregory R. Boyce Protea Biosciences, Morgantown, WV, USA

Stephane Camuzeaux Cancer Proteomics Laboratory, Institute for Women's Health, University College London, London, UK

Pierre Chaurand Department of Chemistry, Université de Montréal, Montreal, QC, Canada

Rainer Cramer Department of Chemistry, University of Reading, Reading, UK

Diego d'Esposito Centro di Spettrometria di Massa Proteomica e Biomolecolare, Istituto di Scienze dell'Alimentazione, Avellino, Italy

Martin Dufresne Department of Chemistry, Université de Montréal, Montreal, QC, Canada

Simona Francese Biomedical Research Centre, Sheffield Hallam University, Sheffield, UK

Carsten Hopf Institute of Medical Technology, University of Heidelberg and Mannheim University of Applied Sciences, Center for Applied Research in Biomedical Mass Spectrometry (ABIMAS), Mannheim, Germany

Instrumental Analytics and Bioanalytics, Mannheim University of Applied Sciences, Mannheim, Germany

T.W. Jaskolla Dr. Franz Köhler Chemie, Bensheim, Germany

Markus Kostrzewa Bruker Daltonik GmbH, Bremen, Germany

Maria Fiorella Mazzeo Centro di Spettrometria di Massa Proteomica e Biomolecolare, Istituto di Scienze dell'Alimentazione, Avellino, Italy

David C. Muddiman Department of Chemistry, W.M. Keck FTMS Laboratory for Human Health Research, North Carolina State University, Raleigh, NC, USA

Bogdan Munteanu Institute of Medical Technology, University of Heidelberg and Mannheim University of Applied Sciences, Center for Applied Research in Biomedical Mass Spectrometry (ABIMAS), Mannheim, Germany

Instrumental Analytics and Bioanalytics, Mannheim University of Applied Sciences, Mannheim, Germany

Milad Nazari Department of Chemistry, W.M. Keck FTMS Laboratory for Human Health Research, North Carolina State University, Raleigh, NC, USA

Peter Nemes Department of Chemistry, W. M. Keck Institute for Proteomics Technology and Applications, The George Washington University, Washington, DC, USA

Nathan Heath Patterson Department of Chemistry, Université de Montréal, Montreal, QC, Canada

Arthur B. Pranada Department of Medical Microbiology, MVZ Dr. Eberhard & Partner Dortmund (ÜBAG), Dortmund, Germany

Pavel Ryumin Department of Chemistry, University of Reading, Reading, UK

Gerold Schwarz Bremen, Germany

Bindesh Shrestha Department of Chemistry, W. M. Keck Institute for Proteomics Technology and Applications, The George Washington University, Washington, DC, USA

Rosa Anna Siciliano Centro di Spettrometria di Massa Proteomica e Biomolecolare, Istituto di Scienze dell'Alimentazione, Avellino, Italy

Prasath Somasundaram AG Systematic Proteome Research & Bioanalytics, Institute for Experimental Medicine, Christian-Albrechts-Universität zu Kiel, Kiel, Germany

Andreas Tholey AG Systematic Proteome Research & Bioanalytics, Institute for Experimental Medicine, Christian-Albrechts-Universität zu Kiel, Kiel, Germany

Aurélien Thomas Unit of Toxicology, CURML, University of Lausanne, Lausanne, Switzerland

John F. Timms Cancer Proteomics Laboratory, Institute for Women's Health, University College London, London, UK

Mark W. Towers Waters Corporation, Wilmslow, UK

Joanna Tucher AG Systematic Proteome Research & Bioanalytics, Institute for Experimental Medicine, Christian-Albrechts-Universität zu Kiel, Kiel, Germany

Callee M. Walsh Protea Biosciences, Morgantown, WV, USA

Kanjana Wiangnon Department of Chemistry, University of Reading, Reading, UK

Part I
Fundamental Improvements in MALDI

Employing 'Second Generation' Matrices

U. Bahr and T.W. Jaskolla

Abstract The 'first generation' of matrix-assisted laser desorption/ionization (MALDI) mass spectrometry (MS) matrices was found randomly during the early days of MALDI by empirical testing of hundreds of small molecules with molecular weights of typically about 150–250 g/mol and high absorption at the wavelength of the laser used for irradiation. For the 'second generation' matrices the structures of established matrix molecules were systematically modified by varying the nature, number, and position of their functional groups. The objective was to gain a better understanding of how the physicochemical processes essential for matrix and analyte ion generation are affected by the matrix molecular structure. With the uncovering of key ionization steps, predictions regarding the MALDI performance of in-silico designed matrix compounds came within reach by computational calculations. This marked a milestone in matrix development and provided valuable information for the creation of optimized compounds. The most comprehensive modifications were done on the core structure of the most widely used matrix α-cyano-4-hydroxycinnamic acid (CHCA) as chemical lead. Three derivatives proved to be outstanding and found their way in different fields of application. The Cl-substituted derivative of CHCA, 4-chloro-α-cyanocinnamic acid (ClCCA), was selected as the most potent matrix for the analysis of several substance classes including peptides. Compared to the hitherto favored CHCA, this new matrix is superior in detecting small amounts of in-solution as well as in-gel digested proteins leading to typically higher sequence coverages. Due to its higher protonation efficiency, discrimination of less basic peptides is strongly diminished which enables more uniform peptide detection. In addition to the more sensitive analysis of acidic peptides, the higher sensitivity also allows for the detection of low-abundant peptides such as phosphopeptides, enzymatically digested peptides with higher numbers

---

Dedication In Memory of Franz Hillenkamp

U. Bahr (✉)
Institute for Pharmaceutical Chemistry, University of Frankfurt,
Max-von-Laue Str. 9, Frankfurt 60438, Germany
e-mail: Ute.Bahr@pharmchem.uni-frankfurt.de

T.W. Jaskolla
Dr. Franz Köhler Chemie, Werner-von-Siemens-Str. 22-28, Bensheim 64625, Germany
e-mail: thorsten@jaskolla.com

R. Cramer (ed.), *Advances in MALDI and Laser-Induced Soft Ionization Mass Spectrometry*, DOI 10.1007/978-3-319-04819-2_1

of missed cleavages or less or even nonspecific cleavage sites (e.g., generated by elastase, slymotrypsin or proteinase K). This matrix can also be used for the analysis of substance classes such as lipids in positive ion mode and labile glycans in negative ion mode. Another CHCA derivative, α-cyano-2,4-difluorocinnamic acid (DiFCCA), was successfully applied for the most sensitive production of positive ions from phosphatidylcholines. In negative ion mode, a third derivative, α-cyano-4-phenylcinnamic acid amide (Ph-CCA-NH_2), showed promising results in lipid analysis. Finally, 1,8-bis(dimethylamino)naphthalene (DMAN) as a very strong base is predestined for negative ionization of acidic compounds. This matrix only generates its intact protonated form $[matrix+H]^+$ and is suitable for the analysis of small molecules in positive and negative ion mode.

1 Introduction

The apparent characteristics of a matrix-assisted laser desorption/ionization (MALDI) matrix are sufficiently high absorption at the irradiation laser wavelength, analyte incorporation into an excess of matrix (i.e., co-crystallization resulting in crystal lattice defects and/or generation of lacunas filled with solvent and, eventually, analytes)[1] and the ability of protonation or deprotonation. Hundreds of small organic molecules have been tested especially in the first years after MALDI was invented but only a few are left today and have found broad application.

Different classes of analytes often require different types of matrix molecules, which can be explained by the necessity of matrix–analyte co-crystallization, the way of charge transfer to the analyte and/or the exclusion of impurities. Examples for the latter are matrices such as 2,5-dihydroxybenzoic acid (DHB), which are advantageous in sample purification through desalting crude analyte samples (or more precisely, through the selective exclusion of ions with high charge density such as alkali metal ions). This is also the case for the binary matrix DHBs (a 9:1 mixture of 2,5-dihydroxybenzoic acid and 2-hydroxy-5-methoxybenzoic acid) for the detection of large proteins or 3-hydroxypicolinic acid with the addition of diammonium hydrogen citrate for oligonucleotide analysis. However, today's most widely used matrix is α-cyano-4-hydroxycinnamic acid (CHCA) (Beavis et al. 1992) since this compound offers high sensitivity for protein and peptide analysis which is currently one of the main applications of MALDI MS. Its advantage over other matrices is the formation of small matrix–analyte crystals, uniform in spatial distribution on the sample target as well as crystal size, either by dried-droplet or thin-layer preparation.

Due to these reasons, Jaskolla et al. started systematic matrix design studies in 2008 with CHCA as lead substance and synthesized derivatives of CHCA by systematically varying its functional groups (Jaskolla et al. 2008). Based on these

[1] In the case of liquid MALDI samples analytes are naturally well-incorporated into an excess of matrix (cf. the chapters on liquid MALDI MS).

studies, 4-chloro-α-cyanocinnamic acid (ClCCA) was designed as most potent matrix molecule which exhibits a clearly higher protonation efficiency compared to CHCA, an efficient conversion of its matrix radical cation to the corresponding protonated species, an absorption maximum close to the wavelength of the nitrogen (N_2) laser typically used in many MALDI mass spectrometers (337 nm), sufficient stability upon irradiation and transfer into (ultra-)high vacuum, efficient analyte incorporation due to its carboxylic acid group, an isotopic pattern which does not impede analyte detection and, last but not least, a crystal morphology appropriate for MALDI MS.

With this compound, a considerable improvement in MALDI performance for proteomics analysis was achieved regarding sensitivity, signal-to-noise (S/N) ratio, sequence coverage, and a more uniform peptide mapping with lower bias for peptides with the strongly basic residue arginine (Jaskolla et al. 2008, 2009a; Leszyk 2010; Papasotiriou et al. 2010).

1.1 A Detailed Background for the MALDI Enthusiast (Can Be Skipped by the MALDI Novice)

At the beginning of the rational MALDI matrix design studies by Jaskolla et al., a plethora of measurements were conducted to obtain crucial matrix and MALDI process information. Since the first newly devised matrix candidates were derivatives of CHCA, these were compared to CHCA with respect to the detectable ion species, ion intensities, and sequence coverage in peptide mass fingerprint experiments. With this information as well as the physicochemical properties of the individual derivatives a knowledge database was created from which conclusions between matrix structure and performance could be drawn. It was realized that an acidic functional group is essential for sensitive detection of analytes exhibiting basic groups such as peptides. Upon (matrix) crystallization with crystal lattice formation, structurally different compounds (such as analytes) are usually excluded from the growing crystal, a process typically used for purification purposes. Nevertheless, it was assumed that the growing matrix crystal surface also contains deprotonated matrix molecules which are capable of temporary binding dissolved protonated analytes by ion–ion interactions. Upon proceeding crystallization at the surrounding crystal surface, the bonded analyte ion would be incorporated into the matrix crystal rather than bulk-separated as in crystallization for purification purposes. Such an analyte incorporation would be the more efficient the stronger the matrix–analyte pairing would be. Consequently, ion–ion interactions between matrix and analyte will result in highest analyte intensities which is in agreement with experiments testing matrix derivatives with carboxylic acids, -amides or -esters as functional group of the α-cyanocinnamic acid (CCA) side chain (Jaskolla et al. 2008). Depending on the pH of the sample preparation, only a fraction of the acidic matrix molecules will be deprotonated and available for analyte bonding—especially if an acid such as trifluoroacetic acid (TFA) is added for analyte protonation. However, taking into account typically used MALDI conditions with matrix:analyte ratios of about 10^3–10^8:1 it is obvious that the deprotonated matrix percentage generally outnumbers the analyte amount.

Following this idea, the sulfonic acid derivative of CHCA was synthesized which should be able to incorporate analytes even more efficiently due to a higher percentage of deprotonated matrices resulting from a lower sulfonic acid pK_a (in addition to an also more efficient analyte protonation in the gas phase; see below). Although extraordinarily high sensitivities were achievable during the first few shots of laser irradiation, all tested sulfonic acid CCA derivatives were extremely volatile at pressures of up to 0.1 mbar. Responsible for this lacking stability is the missing planarity of such matrix compounds which impedes intermolecular π–π-stacking as is the case for CCA derivatives with planar carboxylic acid groups. Furthermore, sulfonic acids contain an unbalanced 1:2-ratio of hydrogen bond donor and -acceptor which is less suitable for intermolecular matrix attraction compared to carboxylic acid groups which exhibit a well-balanced 1:1-ratio of hydrogen bond donor and -acceptor. As a result, the latter matrix compounds are stabilized in the solid state by hydrogen-bridged dimerized carboxylic acid groups as was shown for 4-hydroxycinnamic acid (Bryan and Forcier 1980), 3,4-dihydroxycinnamic acid (Garcia-Granda et al. 1987), sinapic acid (Beavis and Bridson 1993), and 2,5-DHB (Haisa et al. 1982). Another disadvantage results from restricted π–π-orbital overlapping between CCA-sp^2-carbon atoms and the tetrahedral sulfonic acid head group compared to the trigonal planar carboxylic acid head group which also exhibits a sp^2-carbon atom. As a result, sulfonic acid derivatives of CCAs comprise less pronounced chromophore systems and blue-shifted absorption profiles compared to their carboxylic acid analogs. Dependent on the chosen auxochrome(s), such compounds exhibit insufficient absorption properties at the commonly used MALDI wavelengths of 355 or 337 nm (see Table 1).

This leads to the important prerequisite of sufficient absorption at the irradiation wavelength. Elimination of the carboxylic acid head group of CCAs not only lowers the efficiency of analyte incorporation/co-crystallization but also results in a blue-shifted absorption profile with insufficient or at least strongly lowered absorption properties dependent on the selected irradiation wavelength and substituents at the phenyl moiety. The same is true for the elimination of the α-cyano group (see Table 1 and the examples of (α-cyano-)4-methyl- and (α-cyano-)4-methoxycinnamic acid).

Upon analysis of the composition of the phenyl moiety, several effects have to be discussed. First, the size of its aromatic system and the possible mesomeric and inductive effects of the substituents affect the overall absorption profile of the CCA derivative. Increasing the chromophore, e.g., by replacing the phenyl residue with a naphthyl group or extending the vinyl side chain, results in distinct bathochromic shifts of the corresponding absorption profile compared to that of CCA.

This is also the case for substitution by functional groups with positive mesomeric (+M)-effects such as hydroxyl or methoxy groups or, to a lesser extent, with positive inductive (+I)-effects. Halogen substituents exhibit a combination of +M- and comparably strong –I-effects. From fluorine to iodine, the +M-effect weakens due

Table 1 Molar extinction coefficients and structures of selected CCA derivatives (cf. Jaskolla 2010)

Abbreviation	Chemical name	Used solvent and concentration	Extinction coefficient [M^{-1} cm^{-1}] at λ (337 nm)	λ (355 nm)	Structure
CHCA	α-Cyano-4-hydroxycinnamic acid	$3 \cdot 10^{-5}$ M in 10% ACN	18,300	7100	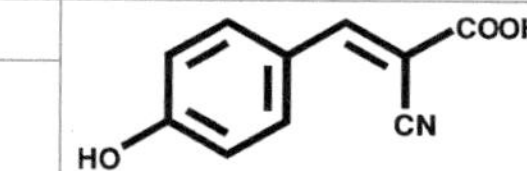
CHCSA	α-Cyano-4-hydroxy-cinnamic sulfonic acid	$3 \cdot 10^{-4}$ M in 0.7% ACN	650	160	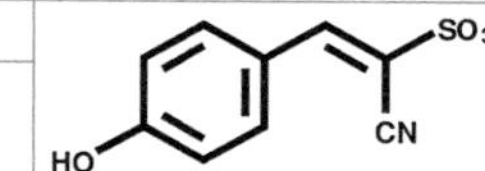
ClCCA	4-Chloro-α-cyanocinnamic acid	$3 \cdot 10^{-4}$ M in 1% ACN	2350	140	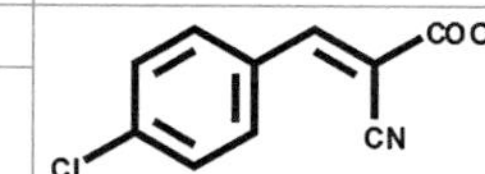
ClCCSA	α-Cyano-4-chloro-cinnamicsulfonic acid	$3 \cdot 10^{-4}$ M in 0.3% ACN	0	0	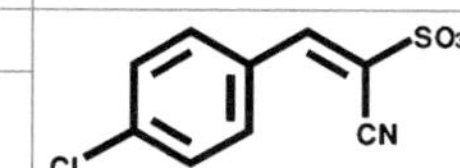
MeCA	4-Methylcinnamic acid	$2 \cdot 10^{-4}$ M in 0.7% ACN	130	60	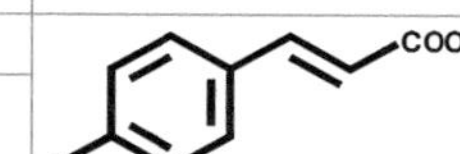
MeOCA	4-Methoxycinnamic acid	$3 \cdot 10^{-5}$ M in 10% ACN	2530	410	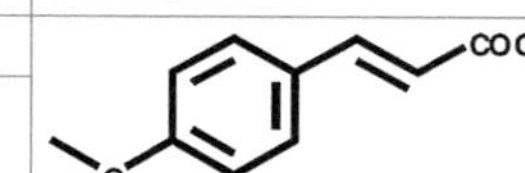
MeCCA	α-Cyano-4-methylcinnamic acid	$3 \cdot 10^{-4}$ M in 0.7% ACN	4950	400	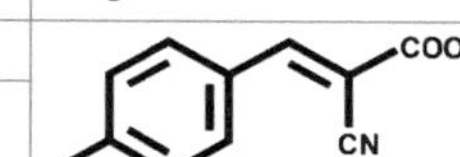

(continued)

Table 1 (continued)

Abbreviation	Chemical name	Used solvent and concentration	Extinction coefficient [M^{-1} cm^{-1}] at λ (337 nm)	λ (355 nm)	Structure
MeOCCA	α-Cyano-4-methoxycinnamic acid	$2 \cdot 10^{-5}$ M in 0.7% ACN	20,300	7850	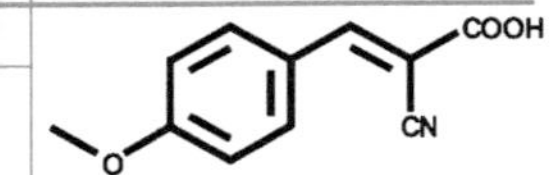
CCA	α-Cyanocinnamic acid	$3 \cdot 10^{-4}$ M in 0.7% ACN	490	23	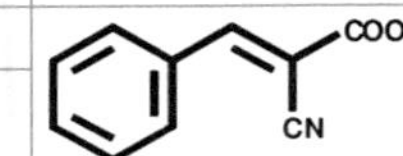
3,4-PhenylCCA	2-Cyano-3-(naphth-2-yl) acrylic acid	$2 \cdot 10^{-5}$ M in 10% ACN	9650	4700	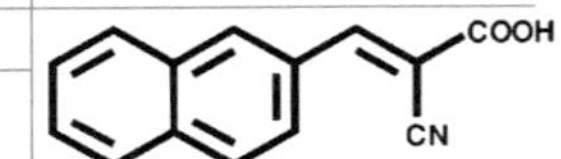
2,3-PhenylCCA	2-Cyano-3-(naphth-1-yl) acrylic acid	$3 \cdot 10^{-5}$ M in 30% ACN	10,770	8100	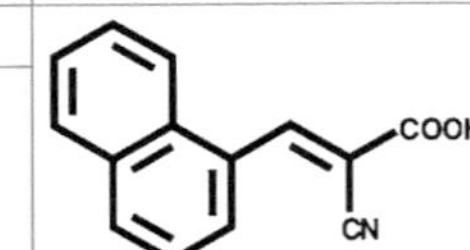
VinylCCA	(2*E*,4*E*)-2-cyano-5-phenyl-2,4-pentadienic acid	$3 \cdot 10^{-5}$ M in 0.7% ACN	30,270	17,930	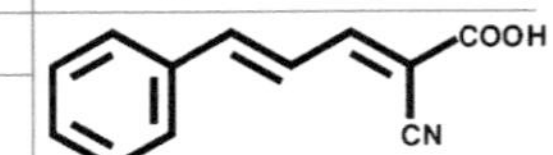
FCCA	α-Cyano-4-fluorocinnamic acid	$3 \cdot 10^{-4}$ M in 0.7% ACN	990	13	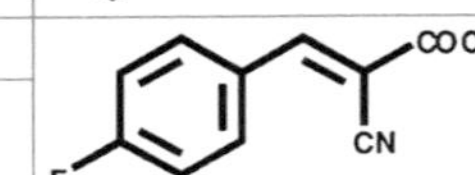
BCCA	4-Bromo-α-cyanocinnamic acid	$3 \cdot 10^{-4}$ M in 0.7% ACN	3460	150	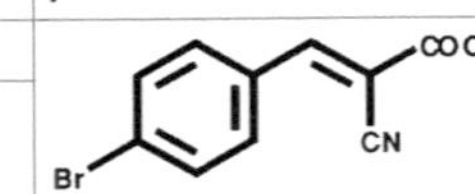

ICCA	α-Cyano-4-iodocinnamic acid	$3 \cdot 10^{-5}$ M in 0.7% ACN	9500 1100	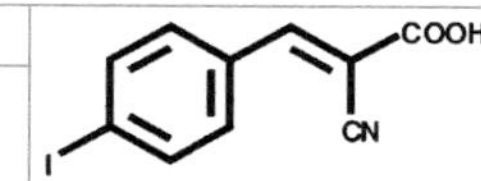
2-FCCA	α-Cyano-2-fluorocinnamic acid	$3 \cdot 10^{-4}$ M in 0.7% ACN	1130 20	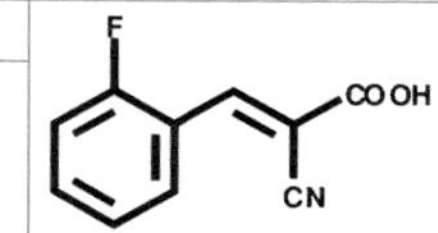
2-ClCCA	α-Cyano-2-chlorocinnamic acid	$3 \cdot 10^{-4}$ M in 0.7% ACN	1150 77	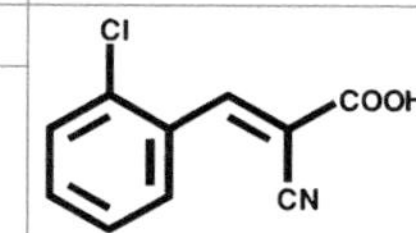
2,3,5-TriClCCA	α-Cyano-2,3,5-trichlorocinnamic acid	$3 \cdot 10^{-4}$ M in 0.7% ACN	910 190	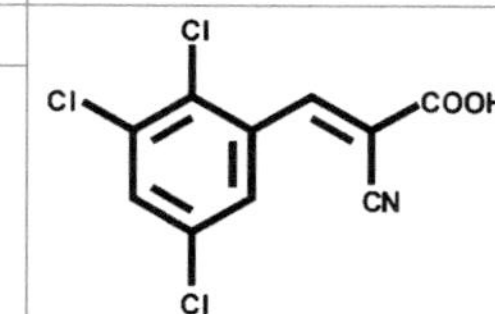
2,3,6-TriClCCA	α-Cyano-2,3,6-trichlorocinnamic acid	$3 \cdot 10^{-4}$ M in 0.7% ACN	280 0	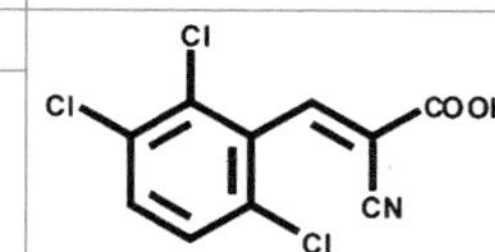
2,4-DiFCCA	α-Cyano-2,4-difluorocinnamic acid	$3 \cdot 10^{-4}$ M in 0.7% ACN	1140 83	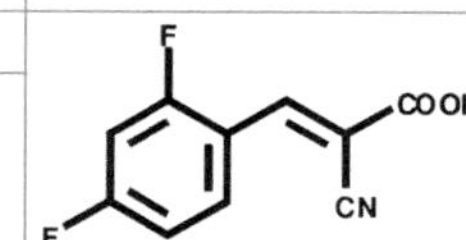
3,5-DiFCCA	α-Cyano-3,5-difluorocinnamic acid	$3 \cdot 10^{-4}$ M in 0.7% ACN	87 0	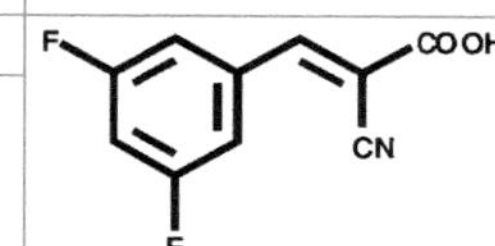

(continued)

Table 1 (continued)

Abbreviation	Chemical name	Used solvent and concentration	Extinction coefficient [M^{-1} cm^{-1}] at λ (337 nm)	λ (355 nm)	Structure
F_3CCCA	α-Cyano-4-(trifluoro-methyl) cinnamic acid	$3 \cdot 10^{-4}$ M in 0.7% ACN	60	0	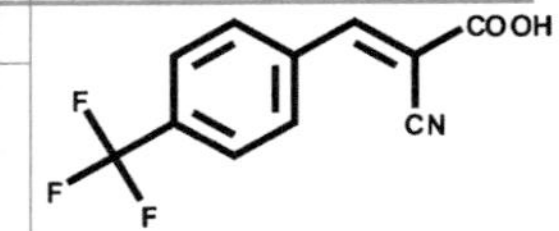
PentaFCCA	α-Cyano-2,3,4,5,6-pentafluorocinnamic acid	$3 \cdot 10^{-5}$ M in 0.44% ACN	0	0	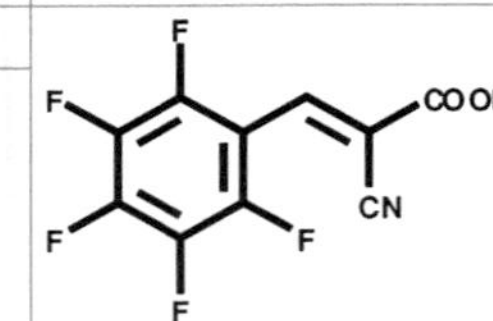
2,3,4,5-TetraClCCA	α-Cyano-2,3,4,5-tetrachlorocinnamic acid	$2 \cdot 10^{-4}$ M in 0.7% ACN	1070	470	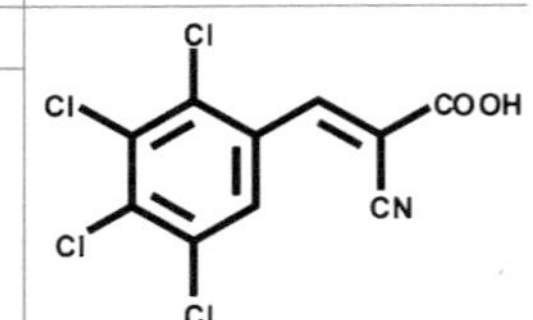
2,3,4,5-TetraBrCCA	α-Cyano-2,3,4,5-tetrabromocinnamic acid	$2 \cdot 10^{-4}$ M in 0.7% ACN	1770	420	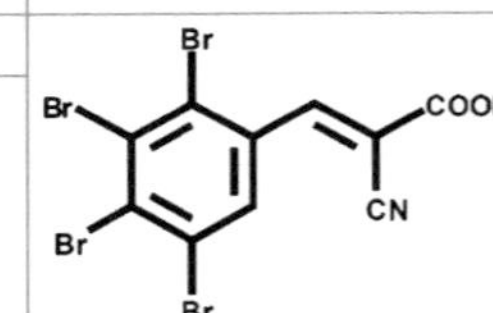
2,3,4,5,6-PentaBrCCA	α-Cyano-2,3,4,5,6-pentabromocinnamic acid	$3 \cdot 10^{-4}$ M in *tert*-BuOH	1620	1360	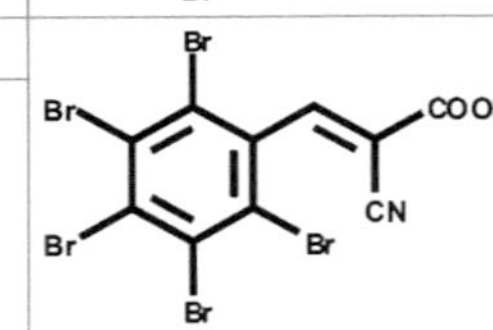

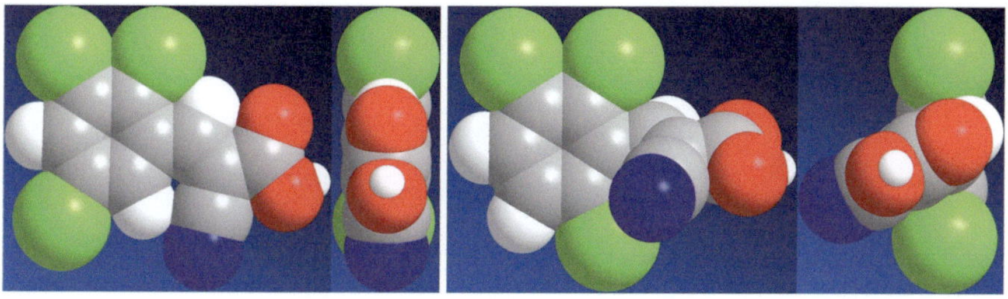

Fig. 1 Planar and out of plane structures of 2,3,5-trichloroCCA (*left*) and 2,3,6-trichloroCCA (*right*) after optimization by DFT-B3LYP/6-311G(2d,2p). Image reproduced from Jaskolla (2010)

to enlarging p-orbitals resulting in diminished overlaps with the p-orbitals from the neighbored carbon. At the same time the electronegativity decreases from fluorine to iodine resulting in diminishing deactivating –I-effects. The overlap of both effects leads to a comparably low absorption at the standard wavelengths for the 4-fluoroCCA derivative (FCCA) but continuously rises upon substitution with higher period halogens.

Substitution at the 2- or 6-position instead of the 4-position changes the morphology of the MALDI sample for the CCA derivatives but has only little consequences for the absorption profile if the substituent does not sterically interfere with the CCA side chain. As an example, the molar absorptivities for 2- and 4-fluoroCCA are comparable, whereas more bulky chlorine leads to a reduction in absorption by switching from the 4- to the 2-position (see Table 1).

Even worse is the combination of bulky substituents at the 2- and the 6-position. In this case the vinyl side chain rotates out of plane which prevents optimal π–π-orbital overlapping between phenyl- and vinyl-group and leads to lacking absorption (see Fig. 1 and Table 1).

With respect to the vinyl side chain carrying the electron-withdrawing cyano- and carboxylic acid group, +M-substituents (e.g., halogens) act as auxochromes at ortho- and para-positions leading to red-shifted absorption profiles. By contrast, halogen substitution at meta-positions causes hypsochromic shifts due to their –I-effect. The influence of the positions of the substituents is demonstrated in Table 1 by the example of α-cyano-2,4-difluoroCCA (DiFCCA, +M-effect at ortho- and para-positions causes red-shift) and α-cyano-3,5-difluoroCCA (–I-effect at meta-positions causes blue-shift).

Elimination of the +M-effect of halogens at o-/p-positions is possible by inserting an additional methylene group between halogen and chromophore. One example for such a combination is para-substitution by a trifluoromethyl group. The resulting compound α-cyano-4-(trifluoromethyl)cinnamic acid (F_3CCCA) does not absorb anymore at 355 nm and exhibits only very low absorption at 337 nm due to the eliminated +M-effect in combination with the distinct –I-effect of the substituent.

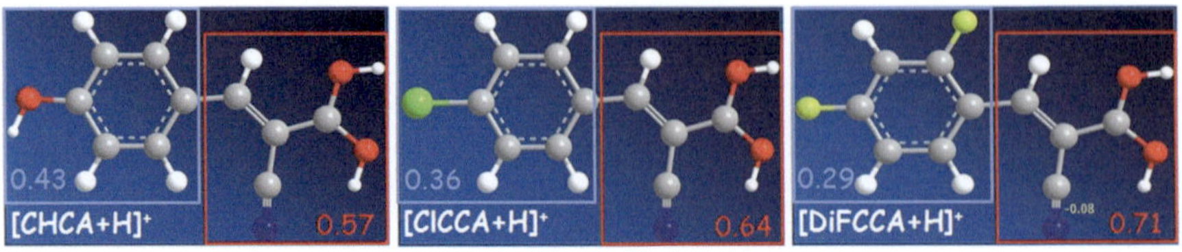

Fig. 2 Charge density distribution of protonated CHCA, ClCCA, and DiFCCA separated into the phenyl moiety and side chain sections. The energetically lowest isomers were calculated by means of DFT(B3LYP)/6-311++G(3df,3pd). The charge density calculations were performed by Mulliken population analysis (UHF/6-311G(d,p)). The sum of all partial charges is 1

What happens if (nearly) all phenyl-hydrogens are substituted against halogens? 2,3,4,5,6-PentafluoroCCA is a low melting crystalline compound without any absorption at 337 and 355 nm due to cumulated –I-effects. 2,3,4,5-Tetrachloro- and 2,3,4,5-tetrabromoCCA as well as 2,3,4,5,6-pentabromoCCA exhibit sufficient absorption at both 337 and 355 nm due to the above discussed weakening –I-effect at higher periods (see Table 1). The pentabromoCCA derivative exhibits two bulky substituents at the 2- and 6-position which cause an angulated geometry as discussed for 2,3,6-trichloroCCA. However, due to the para-bromo substituent this compound still exhibits sufficient absorption. Nevertheless, two disadvantageous effects prevent these compounds to be used as effective MALDI matrices. The first is their extensive isotopic pattern resulting from their high amount of chloro- and bromo-substituents. The corresponding matrix spectra show several matrix clusters which cover *m*/*z*-ranges of up to 30 per ion species due to broad isotopic patterns which impede certain analyte detection. The second disadvantage results from the cumulated electron-withdrawing effects of the halogens, the cyano-, and carboxylic acid group which cause destabilization and induce intense photofragmentation upon irradiation.

Another effect regarding the composition of the phenyl moiety refers to the reactivity of the protonating agent $[matrix+H]^+$. Two factors influence the efficiency of a protonation reaction: the basicity of the analyte (which typically cannot be altered) and that of the proton donating agent (the matrix). The latter point is described by the matrix proton affinity (PA). A low PA equates to a high protonation power due to an energetically unfavorable situation of the $[matrix+H]^+$-ion and enables greater proton transfer and thus more sensitive analyte detection. To decrease the PA of a matrix, its protonated counterpart has to be destabilized which can be realized by confining the extent of delocalization of its positive charge. This can be simply accomplished by inserting electron-withdrawing substituents, which generate positive partial charges in the phenyl moiety. As a result, the positive charge of $[matrix+H]^+$ cannot be delocalized over the whole conjugated system with the result of high charge densities at the remaining delocalization possibilities which is energetically unfavorable in the gas phase provided under MALDI conditions. As can be seen from Fig. 2, the charge density localized at the vinyl side chain increases from 57% for CHCA with an electron-donating hydroxyl group to 64 and 71% in case of one (4-chloroCCA, ClCCA) or two (2,4-difluoroCCA, DiFCCA) electron-withdrawing halogens.

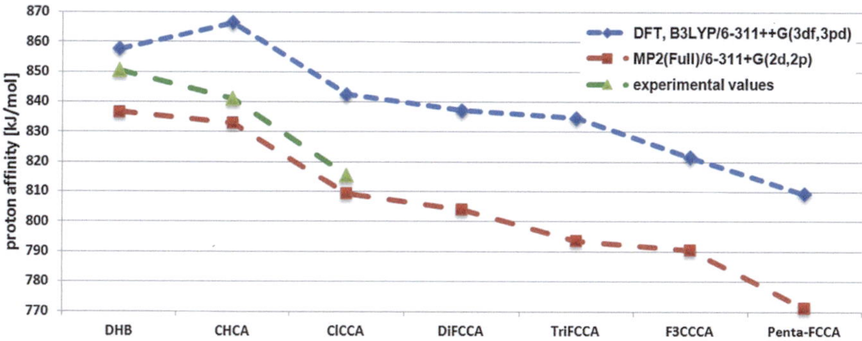

Fig. 3 Calculated and experimentally determined proton affinity values of DHB, CHCA, and several halogenated CCA derivatives. See text for further details

As a result of the restricted charge delocalization, the PA drops with increasing halogenation as can be seen in Fig. 3. The PA values were calculated by means of DFT(B3LYP)6-311++G(3df,3pd) as well as the more accurate MP2(Full)/6-311+G(2d,2p) level of theory. Experimentally determined PA values of DHB (Burton et al. 1997), CHCA (Jørgensen et al. 1998; Porta et al. 2011), and ClCCA (Porta et al. 2011) are between the two calculated values and follow the general trend of the calculated values (see Fig. 3).

Experiments using Cook's Kinetic Method have shown that a lower matrix PA results in a higher proton transfer efficiency (Jaskolla et al. 2008). Solely based on PA, a perhalogenated CCA derivative would be most reactive for proton transfer.

Another important point for designing highly reactive matrices is the formation efficiency of the required matrix species. Ehring et al. have speculated that the protonated matrix species $[matrix+H]^+$ is obtained from hydrogen atom transfer from a neutral matrix to a matrix radical cation generated by photoionization ($matrix + matrix^{\bullet+} \rightarrow [matrix-H]^\bullet + [matrix+H]^+$; Ehring et al. 1992). This assumption is in agreement with findings of Land and Kinsel which support photoionized matrix radical cations to be the essential precursors for analyte protonation with the possible intermediate generation of protonated matrix species (Land and Kinsel 1998).

Starting from the neutral matrix, the question arises which of its hydrogens are accessible for transfer to matrix radical cations. Although it seems reasonable that the carboxylic acid hydrogen will be the most easily one to cleave, substituting the phenyl hydrogens against halogens or other hydrogen-free substituents might significantly lower the hydrogen transfer probability and impede $[matrix+H]^+$-formation. For this purpose, CCA-*d6* (MW = 179 g/mol) was synthesized with only one hydrogen left at the carboxylic acid group and compared to underivatized ^{1}H-CCA (MW = 173 g/mol; see Fig. 4).

The CCA-*d6* peak ratio *m/z* 180:181 corresponds to that of CCA at *m/z* 174:175. Therefore, the peak at *m/z* 181 in the CCA-*d6* spectrum refers to the ^{13}C-isotope of $[matrix+H]^+$ but not to the generation of $[matrix+D]^+$. Consequently, the hydrogen

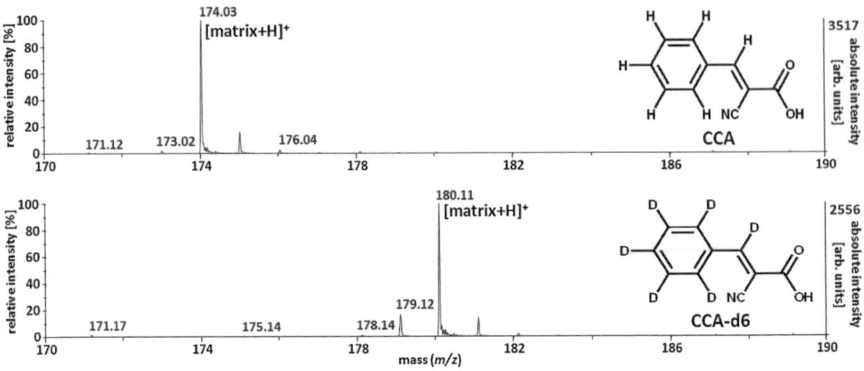

Fig. 4 Comparison between matrix spectra of CCA and CCA-*d6* recorded in positive ion mode with 500 single shots per spectrum. An amount of 10 nmol of each matrix was prepared in 0.5 μL of 70% ACN (c=20 mM), dried and measured at 337 nm using an Applied Biosystems Voyager DE-STR TOF MS. Low laser fluences were chosen for undisturbed isotopic ratios

required for $[matrix+H]^+$ formation originates more or less exclusively from the CCA carboxylic acid group.[2] From this result, elimination of the phenyl- and vinyl-hydrogens should not limit the efficiency of $[matrix+H]^+$ formation.

However, Soltwisch et al. have investigated the influence of the irradiation wavelength and laser fluence on the resulting matrix and analyte ion intensities (Soltwisch et al. 2012). Matrices of interest were, amongst others, CHCA as well as the halogenated counterparts ClCCA, DiFCCA, and TriFCCA. From these data it is evident that there is a shift from the product $[matrix+H]^+$ to the precursor $matrix^{\bullet+}$ with increasing halogenation (see Fig. 5).

As discussed before, reduced amounts of available phenyl hydrogens as a result of matrix halogenation do not directly correlate with restricted $[matrix+H]^+$-formation. Consequently, the apparent impeded conversion of $matrix^{\bullet+}$ to $[matrix+H]^+$ with increasing halogenation seems to result from energetically unfavorable $[matrix+H]^+$ formation enthalpies. To verify this hypothesis, the neutral, radical cationic, protonated, and hydrogen radical abstracted structures of CHCA, ClCCA, DiFCCA, and TriFCCA were energetically minimized using density functional theory (DFT) and the highly extended UB3LYP/6-311++G(3df,3pd) basis set. Starting from optimized geometries, the self-consistent field energies, zero-point energies, vibrational energies and entropies were calculated at 298.15 K for each structure of the four matrices and used for calculating the respective overall free reaction enthalpy ΔG_r of $matrix+matrix^{\bullet+} \rightarrow [matrix-H]^{\bullet}+[matrix+H]^+$. The (zero-point) vibrational energies were corrected by a scaling factor of 0.9877 according to Andersson and Uvdal (2005). The results obtained are given in Table 2.

The formation of protonated matrices according to $matrix+matrix^{\bullet+} \rightarrow [matrix-H]^{\bullet}+[matrix+H]^+$ is endergonic for all investigated cases. Formation of $[CHCA+H]^+$

[2] Note that hydrogens of phenolic hydroxyl groups can be abstracted even more easily as will be discussed later with the example of CHCA.

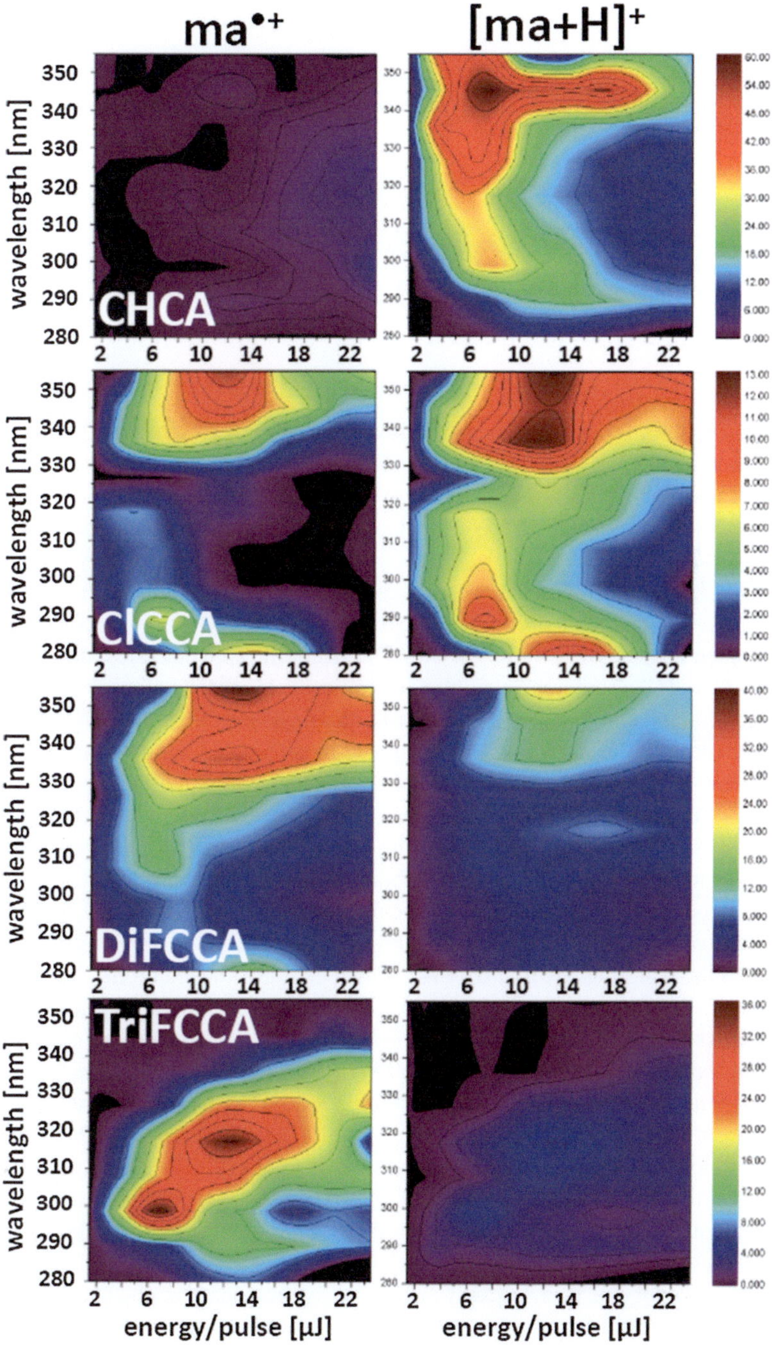

Fig. 5 Heat maps displaying the detected matrix$^{•+}$ (ma$^{•+}$) and [matrix + H]$^+$ ([ma + H]$^+$) intensities in dependence on the irradiation wavelength (*ordinate*) and laser fluence (*abscissa*). The intensities are given in arbitrary units with *red* representing highest intensities. Neat matrix samples (i.e., samples without any additives or analyte) of CHCA, ClCCA, DiFCCA, and TriFCCA were investigated. For more details, see Soltwisch et al. (2012)

Table 2 Calculated free reaction enthalpies ΔG_r for matrix + matrix$^{\bullet+}$ → [matrix − H]$^{\bullet}$ + [matrix + H]$^+$, matrix ionization energies (IE), and the sum of both for several CCA derivatives using DFT and the UB3LYP/6-311++G(3df,3pd) basis set

Matrix	ΔG_r at 298.15 K (kJ/mol)	IE at 298.15 K (kJ/mol)	$\Sigma(\Delta G_r + \text{IE})$ (kJ/mol)
CHCA	13.84	795.88	809.72
ClCCA	86.83	833.93	920.76
DiFCCA	75.11	851.56	926.67
TriFCCA	64.72	864.06	928.78
F_3CCCA	64.66	876.68	941.33

requires the least amount of energy, which is in agreement with the detected efficient conversion of CHCA$^{\bullet+}$ to [CHCA + H]$^+$ (Fig. 5). This is mainly due to the fact that the most stable [CHCA − H]$^{\bullet}$ structure refers to the phenoxy radical whereas the most stable halogenated [CCA − H]$^{\bullet}$ structures are carboxy radicals. Hydrogen abstraction from the carboxylic acid group requires clearly more energy (with ΔG_rs of 1753–1754 kJ/mol) than is necessary for phenols (ΔG_r of only 1666 kJ/mol, see Fig. 6). As a result of this difference, the overall free reaction enthalpies rise for the halogenated matrices (see Table 2). Calculations of the spin densities of the halogenated [CCA − H]$^{\bullet}$-structures reveal that the radical is localized at the carboxylic acid head group only, which explains the nearly identical free reaction enthalpies of the hydrogen abstraction step for the investigated halogenated CCA derivatives.

Generation of [matrix + H]$^+$ by the addition of a hydrogen atom to matrix$^{\bullet+}$ is a strongly exergonic process which releases the more energy the higher the matrix halogenation state is (see blue data in Fig. 6). In this context the charge conditions within the matrix radical cations are important. Here, both the radical and positive charge aspect destabilize the matrix radical cation due to the deficiency in electrons and negative charge. The addition of a hydrogen atom neutralizes the radical and stabilizes the resulting ion [matrix + H]$^+$ which explains the exergonic free reaction enthalpies. The radical as well as the positive charge of the matrix radical cation is delocalized throughout the conjugated system. These delocalizations are the less efficient the more electron-withdrawing (halogen) substituents are present. Therefore, elimination of energetically unfavorable matrix$^{\bullet+}$-states gets more exergonic with higher degrees of halogenation (see Fig. 6). Generation of the F_3CCCA matrix radical cation requires more energy than in the case of TriFCCA. However, the free reaction enthalpy for subsequent [F_3CCCA + H]$^+$ generation does not increase since the protonated F_3CCCA species is more reactive, i.e., energetically higher than that of TriFCCA.

Summation of the free reaction enthalpies of both partial reactions results in the overall free reaction enthalpy ΔG_r given in Table 2. However, this still does not explain why CCA derivatives generate increasingly less [matrix + H]$^+$ ions with increasing halogenation. Since the generation of protonated matrix ions is endergonic, the amount of energy available for this process needs to be considered. For this purpose, the ionization energies (IE) for the generation of the discussed matrix radical cations were calculated (see Table 2). The available energy deposited

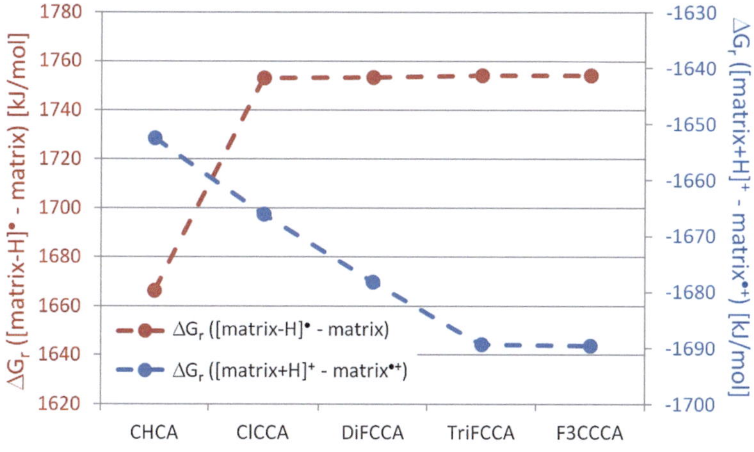

Fig. 6 Free reaction enthalpies for matrix + matrix$^{\bullet+}$ → [matrix − H]$^{\bullet}$ + [matrix + H]$^{+}$, separated in the hydrogen abstraction step matrix → [matrix − H]$^{\bullet}$ (in *red*) and the matrix protonation step matrix$^{\bullet+}$ → [matrix + H]$^{+}$ (in *blue*). The free reaction enthalpies were calculated for the five CCA derivatives CHCA, ClCCA, DiFCCA, TriFCCA, and F_3CCCA. All values are calculated by means of the energetically lowest isomers using DFT and the UB3LYP/6-311++G(3df,3pd) basis set at 298.15 K. For further details, see text

into a specific matrix volume by irradiation (which depends on the laser fluence and the specific matrix extinction coefficient at the irradiation wavelength) will be partially used up for the photoionization step. Consequently, the remaining energy available for secondary reactions such as the formation of protonated matrix ions will be the lower the more energy was consumed for previous photoionization. One has therefore to summarize the photoionization energy and ΔG_r for the subsequent formation of [matrix + H]$^{+}$ which is given in Table 2. Although further factors (e.g., secondary energy loss channels such as fragmentation, fluorescence, or radiationless relaxation) have to be considered as well as somewhat lowered IEs of large clusters compared to the calculated single molecules, the rising energy amounts explain why highly halogenated matrix compounds exhibit restricted conversion efficiencies to the [matrix + H]$^{+}$ species necessary for analyte protonation. This observation is one of the limiting steps in efforts to increase matrix reactivity. For instance, protonated TriFCCA is a highly reactive species but the generated amount is too low for making TriFCCA an efficient MALDI matrix.

A further aspect to discuss refers to the morphology of MALDI samples, which is essentially identical to neat matrix samples. Upon analysis of dozens of CCA derivatives, two usually occurring morphologies have been recognized: Branched needles with typically high surface coverage and small compact crystals with a low surface coverage (see Fig. 7 for the examples of FCCA and ClCCA, and Jaskolla et al. 2008).

High analyte intensities were only detectable in the case of compact crystals. The exact reasons are unknown so far but we assume the surface:volume ratio of the

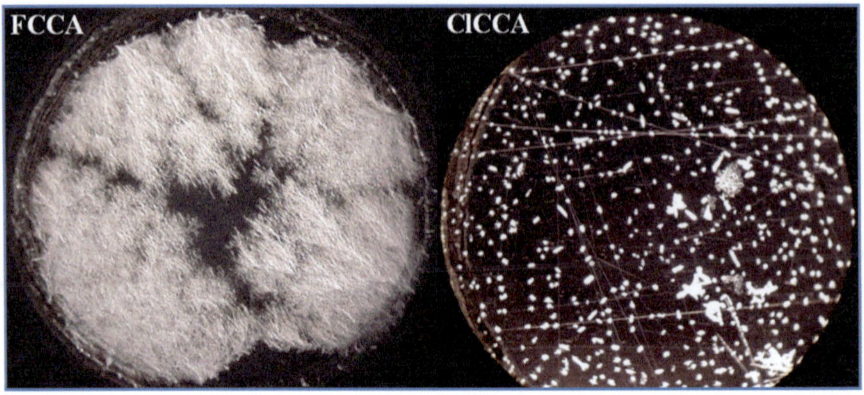

Fig. 7 The two typical MALDI sample morphologies for CCA derivatives using the examples of FCCA (*left*) and ClCCA (*right*). A volume of 0.5 μL of a 20 mM matrix solution in 70% ACN was prepared on a stainless steel target and 'air-dried' under ambient conditions for both CCA derivatives

crystals to be one of the performance indicators. The branched needles exhibit surface:volume ratios, which clearly exceed those of the compact crystals (see Fig. 7). Analytes such as peptides are typically incorporated at the end of the matrix crystallization near the crystal surface (Jaskolla et al. 2009c). Consequently, crystal morphologies with relatively large surfaces will probably exhibit lower analyte concentrations near the surface than in cases of compact morphologies with lower degrees of crystal coverage. Unfortunately, the crystal morphology is neither predictable nor constant. Impurities or additives as well as different surface properties of the sample target might change the morphology in an unpredictable matter. Therefore, sample preparation is a crucial point for improving the analytical performance of matrix compounds (Wiangnon and Cramer 2015).

One last point in evaluating a MALDI matrix is its effective temperature. Matrices with high effective temperatures cause intense analyte fragmentation (Gabelica et al. 2004; Schulz et al. 2006). For ClCCA, the effective temperature is significantly lower than that of CHCA and a little bit higher than that of DHB (Jaskolla et al. 2008).

2 Applications

The expectation that the rationally designed ClCCA matrix enhances the detection of labile compounds was met for labile peptides (Jaskolla et al. 2009a; Leszyk 2010; Openshaw et al. 2010) as well as sialylated glycans and carbohydrates in negative ion mode (Selman et al. 2012). The new matrix ClCCA was also successfully applied for the analysis of chlorinated phospholipids such as dichloramines and chlorimines of phosphatidylethanolamine or phosphatidylserine in positive ion mode (Jaskolla et al. 2009b; Flemmig et al. 2009). In addition, low-molecular metabolites can also be detected with high sensitivity using ClCCA (Schild et al. 2014; Schöner et al. 2014).

With some restrictions regarding the efficiency of $[\text{matrix}+\text{H}]^+$ generation and absorptivity at 355 nm, a second halogenated derivative, DiFCCA, proved to exhibit outstanding properties for the analysis of phosphatidylcholines in positive ion mode (Jaskolla et al. 2009b).

Very recently the CCA lead structure was used again to design a new hydrophobic matrix suitable for negative ionization of lipids (Fülöp et al. 2013). The carboxyl group, which facilitates the incorporation as well as protonation of analytes with basic groups, was replaced by a neutral amide group. To enhance the hydrophobicity of the molecule and to achieve a bathochromic shift in absorption, the electron withdrawing chlorine in ClCCA was replaced by a phenyl group with electron-donating +I-effect. 4-Phenyl-α-cyanocinnamic acid amide (Ph-CCA-NH_2) was identified as the most suitable matrix for analysis and imaging of various lipid classes by MALDI MS in negative ion mode. Compared to other matrices used for lipid analysis, Ph-CCA-NH_2 exhibits superior sensitivity and reproducibility and a relatively low matrix background in the corresponding MALDI spectra.

Another promising new matrix which favors the ionization in negative ion mode is DMAN (Shroff and Svatoš 2009a). DMAN was chosen because it belongs to a class of highly basic compounds called 'proton sponges' which are able to abstract protons from even weak acids. Moreover, this compound has a strong UV absorption between 330 and 350 nm and can be used in MALDI instruments with a nitrogen as well as Nd:YAG laser. A further outstanding feature of DMAN is the absence of all matrix-related ion peaks in the negative ion mode and only one ion signal correlating to protonated DMAN in positive ion mode which makes this compound ideally suited for the analysis of small molecules (<800 Da) such as fatty acids (Shroff and Svatoš 2009b; Shroff et al. 2009). A mixture of DMAN and 9-aminoacridine was applied for the direct lipid analysis of whole bacterial cells (Calvano et al. 2013). However, analyte amounts in the high pmol-range on target (μM analyte concentrations) are typically required (Shroff and Svatoš 2009a, b; Shroff et al. 2009).

In the following, examples for these new matrices in different application fields are given. The structures of these optimized derivatives are shown in Fig. 8.

2.1 Proteomics: Analysis of Peptides

MALDI MS is an important tool for many proteomic studies. In general, protein identification is achieved by peptide mass fingerprinting (PMF) or MS/MS analysis of proteolytic (typically tryptic) peptides followed by protein or DNA sequence database searching. Proteins separated by one- or two-dimensional (1D or 2D) sodium dodecylsulfate (SDS)-polyacrylamide gel electrophoresis (PAGE) are digested directly from excised gel bands. Due to the relatively high ruggedness of

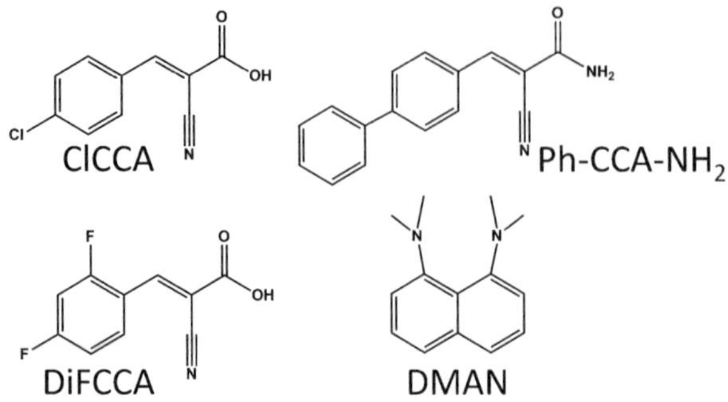

Fig. 8 Matrices discussed in this chapter: 4-chloro-α-cyanocinnamic acid (ClCCA), α-cyano-2, 4-difluorocinnamic acid (DiFCCA), α-cyano-4-phenylcinnamic acid amide (Ph-CCA-NH_2), and 1,8-bis(dimethylamino)naphthalene (DMAN)

MALDI MS, a sample clean-up is often not necessary. Admixtures such as salts and buffers do commonly not disturb the MALDI process (but might change the sample morphology, cf. earlier discussion in Sect. 1). CHCA is still the matrix of choice in MALDI-based proteomic experiments. The simple 'dried-droplet' preparation yields a relatively homogeneous layer of small compact crystals enabling automated high-throughput analysis. Motivation for the continuing search for a better matrix for proteomic analysis is on one hand the detection limit for peptides of about 1 fmol and much higher for peptides from tryptic in-gel digestion, and on the other hand the preferred ionization of strongly basic arginine-containing peptides (Krause et al. 1999) at the cost of small and/or less basic ones. The latter limits the achievable sequence coverage and impedes the identification of peptides with low abundance.

In the following sections comparative experiments between ClCCA and CHCA will be discussed for different kinds of proteomic analysis.

2.1.1 Signal Intensity and S/N Ratio

In-solution digestion of bovine serum albumin (BSA) was used as test sample applying 1 fmol digestion solution onto the MALDI target (Jaskolla et al. 2008; Karas and Jaskolla 2011). Figure 9 exemplifies the enhanced performance of ClCCA compared to CHCA, displaying the sections between *m/z* 1100–1400 of the MALDI spectra for both matrices. Ion intensities (throughout the complete spectrum) are a factor of 2–75 higher for ClCCA, while the S/N ratio increased by an average factor of 22. From an average of three independent experiments and a selected mass tolerance of ±20 ppm for CHCA and ±10 ppm for ClCCA, only seven tryptic peptides were identified out of the CHCA matrix compared to 34 from the ClCCA matrix. The sequence coverage jumped up from 12 to 48%.

Results were overall confirmed in a recent study, in which 200 amol BSA were found to be near the limit of detection (Leszyk 2010). The improvement is particu-

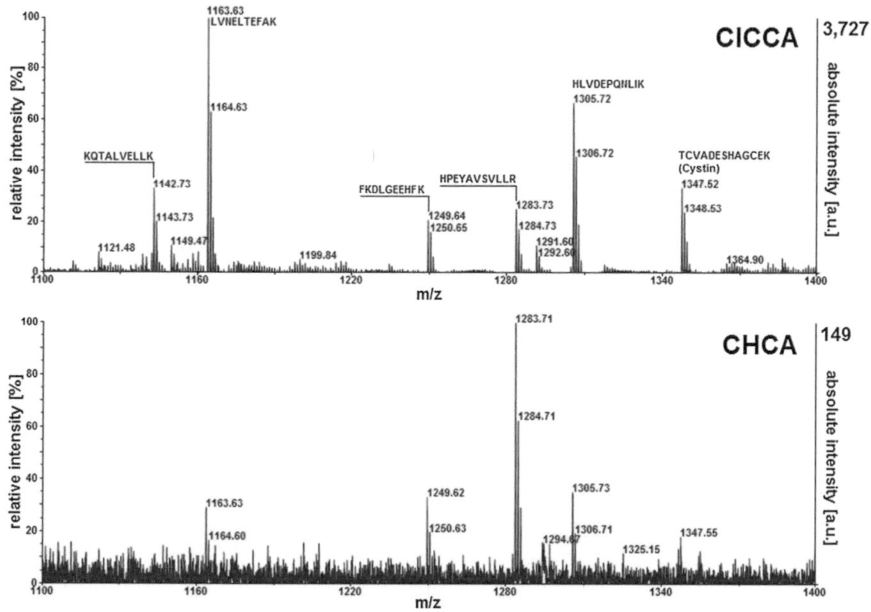

Fig. 9 MALDI mass spectra (*m/z* 1100–1400) of a tryptic in-solution digest of BSA using ClCCA (*upper*) or CHCA (*lower*) as matrix. The sequences of identified peptides are annotated; total sample load was 1 fmol (Karas and Jaskolla 2011)

larly distinct for low protein concentrations (Jaskolla et al. 2009a). Among other proteins and enzymes, tryptic in-solution BSA digests at concentrations of 0.25, 1, and 10 fmol/μL were compared. In all cases, 1 μL analyte solution was applied onto the MALDI target. While the sequence coverages determined at 10 fmol were nearly identical for both matrices, ClCCA exhibited a tenfold higher sequence coverage than CHCA at 0.25 fmol BSA (Jaskolla et al. 2009a).

Ion intensities from tryptic in-gel digestions of BSA showed the same trend. With sensitive silver staining, detection limits are about 1 ng protein/gel band which refers to about 25 fmol BSA (1.65 ng). While protein identification via the MASCOT search engine was possible from faintly visible 25 fmol BSA bands when using ClCCA as matrix (MOWSE score of 77 with 17 identified peptides and 48% sequence coverage) the identification from CHCA failed (no identified peptides). PMF MALDI mass spectra from a tryptic digest of a silver-stained 25 fmol BSA band of a 1D-SDS-PAGE gel can be found in Jaskolla et al. (2009a).

The simple and fast PMF protocol using MALDI with the ClCCA matrix yielded with respect to sequence coverage comparable results to those reported for nanoLC-MS (Winkler et al. 2007).

Comparative analysis with tryptic and less-specific enzymes such as chymotrypsin, elastase, proteinase K and pepsin using the examples of several small to large, acidic, neutral, and basic proteins shows the same enhancement using ClCCA as matrix (see Jaskolla et al. 2008, 2009a; Papasotiriou et al. 2010; and exemplarily Fig. 10).

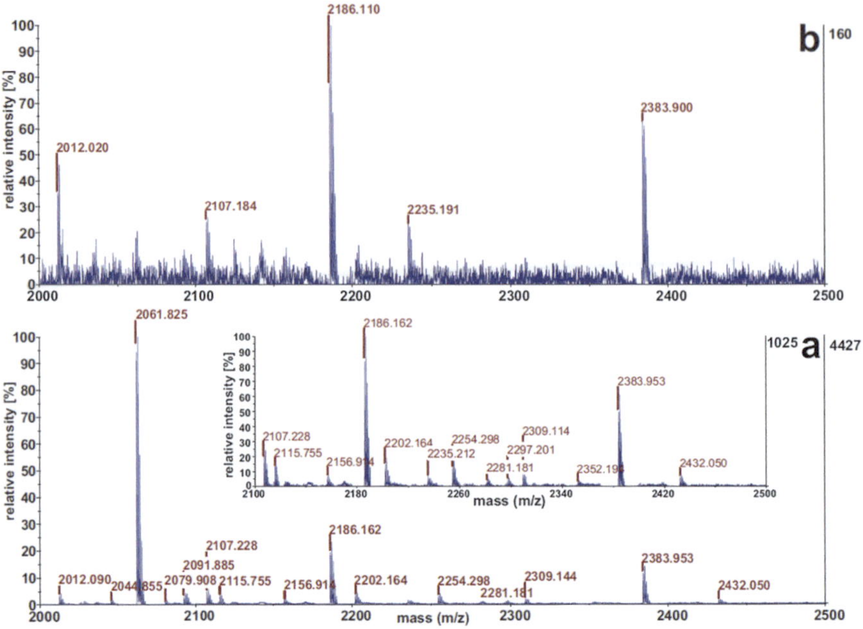

Fig. 10 Positive ion mode MALDI mass spectra of a tryptic in-solution digest of β-casein recorded with (**a**) ClCCA and (**b**) CHCA as matrix. The total analyte load on target was 50 fmol of digested protein. Adapted with permission from Jaskolla et al. (2008). Copyright (2008) National Academy of Sciences, USA

Nonetheless, for de novo sequencing, it has to be mentioned that as a result of the cooler nature of ClCCA its PSD fragmentation efficiency is generally weaker than that of CHCA. However, the peptide fragmentation efficiency can be enhanced by 4-sulfophenylisothiocyanate (SPITC) derivatization, which results in a tenfold increase in fragment ion intensities under PSD and CID conditions for ClCCA compared to CHCA (Leszyk 2010; Openshaw et al. 2010).

2.1.2 Ion Suppression

Besides the higher ion intensities and S/N levels, additional analyte peaks appear in ClCCA spectra (see Figs. 9 and 10) which cannot only be explained by lower noise levels. While in CHCA spectra the most abundant peaks represent peptides with an arginine (R) residue at the C-terminus and only a few low-intensity ion peaks result from cleavages after lysine (K), the ClCCA spectra exhibit a wealth of K-ending peptides and even peptides with only the N-terminus as basic function, e.g., as a result of chymotrypsin or pepsin digestion (Jaskolla et al. 2009a, b, c; Papasotiriou et al. 2010) or tryptic cleavage at the protein's C-terminus (e.g., LVVSTQTALA in the case of BSA; Jaskolla et al. 2008; Jaskolla 2010). These findings were

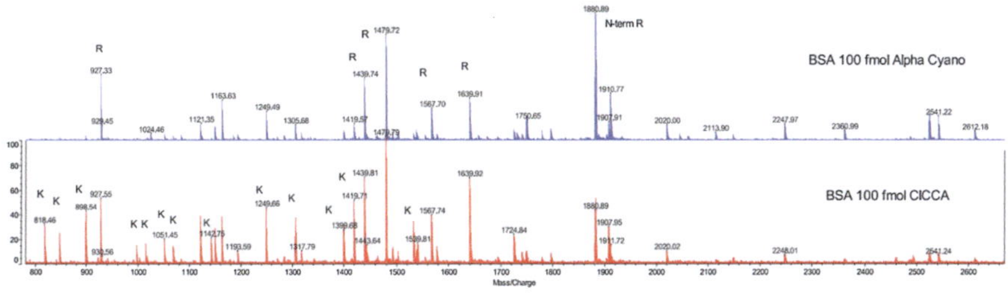

Fig. 11 PMF mass spectra comparison using 100 fmol BSA tryptic digest and CHCA (*upper*) or ClCCA (*lower*) as matrix. Some of the peptides with C-terminal R or K are indicated. The peptide at *m/z* 1880.89 refers to RPCFSALTPDETYVPK (CAM-C) with an arginine at C- and a lysine at N-terminus resulting in increased signal intensity in the case of CHCA (Leszyk 2010). Image used with permission from the *Journal of Biomolecular Techniques* (JBT), © Association of Biomolecular Resource Facilities, www.abrf.org

confirmed by Leszyk (2010) and can be explained by the lower proton affinity of ClCCA which allows for more unbiased protonation of peptides independent of their basicity (see Fig. 11). As a result, the detectable peptides represent much better the actual amino acid distribution of the digested protein, which typically has nearly identical numbers of lysine and arginine residues (Eitner et al. 2010).[3]

2.1.3 Detection of Phosphopeptides

Detection of phosphopeptides in digestion solutions is not only difficult because of the often low relative concentration of these peptide species and the ion suppression due to their acidic nature but they are also labile and can easily loose the phosphate group under MALDI ionization conditions. Because of the cool nature of the matrix ClCCA its performance in analyzing labile peptides such as phosphopeptides was explored (Jaskolla et al. 2008, 2009c; Leszyk 2010). In the mass spectra of an in-solution tryptic digest of 50 fmol β-casein containing about 10% α-casein, substantially more phosphopeptides were detectable with a significant increase in signal intensity and S/N using ClCCA compared to CHCA (see Table 3).

Leszyk analyzed a commercially available phosphopeptide standard containing four synthetic phosphopeptides which appear in tryptic digests of yeast enolase. The mixture comprises a variety of phosphorylation sites and peptides such as singly phosphorylated at Ser, Thr, and Tyr and doubly phosphorylated at Ser. A tenfold improvement in sensitivity and negligible neutral loss of the phosphate group was observed with ClCCA as matrix (Leszyk 2010).

[3] Note that as mentioned in Sect. 1 the sample morphology can have a large effect on the overall MALDI performance. Wiangnon and Cramer have reported that improvements in peptide ion signal intensity and suppression as published earlier for ClCCA are dependent on the use of the MALDI target plate and were not obtained with AnchorChip target plates (Bruker), which lead to markedly different sample morphologies compared to preparations on normal steel target plates (Wiangnon and Cramer 2015).

Table 3 Detectable singly phosphorylated phosphopeptides originating from 50 fmol of a tryptic in-solution digest of β-casein containing α-casein with CHCA or ClCCA as matrix

α-/β-Casein	Sequence	CHCA Intensity (counts)	CHCA S/N	ClCCA Intensity (counts)	ClCCA S/N
β	(K) AVPYPQR (D)	–		2340	256
β	(K) FQSEEQQQTEDELQDK (I)	–		4168	627
β	(K) IEKFQSEEQQQTEDELQDK (I)	–		62	10
α	(K) KISQRYQK (F)	74	19	283	33
α	(K) TVDMESTEVFTK (K)	–		531	69
α	(K) TVDMESTEVFTK (K) Met-Ox.	–		74	10
α	(K) KTVDMESTEVFTK (K)	–		121	16
	(K) TVDMESTEVFTKK (T)				
α	(K) NMAINPSKENLCSTFCK (E) Cys unmod.	39	11	218	29

The sum spectra of five independent mass spectra with 500 shots each were analyzed (modified from Jaskolla et al. 2008; Copyright (2008) National Academy of Sciences, USA)

2.2 *Analysis of Lipids*

MALDI MS is increasingly used for lipid analysis because of its simple and fast sample preparation, its robustness against impurities present in crude tissue or body fluid extracts and its high sensitivity (Fuchs and Schiller 2009). Because lipid species have variegated structures, positive and negative ions need to be measured to cover the complete lipid range. Different matrix compounds are established in the field of lipid analysis with the most frequently used ones being DHB in negative and positive ion mode and 9-aminoacridine (9-AA) in negative ion mode. In addition, *p*-nitroaniline (PNA), 2-mercaptobenzothiazole (MBT), and 2-(2-aminoethylamino)-5-nitropyridine (AAN) are used for the analysis of negative lipid ions. Applications using newly designed matrices in positive and negative ion mode will be described separately.

2.2.1 Positive Ionization

Until recently, mass spectrometric detection of phosphatidylethanolamine (PE) and phosphatidylserine (PS) chloramines and -imines was only possible by ESI MS whereas MALDI MS failed to detect these substances. By using ClCCA as matrix the MALDI MS detection of these species became possible as Na^- adducts (Jaskolla et al. 2009b; Flemmig et al. 2009). It can be assumed that the generation of sodiated chloramines results from protonation of the neutral [PEchloramine − H^+ + Na^+] species and, thus, can be attributed again to the high protonation power of ClCCA. [PEchloramine − H^+ + Na^+] can be protonated at its negatively charged phosphate group with generation of the detectable species [PEchloramine + Na^+]. However, direct protonation of neutral PEchloramine leading to the not observable species [PEchloramine + H]$^+$ is even beyond the protonation power of ClCCA due to the two electron-withdrawing chlorine substituents at the amino group which is the most basic proton acceptor of neutral PEchloramine. Further information and calculation results regarding PAs and sodium cation affinities (SCAs) of these species are provided in Jaskolla et al. (2009b).

DiFCCA was also compared to the established matrices DHB, 9-AA, PNA, MBT, and AAN (Teuber et al. 2010). An organic extract from hen egg yolk was used as a test sample comprising, amongst others, the lipid classes of triacylglycerols (TAGs), phosphatidylcholines (PCs), PEs, phosphatidylinositols (PIs), and sphingomyelins (SMs). Analysis using the different matrices at 337 nm wavelength pointed out that lipids such as TAGs and cholesterols with basicities too low for protonation by currently known matrix compounds are exclusively registered as sodium adducts. Here, DHB showed the highest intensity followed by PNA due to their comparably low SCAs (Zang et al. 2002). Detection of lipids with basic residues such as PCs and PEs suffers from this behavior because the analyte intensity is distributed over protonated and sodiated species. Moreover, detection may be hampered by overlapping peaks. ClCCA and DiFCCA have SCAs higher and PAs clearly lower than DHB which results in efficient formation of protonated species (see above discussion regarding the formation of sodiated chloramines and -imines). For PCs, DiFCCA exhibits the lowest detection limit of about 200 fmol and highest S/N of all tested matrices. The detection of protonated SMs which account for only about 1.5% (m/m) of all phospholipids in hen egg yolk (Fuchs et al. 2007) was possible only with this matrix (shown in the example of SM 16:0 in Teuber et al. 2010). From these experiments it can be concluded that DiFCCA is recommendable for protonable lipids.

2.2.2 Negative Ionization

The above mentioned matrices DHB, 9-AA, 2,6-dihydroxyacetophenone (DHA), AAN, and MBT have also been used for negative lipid analysis. DHB tends to form large crystals and therefore causes poor spot-to-spot reproducibility. The analyte sensitivity is poor and accompanied by strong background noise (Schiller et al. 2007). 9-AA has a high ion intensity variability especially in automated data acquisition and an absorbance minimum at 355 nm which leads to a high ionization energy threshold (Fülöp et al. 2013).

The newly designed matrix Ph-CCA-NH_2 was tested for its applicability to the analysis of a broad range of lipid classes extracted from rat brain (Fülöp et al. 2013). The ion intensities of selected lipid species out of the brain extract were compared using automated measurements to avoid experimenter bias. In the mass range between *m/z* 400 and 2000 the most abundant peaks were identified as hydroxylated and non-hydroxylated sulfatides (SM4s isoforms). With low intensity, phospholipids such as PIs, PSs, negatively charged 1,2-diacyl-PEs, and plasmalogenic PE species as well as gangliosides (GM1) were detected.

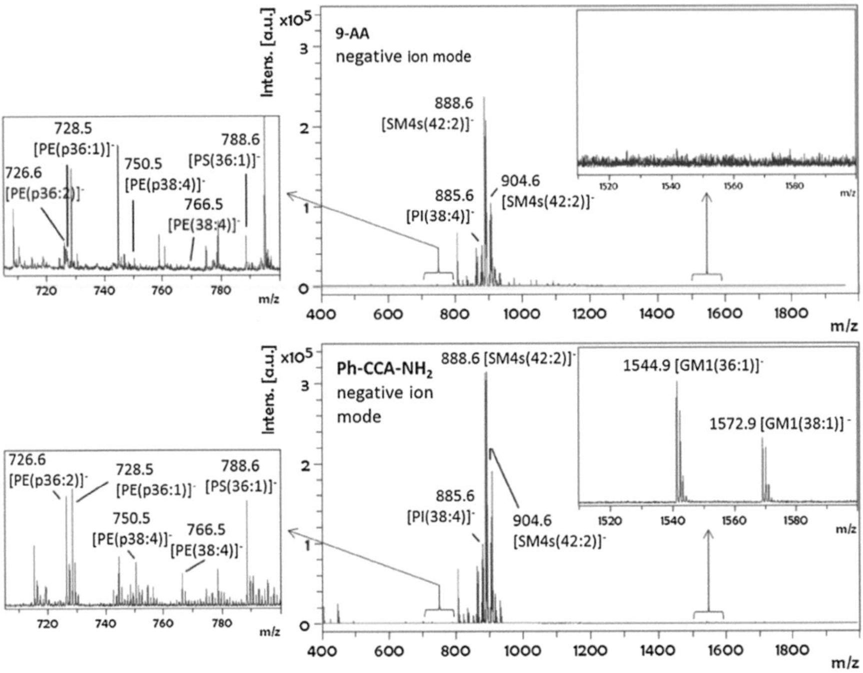

Fig. 12 Negative ion mode MALDI mass spectra recorded at 355 nm showing the detectable lipids from a Sprague–Dawley rat brain total lipid extract. The *upper* spectrum was recorded with 9-AA, the *lower* one with Ph-CCA-NH_2. Reprinted with permission from Fülöp et al. (2013). Copyright (2013) American Chemical Society

Figure 12 displays the comparison between the 9-AA and Ph-CCA-NH_2 matrices. While SM4s are of comparable intensities with both matrices, some of the PE and PS species show higher intensities with the new CCA matrix derivative. Gangliosides were only be detectable with Ph-CCA-NH_2. Furthermore, the matrix ion background was inspected. The matrix spectrum of 9-AA showed intense matrix cluster ions at a laser intensity near ion detection threshold. Fewer background signals appeared when Ph-CCA-NH_2 was used and most of them disappeared in the presence of a lipid extract. This feature was utilized for negative ion detection of sulfatides and other lipids present in rat brain cryosections by MALDI MS imaging (MALDI MSI, Fülöp et al. 2013). While sulfatides could also be detected with 9-AA, imaging and localizing of other lipids such as PEs, PIs, and some phosphatidylglycerol species were only possible with Ph-CCA-NH_2.

Even lower matrix background was observed by combining DMAN matrix (described in detail in Sects. 2.4 and 3.4) with 9-AA (Calvano et al. 2013). This binary matrix system was used for lipid fingerprinting of intact gram-positive (*Lactobacillus sanfranciscensis*) and gram-negative (*Escherichia coli*) bacteria. About 50 major membrane constituents could be identified comprising free fatty acids, mono-, di- and triglycerides, phospholipids, glycolipids, and cardiolipins with m/z values between 200 and 1600. Compared to the tested single matrices, the reduced chemical noise and the absence of matrix-cluster ions increased the achievable analyte S/N and makes this mixture effective for lipidome characterization.[4]

2.3 Analysis of Glycans and Glycopeptides

MALDI MS has found broad application in the field of glycan and glycopeptide analysis (Harvey 1999, 2009; Selman et al. 2010). Standard matrices in this field are DHB and CHCA. Other matrices frequently applied include sinapic acid, 3-hydroxypicolinic acid, DHA, and 2,4,6-trihydroxyacetophenone (THAP). Selman et al. compared the most often used matrix DHB with the new matrix ClCCA for analysis of glycopeptides originating from tryptic digestions of immunoglobulin Gs (IgGs), bovine fetuin, and human apo-transferrin (Selman et al. 2012). Positive as well as negative ions detected in either a reflector or linear TOF mass analyzer were evaluated. Due to the homogeneous distribution of ClCCA microcrystals (cf. Fig. 7), this matrix provides a higher shot-to-shot reproducibility than DHB and facilitates automated high-throughput profiling of glycopeptides. Resulting from higher resolution, mass accuracy, and ion intensity, the limit of detection for glycopeptides was a factor of 2–5 lower when using ClCCA compared to DHB which was in absolute amounts spotted onto the target approximately 220 amol in negative and 44 amol in positive ion reflector mode.

Special attention was paid to the detection of sialylated glycopeptides. It is known that labile substituents such as sialic acids may be lost by in-source and metastable decay and that this loss is strongly dependent on the chosen MALDI matrix. 'Hot' matrices such as CHCA transfer comparatively large amounts of energy to the analyte and lead to an almost complete loss of sialic acids. DHB and ClCCA as 'cooler' matrices enable the analysis of intact sialylated (*N*-acetylneuraminic acid) glycans and glycopeptides in particular when negative ions are registered in a linear TOF (Selman et al. 2012). Negative ion mode MALDI-TOF mass spectra revealed similar abundant peaks from DHB on an AnchorChip target and ClCCA on a common steel target (see Fig. 13).

In a recent study, large-scale IgG1 and IgG2&3 Fc N-glycosylation profiling by negative ion MALDI-TOF MS was performed with ClCCA as matrix and compared to hydrophilic interaction liquid chromatography (HILIC) of enzymatically released and fluorescently labeled glycans (Baković et al. 2013). The age and sex specificity of sialylation, galactosylation, fucosylation, and the occurrence of bisecting GlcNAc in a large number of human individuals was analyzed. Although significantly higher signal levels of galactosylation, bisecting *N*-acetylglucosamine and sialylation were obtained by HILIC analysis glycosylation profiling was comparable between MALDI-TOF MS and HILIC.

[4] Note that the combination of both substances leads to homogenous crystallization, a prerequisite for MALDI imaging.

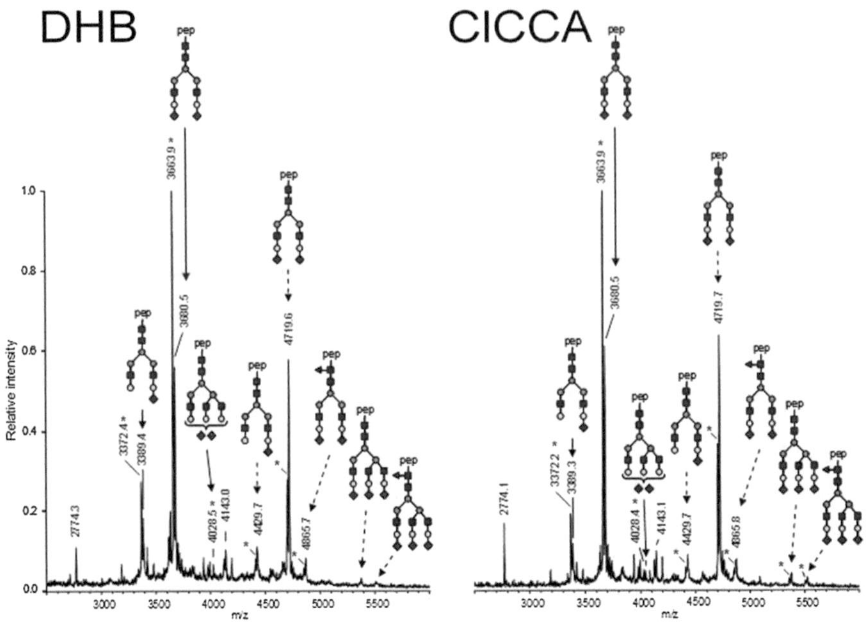

Fig. 13 Negative ion reflectron mode MALDI mass spectra of a tryptic digest of human apotransferrin recorded with DHB spotted on an AnchorChip target (*left*) and ClCCA spotted on a polished steel target (*right*). Identified glycopeptides are marked by *arrows*, post source decay fragments by *asterisks*. *Blue square*, *N*-acetylglucosamine; *green circle*, mannose; *yellow circle*, galactose; *red triangle*, fucose; *purple diamond*, *N*-acetylneuraminic acid Reprinted with permission from Selman et al. (2012). Copyright (2012) John Wiley and Sons.

2.4 Analysis of Small Molecules

MALDI MS of small molecules is a challenge as typical MALDI matrices yield intense matrix signals up to *m*/*z* 500 and sometimes even above (van Kampen et al. 2011). Interferences from the matrix limit the detection and quantification of small molecules. Therefore, matrices without interfering signals at the investigated *m*/*z* range improve analyte detection. Recently, Porta et al. synthesized two new α-cyanocinnamic acid derivatives, (*E*)-2-cyano-3-(naphthalen-2-yl)acrylic acid (NpCCA) and (2*E*)-3-(anthracen-9-yl)-2-cyanoprop-2-enoic acid (AnCCA, Porta et al. 2011). These substances differ from CHCA in the size of the aromatic ring system with a naphthyl (NpCCA) or anthracenyl group (AnCCA) instead of a phenyl moiety but keep much of the chemical functionality of CHCA. The concept was that together with CHCA and ClCCA a series of chemically similar matrix compounds are available with 'tunable' matrix background. Out of this series an appropriate matrix might be selectable without mass spectral interference for the investigated analyte(s). A total of 47 analytes, mostly pharmaceuticals, were selected to compare the four matrix substances with regard to the achievable S/N in selected reaction monitoring mode. Compared with the standard matrix CHCA,

ClCCA yielded an at least 30% increased S/N value for 60% of the compounds and NpCCA and AnCCA led to a higher S/N value for about 30% of the low-molecular weight analytes. The results again show that the matrix with the lowest proton affinity (AnCCA>CHCA>NpCCA>ClCCA) provides the highest absolute analyte ion signals.

Another promising new matrix which favors the ionization in negative mode is DMAN (Shroff and Svatoš 2009a). DMAN was chosen because it belongs to a class of highly basic compounds called 'proton sponges'[5] which are able to abstract protons from even very weak acids. Moreover, DMAN has a strong UV absorption between 330 and 350 nm and can be used in MALDI instruments with nitrogen as well as Nd:YAG lasers.

This matrix was used for studying a wide range of metabolites such as fatty acids, amino acids, fatty acid-amino acid conjugates, plant and animal hormones, vitamins, and short peptides (Shroff and Svatoš 2009a, b; Shroff et al. 2009). In negative ion mode only deprotonated analyte signals were measured with detection limits in the low picomole/femtomole range. As can be seen in Fig. 14 for a selection of fatty acid spectra no matrix-related signals are detectable; only deprotonated ions are registered. It is assumed that the absence of matrix ions in the entire low mass region (<1000 Da) can be attributed to the fact that this highly basic matrix gets protonated by the acidic analyte forming an extremely stable hydrogen chelated cation which is observable in the positive ion mass spectrum.

To study the metabolite distribution in roots and root nodules of *Medicago truncatula* during nitrogen fixation, MALDI mass spectrometry imaging (MSI) was used (Ye et al. 2013). The detection of metabolites with DHB in positive ion mode was complemented by negative ion detection with DMAN. A large array of organic acids, amino acids, sugars, lipids, flavonoids and their conjugates were found.

3 Materials and Protocols

3.1 Analysis of Peptides

3.1.1 Chemicals

The matrix compound ClCCA is commercially available from Sigma-Aldrich, Taufkirchen, Germany, in MALDI matrix quality at ≥95% (order no. 94141) and 97% (order no. 741140) purity. Acetonitrile (ACN) should be used in HPLC grade, TFA in the highest available purity and water in ultrahigh purity.

[5] 'Proton sponges' can contaminate MS instruments with the effect of reduced signal intensities in subsequent positive ion mode measurements. Therefore, 'proton sponges' need to be carefully employed in mass spectrometry.

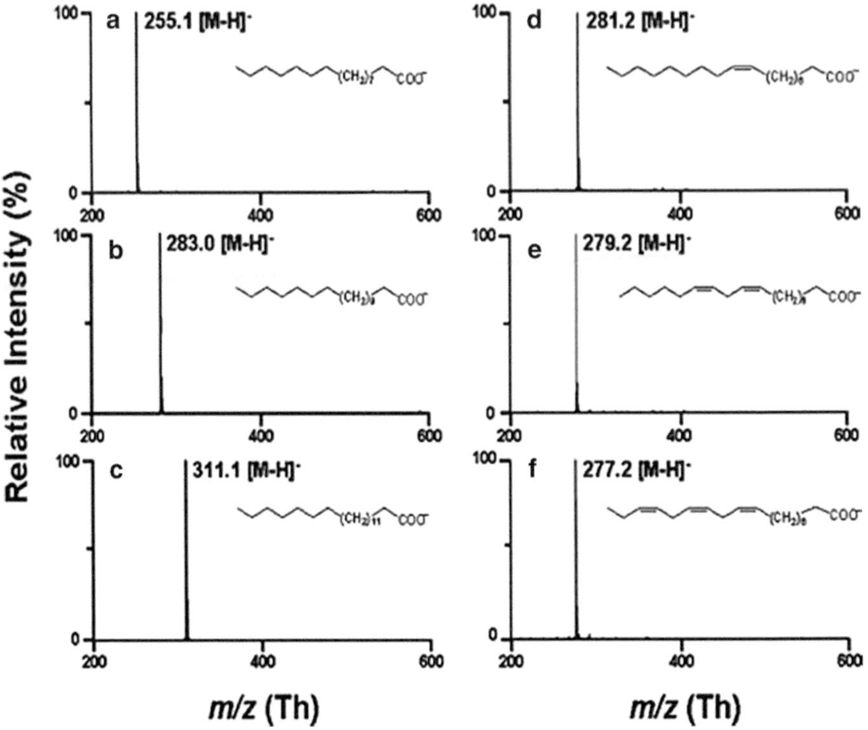

Fig. 14 MALDI mass spectra of several saturated and desaturated fatty acids using DMAN as matrix in negative ion mode. The analyte amount on target was 100 pmol in all cases. (**a**) Palmitic acid; (**b**) stearic acid, (**c**) arachidic acid, (**d**) oleic acid, (**e**) linoleic acid, and (**f**) linolenic acid. Reprinted with permission from Shroff and Svatoš (2009b). Copyright (2009) John Wiley and Sons

3.1.2 'Dried-Droplet' Sample Preparation

For preparation of the ClCCA matrix solution, a few milligrams of ClCCA are dissolved in 70% ACN/0.1% TFA (v/v) to obtain a concentration of 5 mg/mL. Analyte solutions of in-solution digests, phosphopeptides and in-gel digests can be prepared at concentrations down to 1 nM in 30% ACN/0.1% TFA. For digestion conditions, see the literature.

A volume of 1–2 μL of the diluted analyte solution and 0.5–1 μL of the matrix solution are typically mixed directly on a polished stainless steel target. Optionally, 1 μL of an aqueous 10 mM $NH_4H_2PO_4$ can be added to suppress matrix-cluster ions (Smirnov et al. 2004). The sample is then dried, typically in a stream of ambient air.

Comments:

It is recommended to use 0.5 μL of 20 mM ClCCA solutions (about 4 mg/mL) for 'dried-droplet' preparations. Analytes should be diluted to nanomolar concentrations. The ACN percentage of the combined analyte/matrix solution on the MALDI target should not be below 50% (v/v). A somewhat higher percentage of organic solvent is beneficial for the formation of uniform crystal morphology upon drying, comparable to CHCA. Using a lower proportion of organic solvent can sporadically lead to the erratic formation of needle-like crystals with nonuniform analyte intensities (Jaskolla et al. 2009a).

3.1.3 MALDI MS Measurement

MALDI-(Q/IT)-TOF or MALDI orbitrap mass spectrometers can be used for mass analysis. The achievable performances are higher if 337 nm N_2 lasers are used but ClCCA also works well at 355 nm (Porta et al. 2011; Selman et al. 2012; Jaskolla and Karas 2013). To maintain an optimal S/N value, the laser irradiance should be about 10% above the threshold for ion detection. For each mass spectrum at least 100–500 single laser shots should be accumulated.

Comments:

Compared to CHCA the ClCCA matrix exhibits a hypsochromic shift due to the previously discussed electron-withdrawing chloro-substituent resulting in reduced absorption at 355 nm (see Table 1). As a consequence, highest performance is achievable at N_2 lasers emitting at 337 nm. Also, 10–15% more laser power is necessary for threshold signals than what is typically required for CHCA. ClCCA has a somewhat higher vapor pressure than CHCA and therefore is more volatile under vacuum. The matrix spot shows noticeable disappearance by optical inspection and loss in signal intensities after 1 day under vacuum. The consequence is that samples have to be analyzed within the first several hours after being placed in vacuum or the plate should be ejected if samples cannot be analyzed on the same day. Under atmospheric pressure the matrix crystals are stable for at least several days (Leszyk 2010).

3.2 Analysis of Lipids

3.2.1 Chemicals

Both ClCCA and DiFCCA are commercially available as MALDI matrix (Sigma-Aldrich). All solvents should be obtained in the highest commercially available purity. The matrix Ph-CCA-NH_2 is not yet commercially available but its synthesis is described in Fülöp et al. (2013).

3.2.2 Sample Preparation and Mass Spectrometry

Because of the poor solubility of many lipids in commonly used matrix solvents, sample preparation on a stainless steel target is performed by applying 0.5–1 μL of lipid (extract) solution in an organic solvent such as chloroform and methanol followed by air-drying. The organic solvent might cause sample spreading over the target surface. During sample deposition it might therefore be helpful to position the pipette tip close to the target surface followed by slow solvent delivery with immediate evaporation. In the next step the same or a lower volume of matrix solution is added and air-dried. Solution concentrations for the different matrices are:

- 30 mM (6.21 mg/mL) ClCCA in 70/30 ACN/1.5% TFA (v/v)
- 10 mg/mL DiFCCA in 70% ACN (v/v)
- 5 mg/mL Ph-CCA-NH_2 in 80% ACN or 90% acetone (v/v)

ClCCA and DiFCCA are preferably used in MALDI instruments with a nitrogen laser (337 nm) whereas Ph-CCA-NH_2 requires an optimum irradiation wavelength of 355 nm (Nd:YAG laser).

Comment:

Although the absorption of Ph-CCA-NH_2 at 337 nm is sufficiently high, only low ion intensities of lipids are typically registered. Therefore, it is suggested to use 9-AA as matrix for the detection of negatively charged lipids when MALDI instruments are used that are equipped with a nitrogen laser (Fülöp et al. 2013).

3.3 Analysis of Glycans and Glycopeptides

3.3.1 Chemicals

As before ClCCA is available in matrix quality from Sigma-Aldrich.

3.3.2 Sample Preparation and Mass Spectrometry

A volume of 1–3 μL of aqueous glycopeptide or glycan solution is spotted onto a polished steel target plate and allowed to dry at room temperature. Subsequently, 1 μL of ClCCA matrix solution (5 mg/mL in 50–70% ACN) is applied on top of each sample and allowed to dry. MALDI-TOF instruments with 355 and 337 nm lasers can be used in positive and negative ion mode.

3.4 Analysis of Small Molecules

3.4.1 Chemicals

DMAN was purchased from Sigma-Aldrich. All solvents should be of HPLC grade.

3.4.2 Sample Preparation and Mass Spectrometry

Approximately 2 mg/mL DMAN is dissolved in ethanol. Standard samples or purified extracts can also be dissolved in ethanol. Identical volumes of matrix and analyte solution are premixed in an Eppendorf tube, and 1 μL of the resulting mixture is spotted on a MALDI target plate and allowed to dry, e.g., under a gentle stream of argon as described in (Shroff and Svatoš 2009a). MALDI-TOF instruments are typically used with 337 nm (N_2 between laser) or 355 nm (Nd:YAG laser) in negative ion mode.

Health and safety note:

DMAN is a strong base and could be harmful if swallowed. Use gloves and eye protection when working with this substance. In case of contact with the eyes, rinse immediately with plenty of water and seek medical advice (Shroff and Svatoš 2009a*).*

References

Andersson MP, Uvdal P (2005) New scale factors for harmonic vibrational frequencies using the B3LYP density functional method with the triple-ξ basis set 6-311+G(d,p). J Phys Chem A 109:2937–2941

Baković MP, Selman MHJ, Hoffmann M et al (2013) High-throughput IgG Fc N-glycosylation profiling by mass spectrometry of glycopeptides. J Proteome Res 12:821–831

Beavis RC, Bridson JN (1993) Epitaxial protein inclusion in sinapic acid crystals. J Phys D Appl Phys 26:442–447

Beavis RC, Chaudhary T, Chait BT (1992) α-Cyano-4-hydroxycinnamic acid as a matrix for matrix assisted laser desorption mass spectrometry. Org Mass Spectrom 27:156–158

Bryan RF, Forcier PG (1980) Crystal structure basis for the absence of thermal mesomorphism in p-hydroxy-*trans*-cinnamic acid. Mol Cryst Liq Cryst 60:157–165

Burton RD, Watson CH, Eyler JR (1997) Proton affinities of eight matrices used for matrix-assisted laser desorption/ionization. Rapid Commun Mass Spectrom 11:443–446

Calvano CD, Monopoli A, Ditaranto N et al (2013) 1,8-Bis(dimethylamino)naphthalene/9-aminoacridine: a new binary matrix for lipid fingerprinting of intact bacteria by matrix assisted laser desorption ionization mass spectrometry. Anal Chim Acta 798:56–63

Ehring H, Karas M, Hillenkamp F (1992) Role of photoionization and photochemistry in ionization processes of organic molecules and relevance for matrix-assisted laser desorption/ionization mass spectrometry. Org Mass Spectrom 27:427–480

Eitner K, Koch U, Gawęda T et al (2010) Statistical distribution of amino acid sequences: a proof of darwinian evolution. Bioinformatics 26:2933–2935

Flemmig J, Spalteholz H, Schubert K et al (2009) Modification of phosphatidylserine by hypochlorous acid. Chem Phys Lipids 161:44–50

Fuchs B, Schiller J (2009) Application of MALDI TOF mass spectrometry in lipidomics. Eur J Lipid Sci Technol 11:83–98

Fuchs B, Schiller J, Süß R et al (2007) A direct and simple method of coupling matrix-assisted laser desorption and ionization time-of-flight mass spectrometry (MALDI-TOF MS) to thin-layer chromatography (TLC) for the analysis of phospholipids from egg yolk. Anal Bioanal Chem 389:827–834

Fülöp A, Porada MB, Marsching C et al (2013) 4-Phenyl-α-cyanocinnamic acid amide: screening for a negative ion matrix for MALDI-MS imaging of multiple lipid classes. Anal Chem 85:9156–9163

Gabelica V, Schulz E, Karas M (2004) Internal energy build-up in matrix assisted laser desorption/ionization. J Mass Spectrom 39:579–593

Garcia-Granda S, Beurskens G, Beurskens PT et al (1987) Structure of 3,4-dihydroxy-transcinnamic acid (caffeic acid) and its lack of solid-state topochemical reactivity. Acta Crystallogr C43:683–685

Haisa M, Kashino S, Hanada SI et al (1982) The structures of 2-hydroxy-5-methylbenzoic acid and dimorphs of 2,5-dihydroxybenzoic acid. Acta Crystallogr B38:1480–1485

Harvey DJ (1999) Matrix-assisted laser desorption/ionization mass spectrometry of carbohydrates. Mass Spectrom Rev 18:349–450

Harvey DJ (2009) Analysis of carbohydrates and glycoconjugates by matrix-assisted laser desorption/ionization mass spectrometry: an update for 2003–2004. Mass Spectrom Rev 28:273–361

Jaskolla TW (2010) Analyse und Optimierung der Matrixeigenschaften in der MALDI Massenspektrometrie. Shaker Verlag, Aachen

Jaskolla TW, Karas M (2013) Use of halogenated derivatives of the cyanocinnamic acid as matrices in MALDI mass spectrometry. US Patent 2013/0040395

Jaskolla TW, Lehmann WD, Karas M (2008) 4-Chloro-α-cyanocinnamic acid is an advanced, rationally designed MALDI matrix. Proc Natl Acad Sci U S A 105:12200–12205

Jaskolla TW, Papasotiriou DG, Karas M (2009a) Comparison between the matrices α-cyano-4-hydroxycinnamic acid and 4-chloro-α-cyanocinnamic acid for trypsin, chymotrypsin, and pepsin digestions by MALDI-TOF mass spectrometry. J Proteome Res 8:3588–3597

Jaskolla TW, Fuchs B, Karas M et al (2009b) The new matrix 4-chloro-α-cyanocinnamic acid allows the detection of phosphatidylethanolamine chloramines by MALDI-TOF mass spectrometry. J Am Soc Mass Spectrom 20:867–874

Jaskolla TW, Karas M, Roth U et al (2009c) Comparison between vacuum sublimed matrices and conventional dried droplet preparation in MALDI-TOF mass spectrometry. J Am Soc Mass Spectrom 20:1104–1114

Jørgensen TJD, Bojesen G, Rahbek-Nielsen H (1998) The proton affinities of seven matrix-assisted laser desorption/ionization matrices correlated with the formation of multiply charged ions. Eur Mass Spectrom 4:39–45

Karas M, Jaskolla TW (2011) Use of cyanocinnamic acid derivatives as matrices in MALDI mass spectrometry. US Patent 2011/0121166A1

Krause E, Wendschuh H, Jungblut PR (1999) The dominance of arginine-containing peptides in MALDI derived tryptic mass fingerprints of proteins. Anal Chem 71:4160–4165

Land CM, Kinsel GR (1998) Investigation of the mechanism of intracluster proton transfer from sinapinic acid to biomolecular analytes. J Am Soc Mass Spectrom 9:1060–1067

Leszyk JD (2010) Evaluation of the new MALDI matrix 4-chloro-α-cyanocinnamic acid. J Biomol Tech 21:81–91

Openshaw ME, Eagle G, Yamazaki Y, et al (2010) Evaluation of 4-chloro-alpha-cyanocinnamic acid (Cl-CCA) using MALDI-TOF and MALDI-QIT-TOF. In: Proceedings of the 58th annual ASMS conference on mass spectrometry and allied topics (Poster TP-581), Salt Lake City, UT, 23–27 May 2010

Papasotiriou DG, Jaskolla TW, Markoutsa S et al (2010) Peptide mass fingerprinting after less specific in-gel proteolysis using MALDI-LTQ-Orbitrap and 4-chloro-α-cyanocinnamic acid. J Proteome Res 9:2619–2629

Porta T, Grivet C, Knochenmuss R et al (2011) Alternative CHCA-based matrices for the analysis of low molecular weight compounds by UV-MALDI-tandem mass spectrometry. J Mass Spectrom 46:144–152

Schild H-A, Fuchs SW, Bode HB et al (2014) Low-molecular-weight metabolites secreted by *Paenibacillus larvae* as potential virulence factors of american foulbrood. Appl Environ Microbiol 80:2484–2492

Schiller J, Süß R, Fuchs B et al (2007) The suitability of different DHB isomers as matrices for the MALDI-TOF MS analysis of phospholipids: which isomer for what purpose? Eur Biophys J 36:517–527

Schöner TA, Fuchs SW, Reinhold-Hurek B et al (2014) Identification and biosynthesis of a novel xanthomonadin-dialkylresorcinol-hybrid from *Azoarcus* sp. BH72. PLoS One 9:e90922

Schulz E, Karas M, Rosu F et al (2006) Influence of the matrix on analyte fragmentation in atmospheric pressure MALDI. J Am Soc Mass Spectrom 17:1005–1013

Selman MHJ, McDonnell LA, Palmblad M et al (2010) Immunoglobulin G glycopeptide profiling by matrix-assisted laser desorption ionization fourier transform ion cyclotron resonance mass spectrometry. Anal Chem 82:1073–1081

Selman MHJ, Hoffmann M, Zauner G et al (2012) MALDI-TOF-MS analysis of sialylated glycans and glycopeptides using 4-chloro-α-cyanocinnamic acid matrix. Proteomics 12:1337–1348

Shroff R, Svatoš A (2009a) Proton sponge: a novel and versatile MALDI matrix for the analysis of metabolites using mass spectrometry. Anal Chem 81:7954–7959

Shroff R, Svatoš A (2009b) 1,8-Bis(dimethylamino)naphthalene: a novel superbasic matrix for matrix-assisted laser desorption/ionization time-of-flight mass spectrometric analysis of fatty acids. Rapid Commun Mass Spectrom 23:2380–2382

Shroff R, Rulíšek L, Doubský J et al (2009) Acid-base-driven matrix-assisted mass spectrometry for targeted metabolomics. Proc Natl Acad Sci U S A 106:10092–10096
Smirnov IP, Zhu X, Taylor T et al (2004) Suppression of α-cyano-4-hydroxycinnamic acid matrix clusters and reduction of chemical noise in MALDI-TOF mass spectrometry. Anal Chem 76:2958–2965
Soltwisch J, Jaskolla TW, Hillenkamp F et al (2012) Ion yields in UV-MALDI mass spectrometry as a function of excitation laser wavelength and optical and physico-chemical properties of classical and halogen-substituted MALDI matrixes. Anal Chem 84:6567–6576
Teuber K, Schiller J, Fuchs B et al (2010) Significant sensitivity improvements by matrix optimization: a MALDI-TOF mass spectrometric study of lipids from hen egg yolk. Chem Phys Lipids 163:552–560
van Kampen JJA, Burgers PC, de Groot R et al (2011) Biomedical application of MALDI mass spectrometry for small-molecule analysis. Mass Spectrom Rev 30:101–120
Wiangnon K, Cramer R (2015) Sample preparation: a crucial factor for the analytical performance of rationally designed MALDI matrices. Anal Chem 87:1485–1488
Winkler C, Denkler K, Wortelkamp S et al (2007) Silver- and coomassie-staining protocols: detection limits and compatibility with ESI-MS. Electrophoresis 28:2095–2099
Ye H, Gemperline E, Venkateshwaran M et al (2013) MALDI mass spectrometry-assisted molecular imaging of metabolites during nitrogen fixation in the *Medicago truncatula–Sinorhizobium meliloti* symbiosis. Plant J 75:130–145
Zang J, Knochenmuss R, Stevenson E et al (2002) The gas-phase sodium basicities of common matrix-assisted laser desorption/ionization matrices. Int J Mass Spectrom 213:237–250

Efficient Production of Multiply Charged MALDI Ions

Pavel Ryumin and Rainer Cramer

Abstract This chapter details a newly developed MALDI method which allows the generation of multiply charged ions of peptides and proteins similar to those produced by electrospray ionization (ESI) with high sensitivity and low sample consumption. A straightforward modification of a commercially available mass spectrometer with an atmospheric pressure (AP) ion source, and the necessary MALDI sample preparation protocol are described. This new method allows the combination of MALDI with mass analyzers of limited *m/z* range (e.g., quadrupoles, ion traps) for MS analysis of large biomolecules and ETD/ECD fragmentation for enhanced MS/MS analyses. In combination with ion mobility spectrometry (IMS) the signal-to-noise ratio of the multiply charged analyte ions can be significantly increased by filtering out the singly charged MALDI chemical noise ions.

1 Introduction

Analysis of organic molecules, specifically biomolecules, is an important application where mass spectrometry has become a tool of choice helped by the discovery of soft ionization techniques such as electrospray ionization (ESI) by Fenn et al. (1989) and matrix-assisted laser desorption/ionization (MALDI) by Karas et al. (1987) and Karas and Hillenkamp (1988).

Both ESI and MALDI have been extensively developed in conjunction with corresponding mass analyzers which have progressively improved in sensitivity and versatility. Due to their different nature, in particular with respect to ion formation and the resulting ion charge states, some mass analyzers are better suited for one ionization method than the other. Compared to conventional MALDI, ESI is a continuous ionization technique performed at atmospheric pressure (AP). More importantly, a distinctive feature of ESI is the predominantly multiply charged ion yield, which is well suited for many mass analyzers and ion transport systems that perform best at low *m/z* ranges. In contrast, conventional MALDI provides

P. Ryumin • R. Cramer (✉)
Department of Chemistry, University of Reading, Whiteknights, Reading RG6 6AD, UK
e-mail: p.ryumin@reading.ac.uk; r.k.cramer@reading.ac.uk

R. Cramer (ed.), *Advances in MALDI and Laser-Induced Soft Ionization Mass Spectrometry*, DOI 10.1007/978-3-319-04819-2_2

predominantly singly charged ions, resulting in simpler and easier-to-interpret spectra but suffering from the above mentioned technical challenges such as ion transmission of molecular ions with large *m/z* values from ambient or sub-ambient pressures to the high vacuum analyzer region.

Multiply charged MALDI ions would allow taking advantage of the ion transport systems and mass analyzers developed for ions with low *m/z* values such as in ESI. Thus, using the same mass separation technology as for ESI would immediately widen the range of MS performance for MALDI and also significantly reduce the cost of instrumentation. For instance, in Fourier transform (FT)-based mass analyzers such as Orbitraps or FT-ICR instruments the measured oscillation frequency depends on the *m/z* ratio, providing higher resolving power for the same acquisition time (Makarov 2000; Marshall et al. 1998). In ion detection, the gain of microchannel plate detectors (Xian et al. 2012) depends on the *m/z* value of the ion, typically leading to the discrimination of heavier molecules.

Another major advantage of a higher charge state is that multiply charged molecules such as peptides are also more amenable to fragmentation both by collision-induced dissociation (CID) and electron transfer dissociation (ETD)/electron capture dissociation (ECD). According to a plethora of publications (e.g., Cramer and Corless 2001; Huang et al. 2005; Good et al. 2007), CID fragmentation spectra of doubly charged peptide precursor ions provide different information compared to the spectra obtained by fragmentation of singly charged ions. The recently developed ECD by Zubarev et al. and ETD by Syka et al. preserve labile post-translational modifications and therefore provide complementary information to CID. At the core of the ECD/ETD fragmentation lays the capture of an electron by the precursor cation, reducing its charge state and demanding at least two charges at the precursor ion for obtaining fragment ions that can then be detected.

As MALDI is relatively tolerant to contaminants, has high spatial resolution for imaging applications and is a time-controlled, pulsed desorption/ionization event, which has the potential to significantly reduce sample wastage/consumption, there are some distinct analytical advantages compared to ESI. The possible combination of these and the ability to generate predominantly multiply charged ion yields similar to ESI are of great interest and have inspired the MS community to develop methods which would deliver such powerful combination.

Recently one promising technique has been developed (Cramer et al. 2013) allowing the generation of persistent yields of highly charged ions for peptide and protein analysis by utilizing an AP-UV-MALDI source and glycerol-based liquid sample preparations, leading to durable multiply charged ion signals with femtomole detection limits and very low sample consumption.

2 Applications

Conventional MALDI has been successfully applied to the analysis of peptides (Karas et al. 1987), polymers (Bahr et al. 1992), and large proteins (Karas and Hillenkamp 1988; Tanaka et al. 1988), to name but a few. Recently, it has particularly become important in the field of clinical microbiology (Fournier et al. 2013), lipidomics (Köfeler et al. 2012), and MALDI imaging (Angel and Caprioli 2012).

Various studies have been reported on the generation of multiply charged ions in MALDI sources, with increasing frequency in recent years. In 2001, Cramer and Corless were able to generate up to 25% of the ion yield of [Glu^{1}]-fibrinopeptide B as doubly charged ion species using intermediate vacuum and an infrared (IR) laser in their studies of CID fragmentation pathways for ESI- and MALDI-generated ions (Cramer and Corless 2001). For UV-MALDI, the doubly charged ion production was reported to be less than 10% of the total analyte ion current. In 2003, Frankevich et al. were able to generate multiply charged ions from an electron-free PEEK substrate sufficient enough to acquire ECD fragmentation spectra of the peptide bradykinin (Frankevich et al. 2003). Later, Kononikhin et al. were able to generate multiply charged MALDI ions of proteins from electrosprayed sample layers using on-target protein amounts above 200 pmol (Kononikhin et al. 2005), while König et al. have shown the generation of predominantly multiply charged peptide ions at the picomole level using an IR-MALDI source at atmospheric pressure and glycerol as matrix (König et al. 2007). The latter group was also able to couple a commercial ion trap and a modified MALDI source for the analysis of small proteins such as ubiquitin, cytochrome C, and myoglobin. However, as IR-MALDI requires a more complex and expensive setup, and the reported detection limits were relatively high, the proposed method was not widely adopted. Finally, Trimpin et al. have introduced the so-called laserspray ionization technique for the generation of multiply charged ions and were able to perform ETD fragmentation but the reported laser fluence has been substantially higher than in MALDI and leads to rapid sample consumption (Trimpin et al. 2010).

The approach described in this book chapter has so far been applied to the ionization of peptides and proteins. The detection limits of down to 50 fmol were reported with laser energies as low as 1–10 μJ (Cramer et al. 2013). It was shown that one important aspect for this promising methodology was the employment of (glycerol-based) liquid MALDI samples. Previously, the liquid nature of glycerol-based liquid MALDI matrices/samples could be utilized for high-throughput proteomics analysis (Cramer and Corless 2005), for accurate tuning and quantitation in UV-MALDI MS (Palmblad and Cramer 2007) and also for varying the solvent properties, e.g., the pH level for advanced analyte and reaction monitoring (Towers and Cramer 2007).

3 Materials and Protocols

3.1 AP-MALDI Source Design

Here we describe the modification steps we undertook to build an AP-MALDI source capable of producing multiply charged ions. In our setup, a Waters Synapt G2-Si qToF mass spectrometer has been adapted to be equipped with a home-built AP-MALDI ion source but previously similar sources were developed for other instruments such as a

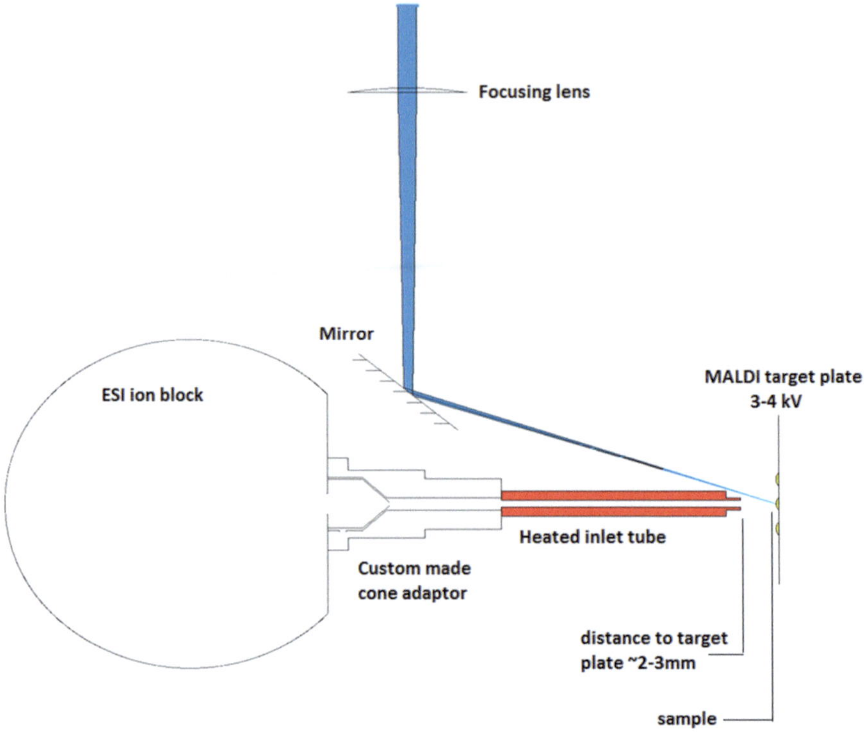

Fig. 1 Schematic view of an ion source for the production of multiply charged MALDI ions using liquid AP-MALDI. The laser beam is guided onto a MALDI target plate, which is mounted in front of a heated ion transfer tube. A high voltage potential is applied between the target plate and ion transfer tube

Q-Star Pulsar i instrument (AB Sciex, Toronto, Canada; see Cramer et al. 2013). Thus, this technology can be easily implemented on a variety of available MS instruments, in particular on instruments that already provide AP ion sources and a heated ion transfer capillary or tube (see below) as found in many of the orbitrap instruments.

3.1.1 Mass Spectrometer Modification

A schematic diagram of the instrument modification is shown in Fig. 1. The standard ESI source enclosure of the instrument is removed, leaving the electrospray ion source block as an ion guide from the AP region to the mass spectrometer. The resistance interlock and the mechanical microswitch interlock are overridden with custom-made interlock defeats.[1] The standard cone adaptor is replaced with a custom-made adaptor,

[1] Note that any instrument modification and development work (including defeating interlocks and setting up lasers) will have Health and Safety implications and can invalidate the warranty and certain service/maintenance contracts for commercial instrumentation. For the former, please consult your local Health and Safety officer and the pertinent local rules and regulations.

which has a 1/8″-Swagelok thread for mounting a custom-made heated ion transfer tube. This heated ion transfer tube (details of the manufacturing are provided below) is easily exchangeable, attached to the MS and helps the charged droplets to desolvate. A low-voltage DC power supply (0–30 V; 0–5 A) is used to power the heating element.

In front of the heated ion transfer tube a MALDI target plate is mounted on an *X–Y* translational stage within a distance of 3 mm from the end of the ion transfer tube. In order to promote the ionization process as well as create additional force which drives the charged droplets inside the mass spectrometer a high voltage potential is applied between the target plate and the ion transfer tube. The target plate is typically insulated from the rest of the setup, e.g., by a home-made PEEK target plate holder.

To accommodate the optical system the top cover of the instrument is removed and an optical breadboard is mounted to the instrument frame. In our current setup, a 337-nm (nitrogen) laser (MNL 103 LD from LTB Lasertechnik GmbH, Berlin, Germany) with a laser pulse duration of 3 ns, 100 μJ pulse energy and a laser pulse repetition rate of 2–30 Hz is used to irradiate the sample.[2] A laser attenuator as well as guiding mirrors and a focusing lens with a focal length of 150 mm are used to moderate the laser power and subsequently steer and focus the laser beam onto a small (~200 μm in diameter) spot on the target plate. The on-target laser focus/spot location should be aligned with the ion transfer tube axis.

3.1.2 Procedure for Manufacturing a Heated Ion Transfer Tube

Materials

- Stainless steel tubing with an inner diameter of ~1.6 mm and an outer diameter of 1/8″
- Tube cutting tool
- Deburring tool
- High-temperature chemical set cement, e.g., OMEGABOND® 600 Powder purchased from Omega Engineering Ltd, Manchester, United Kingdom
- 35SWG 80/20 nichrome wire purchased from RS components, Weldon, United Kingdom
- Reagent grade methanol for cleaning

Protocol

- Cut the stainless steel tubing to a length of 60 mm
- Deburr rough ends
- Submerge in methanol and clean in an ultrasonic bath for 30 min
- Prepare the thermal cement mixture according to the manufacturer's instructions
- Coat the outer wall of the capillary with the thermal cement, thus providing electrical insulation

[2] Note that other UV lasers have been successfully used (e.g., Cramer et al. 2013).

- Leave it to dry for 24 h under ambient conditions
- Evenly wind the nichrome wire around the cement-coated tube making sure that the total resistance will be ~10 Ω
- Coat the wound tube once more with the thermal cement
- Leave it to dry for 24 h under ambient conditions

3.2 Sample Preparation for Liquid AP-MALDI

Materials

- 2,5-Dihydroxybenzoic acid (DHB) in MALDI matrix purity (e.g., cat. no. 85707 from Sigma-Aldrich, Poole, UK)
- α-Cyano-4-hydroxycinnamic acid (CHCA) in MALDI matrix purity (e.g., cat. no. 70990 from Sigma-Aldrich)
- Acetonitrile (ACN; e.g., cat. no. BIO-01204101 from Greyhound, Birkenhead, United Kingdom)
- Glycerol (e.g., cat. no. G9012 from Sigma-Aldrich)
- Trifluoroacetic acid (TFA; e.g., cat. no. 85183 from Sigma-Aldrich)

Glycerol/DHB-Based Liquid Matrix Preparation Protocol

- Dissolve 10 mg of DHB in 100 μL of 70% ACN
- Whirl-mix for 30 s
- Sonicate in an ultrasonic bath for 15 min
- Add 60% glycerol by volume
- Whirl-mix for 30 s
- If needed, help the mixing process with a clean pipette tip, e.g., by breaking phase layers if phase separation has occurred
- Whirl-mix for 30 s
- Sonicate in an ultrasonic bath for 15 min

Glycerol/CHCA-Based Liquid Matrix Preparation Protocol

- Dissolve 10 mg of CHCA in 1 mL of 70% ACN
- Whirl-mix for 30 s
- Sonicate in an ultrasonic bath for 15 min
- Add 60% glycerol by volume
- If needed, help the mixing process with a clean pipette tip, e.g., by breaking phase layers if phase separation has occurred
- Whirl-mix for 30 s
- Sonicate in an ultrasonic bath for 15 min

Analytes are typically dissolved in 0.1% TFA solution.

A volume of 0.5 µL of analyte solution and 0.5 µL of matrix solution are spotted onto a stainless steel MALDI target plate. The spotted sample is then left to mix under ambient conditions for 5 min and will remain liquid during the subsequent MS analysis.[3]

3.3 MS Data Acquisition

3.3.1 Mass Spectrometer Settings

In our lab, default Synapt G2-Si acquisition parameters for the ESI mode are typically used to acquire the data. All data are collected in the sensitivity mode. The *m/z* range window is set to 50–2000 for small peptides and to 50–5000 for protein data acquisition. Each spectrum is normally obtained from the combination of 60 scans with a scan duration set to 1 s.

3.3.2 Ion Source Parameters

The MALDI target plate is typically placed in ~3 mm distance from the heated ion transfer tube entrance. The laser is guided onto the target plate under the smallest practically possible incidence angle, which is ~30° in our setup. The power for heating the nichrome wire, and thus the ion transfer tube, has to be optimized. In our lab, it is set to ~30 W, resulting in a tube wall outer temperature of ~300–350 °C. With the described setup it has been found that increasing the applied high voltage potential between the MALDI target and ion transfer tube leads to an increase in the ion yield. Thus, it should be set to the maximum safe setting, which does not produce arcing (~3.5 kV in our setup). During initial trials using the above described experimental setup, it was observed that at the scan acquisition time of 1 s the laser pulse repetition rate of 10 Hz is sufficient to produce a stable continuous ion yield. Therefore, this setting was used for all data acquisitions. The laser energy per shot varies in the range of 5–30 µJ with a typical value of ~15 µJ for obtaining good spectra of multiply charged peptide ions for the above described setup. One of the particular advantages of the liquid matrices is the self-healing property after laser ablation which provides high signal durability over thousands of laser shots. This feature facilitates the optimization of the ion yield by varying essential parameters such as the distance between the focal lens and the target while monitoring the ion

[3]The sample remains liquid under ambient conditions for weeks using this sample preparation protocol. For the DHB-based sample preparation, it was observed that for relatively high amounts of analyte (e.g., 5 pmol on target) the deposited sample may be used for at least 2 weeks, providing good analyte signal.

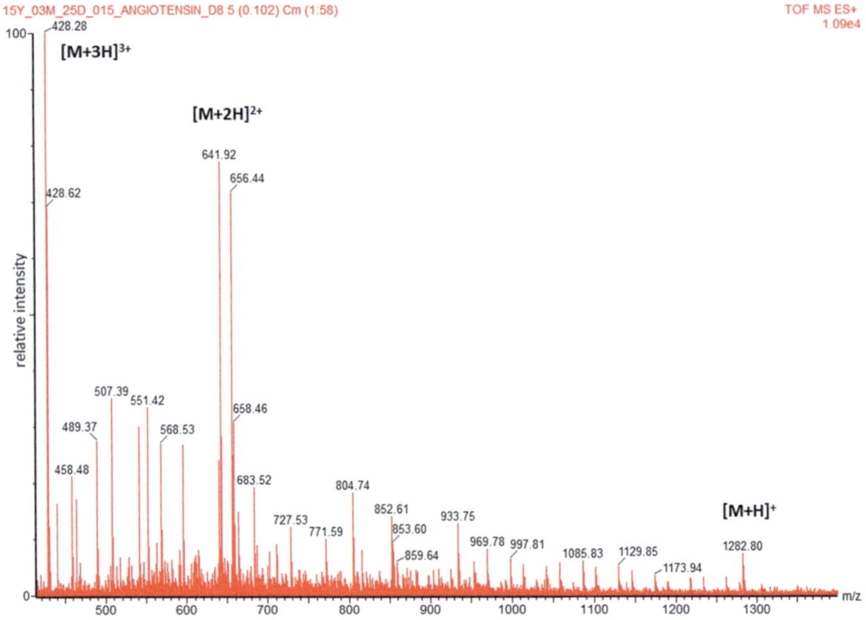

Fig. 2 AP-UV-MALDI spectrum of [Val-5]-Angiotensin I using the glycerol/CHCA-based liquid matrix. A total of 500 fmol of the analyte was deposited on the target. The triply (*m/z* 428.28), doubly (*m/z* 641.92), and singly (*m/z* 1282.80) charged analyte ion species are observed

signal response on the mass spectrometer. For the above parameter and our setup, the optimal laser spot diameter was found to be ~200 μm.[4]

3.3.3 Example Data

Under the above described conditions we were able to acquire spectra of multiply charged [Val-5]-Angiotensin 1 (see Fig. 2). The analyte solution was spotted alongside with the glycerol/CHCA-based matrix solution as described previously. Even for a small peptide the most abundant ion signal and base peak of the spectrum originates from the triply charged analyte ion. The singly charged chemical noise ions can be filtered out by applying IMS filtering, leading to the much improved spectrum shown in Fig. 3.

In Fig. 4 a spectrum of multiply charged ubiquitin ions is shown, using the glycerol/DHB-based liquid matrix preparation protocol. Again, 500 fmol of analyte

[4]For this the MALDI target plate was coated with a saturated solution of CHCA in acetone, which immediately dries, leaving a thin layer of CHCA crystals behind. The thin film was then irradiated by the laser at the maximum energy for a sufficiently long period of time, making sure that the matrix on this spot was completely ablated. The size of the ablation spot was then measured under a microscope.

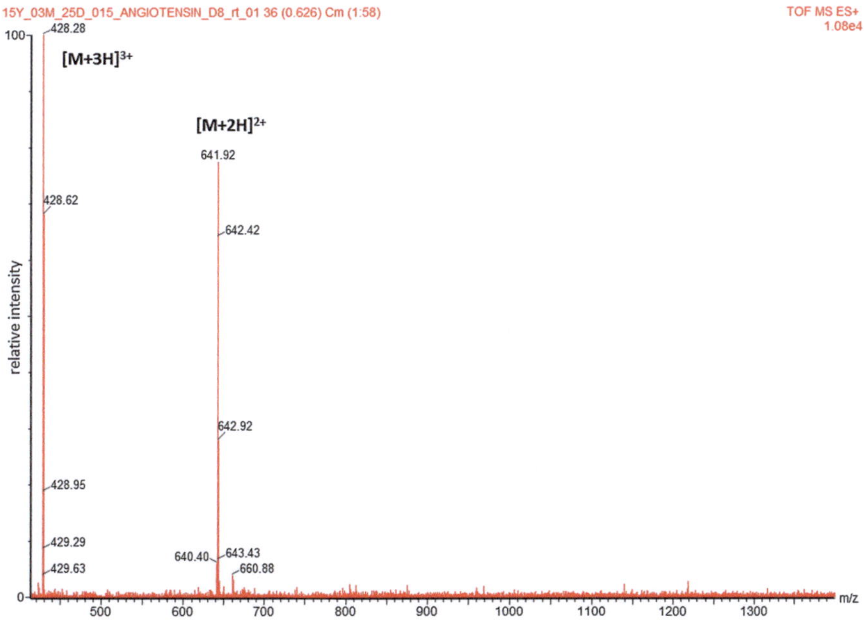

Fig. 3 AP-UV-MALDI spectrum of [Val-5]-Angiotensin I from Fig. 2 with IMS filtering applied, eliminating singly charged ion signals

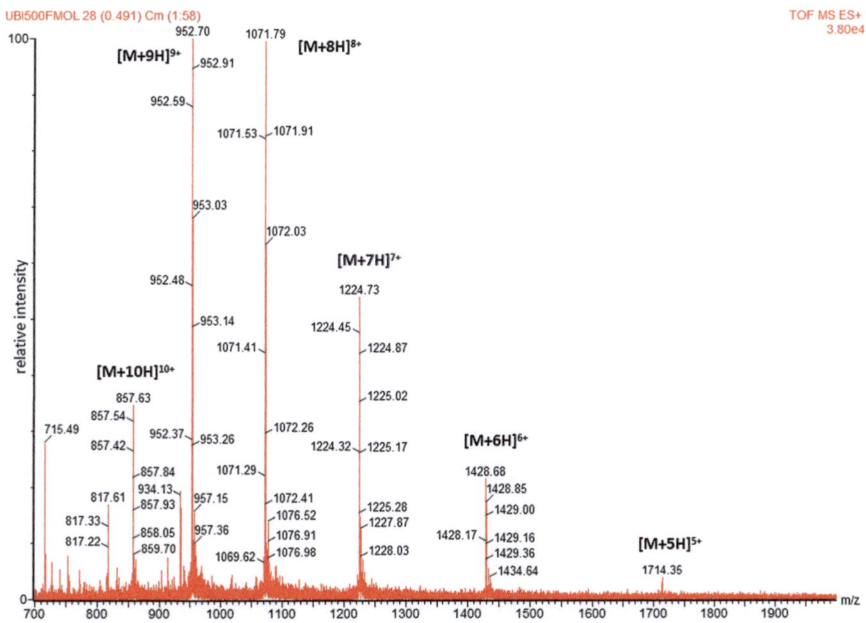

Fig. 4 AP-UV-MALDI spectrum of ubiquitin using the glycerol/DHB-based liquid matrix. A total of 500 fmol of the analyte was deposited on the target. Analyte ion charge states from 5 to 10 can easily be observed

was spotted on the target. This spectrum is similar to the ubiquitin spectra typically generated by ESI on this instrument.

Further information about the data supporting these findings and requests for access to the data can be directed to Professor Rainer Cramer.

Acknowledgement This work is supported by the EPSRC through grant EP/L006227/1.

References

Angel PM, Caprioli RM (2012) Matrix-assisted laser desorption ionization imaging mass spectrometry: in situ molecular mapping. Biochemistry 52(22):3818–3828. doi:10.1021/bi301519p

Bahr U, Deppe A, Karas M et al (1992) Mass spectrometry of synthetic polymers by UV-matrix-assisted laser desorption/ionization. Anal Chem 64(22):2866–2869. doi:10.1021/ac00046a036

Cramer R, Corless S (2001) The nature of collision-induced dissociation processes of doubly protonated peptides: comparative study for the future use of matrix-assisted laser desorption/ionization on a hybrid quadrupole time-of-flight mass spectrometer in proteomics. Rapid Commun Mass Spectrom 15(22):2058–2066. doi:10.1002/rcm.485

Cramer R, Corless S (2005) Liquid ultraviolet matrix-assisted laser desorption/ionization—mass spectrometry for automated proteomic analysis. Proteomics 5(2):360–370. doi:10.1002/pmic.200400956

Cramer R, Pirkl A, Hillenkamp F, Dreisewerd K (2013) Liquid AP-UV-MALDI enables stable ion yields of multiply charged peptide and protein ions for sensitive analysis by mass spectrometry. Angew Chem Int Ed Engl 52(8):2364–2367. doi:10.1002/anie.201208628

Fenn JB, Mann M, Meng CK et al (1989) Electrospray ionization for mass spectrometry of large biomolecules. Science 246(4926):64–71

Fournier P-E, Drancourt M, Colson P et al (2013) Modern clinical microbiology: new challenges and solutions. Nat Rev Microbiol 11(8):574–585. doi:10.1038/nrmicro3068

Frankevich V, Zhang J, Dashtiev M, Zenobi R (2003) Production and fragmentation of multiply charged ions in 'electron-free' matrix-assisted laser desorption/ionization. Rapid Commun Mass Spectrom 17(20):2343–2348. doi:10.1002/rcm.1194

Good DM, Wirtala M, McAlister GC, Coon JJ (2007) Performance characteristics of electron transfer dissociation mass spectrometry. Mol Cell Proteomics 6(11):1942–1951. doi:10.1074/mcp.M700073-MCP200

Huang Y, Triscari JM, Tseng GC et al (2005) Statistical characterization of the charge state and residue dependence of low-energy CID peptide dissociation patterns. Anal Chem 77(18):5800–5813. doi:10.1021/ac0480949

Karas M, Hillenkamp F (1988) Laser desorption ionization of proteins with molecular masses exceeding 10,000 daltons. Anal Chem 60(20):2299–2301. doi:10.1021/ac00171a028

Karas M, Bachmann D, Bahr U, Hillenkamp F (1987) Matrix-assisted ultraviolet laser desorption of non-volatile compounds. Int J Mass Spectrom Ion Process 78:53–68. doi:10.1016/0168-1176(87)87041-6

Köfeler HC, Fauland A, Rechberger GN, Trötzmüller M (2012) Mass spectrometry based lipidomics: an overview of technological platforms. Metabolites 2(4):19–38. doi:10.3390/metabo2010019

König S, Kollas O, Dreisewerd K (2007) Generation of highly charged peptide and protein ions by atmospheric pressure matrix-assisted infrared laser desorption/ionization ion trap mass spectrometry. Anal Chem 79(14):5484–5488. doi:10.1021/ac070628t

Kononikhin A, Nikolaev E, Frankevich V, Zenobi R (2005) Letter: multiply charged ions in matrix-assisted laser desorption/ionization generated from electrosprayed sample layers. Eur J Mass Spectrom 11(1):257–259. doi:10.1255/ejms.729

Makarov A (2000) Electrostatic axially harmonic orbital trapping: a high-performance technique of mass analysis. Anal Chem 72(6):1156–1162. doi:10.1021/ac991131p

Marshall AG, Hendrickson CL, Jackson GS (1998) Fourier transform ion cyclotron resonance mass spectrometry: a primer. Mass Spectrom Rev 17(1):1–35. doi:10.1002/(SICI)1098-2787(1998)17:1<1::AID-MAS1>3.0.CO;2-K

Palmblad M, Cramer R (2007) Liquid matrix deposition on conductive hydrophobic surfaces for tuning and quantitation in UV-MALDI mass spectrometry. J Am Soc Mass Spectrom 18(4):693–697. doi:10.1016/j.jasms.2006.11.013

Tanaka K, Waki H, Ido Y et al (1988) Protein and polymer analyses up to m/z 100,000 by laser ionization time-of-flight mass spectrometry. Rapid Commun Mass Spectrom 2(8):151–153. doi:10.1002/rcm.1290020802

Towers M, Cramer R (2007) Liquid matrices for analyses by UV-MALDI mass spectrometry. Spectroscopy 22(11):29–37

Trimpin S, Inutan ED, Herath TN, McEwen CN (2010) Laserspray ionization, a new atmospheric pressure MALDI method for producing highly charged gas-phase ions of peptides and proteins directly from solid solutions. Mol Cell Proteomics 9(2):362–367. doi:10.1074/mcp.M900527-MCP200

Xian F, Hendrickson CL, Marshall AG (2012) High resolution mass spectrometry. Anal Chem 84(2):708–719. doi:10.1021/ac203191t

Part II
Applications of Liquid MALDI Mass Spectrometry

Ionic Liquids and Other Liquid Matrices for Sensitive MALDI MS Analysis

Mark W. Towers and Rainer Cramer

Abstract Although liquid matrix-assisted laser desorption/ionization (MALDI) has been used in mass spectrometry (MS) since the early introduction of MALDI, its substantial lack of sensitivity compared to solid (crystalline) MALDI was for a long time a major hurdle to its analytical competitiveness. In the last decade, this situation has changed with the development of new sensitive liquid matrices, which are often based on a binary matrix acid/base system. Some of these matrices were inspired by the recent progress in ionic liquid research, while others were developed from revisiting previous liquid MALDI work as well as from a combination of these two approaches. As a result, two high-performing liquid matrix classes have been developed, the ionic liquid matrices (ILMs) and the liquid support matrices (LSMs), now allowing MS measurements at a sensitivity level that is very close to the level of solid MALDI and in some cases even surpasses it. This chapter provides some basic information on a selection of highly successful representatives of these new liquid matrices and describes in detail how they are made and applied in MALDI MS analysis.

1 Introduction

The key to high-quality mass spectrometry (MS) data from a matrix-assisted laser desorption/ionization (MALDI) experiment at high sensitivity is sample preparation and the most important factor of the preparation is the matrix and how it interacts with the analyte. As might be expected the matrix chosen will have a considerable effect on the ionization efficiencies of the various analytes (cf. Chapter 'Employing 'Second Generation' Matrices'), pattern of interference from the matrix ions and potential for fragmentation of the analyte during the ionization process. In

M.W. Towers (✉)
Waters Corporation, Altrincham Road, Wilmslow SK9 4AX, UK
e-mail: Mark_Towers@waters.com

R. Cramer
Department of Chemistry, University of Reading, Whiteknights, Reading RG6 6AD, UK
e-mail: r.k.cramer@reading.ac.uk

R. Cramer (ed.), *Advances in MALDI and Laser-Induced Soft Ionization Mass Spectrometry*, DOI 10.1007/978-3-319-04819-2_3

general the only steadfast rule for a matrix is that it needs to absorb the wavelength of the laser light used and promote the formation of analyte ions. However, a number of other qualities are also highly desirable. For instance, if MALDI is to be performed under vacuum the matrix needs to be vacuum stable, the matrix should also not induce chemical reactions that lead to an altered analyte, and in order to simplify the data acquisition and facilitate quantitative measurements (cf. Chapter 'Quantitative MALDI MS Using Ionic Liquid Matrices') the sample spot should ideally be formed of a complete homogenous layer of matrix and analyte. In reality, most matrices do not possess all of these qualities and so compromises must be made based on the application.

The most commonly used matrices are derivatives of weak organic acids, for example α-cyano-4-hydroxycinnamic acid (CHCA) (Beavis et al. 1992) (cf. Chapter 'Employing 'Second Generation' Matrices') and 2,5-dihydroxybenzoic acid (DHB) (Strupat et al. 1991; Karas et al. 1993). These matrices are generally employed using a 'dried-droplet' method where the matrix is dissolved in a suitable solvent and placed on a target with the analyte solution to crystallize. By adjusting the solvents and the drying conditions the spot morphology can be fine-tuned for the desirable outcome. Two notable examples of this are the matrix layer technique for DHB (Garaguso and Borlak 2008) and the thin-layer affinity technique for CHCA (Gobom et al. 2001). However, in practice these more elaborate methods can be time-consuming and difficult for the inexperienced to replicate reproducibly.

Other interesting sample preparation methods can be found in liquid MALDI. The principle behind liquid MALDI is that rather than the matrix and analyte form a co-crystallized layer, the matrix material remains in solution co-dissolved with the analyte. Liquid matrices have been utilized in MALDI for almost as long as MALDI has existed as a technique. In 1988, Tanaka et al. used a suspension of tungsten powder in glycerol for their Nobel-prize winning work ionizing proteins and polymers with masses of up to 100 kDa (Tanaka et al. 1988). Liquid matrices, and thus liquid MALDI, however lost favor due to their substantially lower analytical sensitivity compared to matrices leading to crystalline (solid) MALDI samples. Recently, new liquid matrix systems have been developed which promise to re-address this sensitivity gap.

There are typically two forms of liquid matrices that are applied in liquid MALDI MS, the ionic liquid matrix (ILM) and the liquid support matrix (LSM). For the latter, the matrix material (the component that mainly promotes the desorption/ionization process) is typically dissolved or suspended in a vacuum-stable liquid. For instance, binary matrix systems are frequently dissolved in glycerol but even metal powder chromophores have been used as matrix suspended in glycerol.[1] The more recent ILMs have branched off from research on ionic liquids, which are commonly defined as salts that melt below 100 °C, are non-explosive/-flammable, stable below 250 °C, and possess negligible vapor pressure (Wasserscheid and Keim 2000). Another key reason for research into ionic liquids is the countless combinations

[1] Although metal powder (metal nanoparticles) can efficiently absorb the MALDI laser energy and thus provide the necessary energy for desorption, other liquid MALDI sample components are most likely more important for the ionization process (proton transfer reactions).

possible. Alteration of the cation/anion content gives ionic liquids a high degree of tuneability which makes them highly interesting with regard to the creation of novel tailor-made solvent systems. These properties therefore make ionic liquids also highly attractive as potential liquid matrices for MALDI. However, early studies showed that the classical ionic liquids did not perform well as MALDI matrices (Armstrong et al. 2001), although it was thought that the ionic nature should efficiently support analyte ionization. There could be many reasons for this such as insufficient laser energy absorption, unfavorable analyte incorporation and/or sample morphology, and last but not least factors that lead to reduced (gas phase) charge/proton transfer reactions. This research did however result in a new class of ionic liquids, which are suitable as MALDI matrices and were termed ILMs. These new matrices were formed from combinations of classical crystalline matrices such as DHB and CHCA co-dissolved in an organic solvent along with an organic base such as 3-aminoquinoline (3-AQ); subsequent removal of the solvent resulted in the formation of the ILM (Armstrong et al. 2001; Zabet-Moghaddam et al. 2004a, b).

Previous work on a similar system had been carried out by Sze et al. (1998). In this case the system was an LSM using CHCA and 3-AQ with glycerol as the liquid support. The intriguing point observed here was that the addition of the organic counter base 3-AQ allowed a tenfold or more increase in the solubility of the solid CHCA matrix.[2] One of the more successful combinations found in this study was that of CHCA and 3-AQ in glycerol in a ratio of 1:4:6 by weight. The ratio of CHCA to 3-AQ was found to be a key factor in the ionization performance and liquid characteristic of the matrix. This matrix formulation has subsequently been optimized and explored (Cramer and Corless 2005; Palmblad and Cramer 2007; Towers et al. 2010) and will be discussed in more detail later.

Liquid matrices present a number of potential advantages over the current solid (crystalline) matrices and great strides have been made to bring the overall sensitivity of these matrices to a point where they can under certain conditions be seen to equal or even surpass their crystalline counterparts (Towers et al. 2010). One of their distinct properties is the simple morphology with a high degree of homogeneity (Armstrong et al. 2001; Zabet-Moghaddam et al. 2004b; Cramer and Corless 2005; Palmblad and Cramer 2007; Towers et al. 2010; Mank et al. 2004). This gives them a significant advantage in terms of the potential for high-throughput automated analysis when compared to the crystalline MALDI samples as there is no need to search for 'sweet spots' or to acquire data from multiple positions within a sample to achieve reproducible and comparable ion abundances. The liquid nature also means that the sample surface is inherently self-healing and renewing, leading to long-lasting ion signal yields from a single desorption spot (Palmblad and Cramer 2007; Towers et al. 2010). In contrast to crystalline MALDI samples where individual positions are quickly consumed and positional variations are observed, liquid MALDI samples provide highly improved shot-to-shot reproducibility and a significant increase in sample longevity. These properties lend themselves well to

[2] Note that CHCA and 3-AQ will also form an ILM. Thus, much of the improvement in solubility might come from the formation of an ILM.

quantitative analysis, which has been demonstrated across a broad range of abundances (Mank et al. 2004; Santos et al. 2004; Li and Gross 2004; Bungert et al. 2004) even without the use of an internal standard (Palmblad and Cramer 2007; Towers et al. 2010). In addition, liquid matrices have been reported to be tolerant to contamination and commonly used buffers such as ammonium bicarbonate at levels which would impair the crystallization of solid MALDI samples (Towers et al. 2010), allowing the analysis of biological samples of limited purity which would normally require extensive clean-up (Cramer and Corless 2005).

2 Applications

It has been shown that liquid matrices can improve the analysis of a variety of molecules including peptides (Cramer and Corless 2005; Towers et al. 2010; Zabet-Moghaddam et al. 2006), proteins (Mank et al. 2004), phosphopeptides (Tholey 2006), glycopeptides (Fukuyama et al. 2008), oligosaccharides (Fukuyama et al. 2008; Laremore et al. 2006), phospholipids (Li et al. 2005) and polymers (Armstrong et al. 2001). In general, liquid matrices have been shown to be useful for many applications, including but not limited to quantitation (cf. Chapter 'Quantitative MALDI MS Using Ionic Liquid Matrices'), the analysis of labile biomolecules, and enzyme reaction monitoring (Mank et al. 2004; Towers and Cramer 2007).

Quantitation has historically been very hard to perform by MALDI MS analysis. With solid MALDI samples quantitation based on MS ion signals without comparison to isotope-labeled reference standards[3] requires very careful time-consuming sample preparation with a high number of replicates to overcome the issues of poor shot-to-shot ion signal reproducibility, mainly due to the variance in sample morphology and matrix/analyte homogeneity. The use of an internal standard of known concentration is often needed but if this standard differs too much from the analyte of interest, as it is often the case for non-isotope-labeled reference standards, the distribution and ionization efficiencies between analyte and reference can still vary widely and so their ion signals. Thus, in general MALDI is widely considered to be qualitative but not quantitative. In 2004, however, Li and Gross first demonstrated that ILMs can be utilized for quantitative measurements using a nonlabeled internal standard. Values of R^2>0.992 were achieved for peptides and >0.994 for oligodeoxynucleotides and >0.998 for proteins in a range from 1 to 50 pmol on target with a variety of ILMs (Li and Gross 2004). In 2006, Tholey et al. demonstrated the quantitation of peptides in the low-pmol range (up to 7 pmol) using an ILM without the use of an internal standard, resulting in an R^2 value of 0.995. However, the linear

[3] One approach to avoid the detrimental effects of analyte ion suppression/ionization competition in quantitative MS measurements is based on signal comparison to a co-prepared and co-analyzed isotope-labeled reference, circumventing differences in the desorption/ionization process by assuming that analyte and reference undergo the same processes as their physicochemical properties are virtually the same (apart from a slight shift in mass).

range of quantitation was limited to one order of magnitude and multiple spots were required for each data point (Tholey et al. 2006). In 2007, Palmblad and Cramer demonstrated a significant improvement in the dynamic range achievable for peptides without an internal standard using a glycerol-based LSM incorporating CHCA and 3-AQ. By including an additional matrix dilution step and employing a disposable hydrophobic Teflon™ tape as target surface, very sensitive measurements of low- to high-fmol amounts (up to 1 pmol) of Bradykinin and Angiotensin I were achieved with R^2 values of >0.998 (Palmblad and Cramer 2007). Palmblad and Cramer also demonstrated that the highly stable ion current produced by the glycerol-based LSM can be utilized for tuning a Q-Tof instrument, which is typically undertaken in electrospray mode, thus eliminating the usual double switch of sources for instruments with an addtional MALDI source. Towers et al. in 2010 showed similar results for quantitative measurements substituting the CHCA in the LSM with 4-chloro-α-cyanocinnamic acid (ClCCA). Here, seven peptides could be quantified in a single mixture over a range of three orders of magnitude with an average R^2 value of >0.98 (Towers et al. 2010). More on quantitative MALDI MS using ILMs can be found in Chapter 'Quantitative MALDI MS Using Ionic Liquid Matrices.'

Liquid matrices have shown particular promise in the area of labile (bio)molecules. Labile molecules pose a particular challenge for MS analysis due to their tendency to fragment during the ionization process leading to a loss of critical information. This is particularly true when analyzing oligosaccharides and glycopeptides as the sulfate groups, sialic acid groups or in the case of glycopeptides the whole glycan chain can easily become disassociated from the analyte with the result that the intact molecular ions are rarely detected. Multiple papers have been published on the use of ILMs for the analysis of oligosaccharides (Fukuyama et al. 2008; Laremore et al. 2006; Kolli and Orlando 1996, 1997; Laremore et al. 2007). One paper of particular note is that of Fukuyama et al. (2008), which demonstrated the use of ILMs formed from *p*-coumaric acid with 1,1,3,3-tetramethylguanidium and CHCA with 1,1,3,3-tetramethylguanidium for preferential detection of glycopeptide ions from a digest of ribonuclease B while suppressing the disassociation of sialic acids.

Phosphopeptides are another class of molecules which whilst biologically very important can be difficult to analyze by MALDI MS. The presence of nonphosphorylated peptides has been shown to suppress the ionization of phosphorylated peptides. In addition, due to the labile nature of the phosphopeptides metastable fragmentation can often be seen to occur, demanding 'cooler/softer' matrices such as DHB for their improved detection (Glückmann and Karas 1999). Best results have been achieved using DHB as a crystalline matrix with the addition of phosphoric acid (Kjellström and Jensen 2004). However, DHB-based solid MALDI samples tend to be highly inhomogeneous with a complicated morphology and topology. As discussed liquid matrices can result in highly homogeneous sample preparations, and in 2006, Tholey showed how ILMs formed with DHB and either pyridine or *n*-butylamine (combined with phosphoric acid in positive mode only) could be used to analyze phosphopeptides with an increased sensitivity and improved homogeneity for positive and negative ion mode measurements (Tholey 2006).

Liquid matrices have also shown promise in the area of reaction monitoring. They have been used to successfully enable quantitative measurements of sugar converting enzymes (Bungert et al. 2004). In 2004 Mank et al. (2004) have shown for the first time that ILMs can be used for the direct screening of enzymatic reactions, reporting the desialylation of sialyllactose with sialidase from *Clostridium perfringens* in the presence of diluted aqueous DHB/butylamine ILM. In 2007, Towers and Cramer demonstrated the tuneability of their LSM system for the analysis of enzymatic reactions occurring in the matrix solution by pH-tuning of a dilute glycerol-based LSM typically used for MALDI sample spotting, allowing for proteolytic trypsin digestion of cytochrome C and myoglobin.

Finally, peptide mass fingerprinting is an area where liquid matrices have also been shown to be quite useful. Liquid matrices have been shown to allow for increased sequence coverage (Calvano et al. 2009), higher tolerance to contaminants (Cramer and Corless 2005; Towers et al. 2010), and analytical sensitivities equal to or better than their crystalline counterparts (Towers et al. 2010).

3 Materials and Protocols

The following protocols are suitable for the creation of basic LSMs and ILMs using CHCA or the newly introduced solid matrix ClCCA (Jaskolla et al. 2008) (cf. Chapter 'Employing 'Second Generation' Matrices') in combination with 3-AQ which are quick and easy to prepare. The protocols are based on those published by Sze et al. (1998) and further modified by Cramer and Corless (2005) and Towers et al. (2010). The glycerol content in the described LSMs increases the overall homogeneity of the MALDI sample and improves its ability to tolerate high amounts of buffer compared to the crystalline matrices and ILMs, and to allow for pH-tuning by the addtion of TFA or ammonium bicarbonate. The ILMs, however, although less homogeneous, show greater overall sensitivity.

For a glycerol-based LSM of CHCA and 3-AQ combine CHCA, 3-AQ, a solution of 10 mM ammonium phosphate (AP) 50% methanol (MeOH) and glycerol in an MS-compatible microcentrifuge tube[4] in a ratio of 1:3:5:5 where 1 mg of solid is equal to 1 µL of solvent. The order in which the components are added can have an influence on the speed, in which the matrix components dissolve. It is best practice to add the solid components to the tube first and then the AP/50% MeOH solution. It can be advantageous, although not strictly necessary, to vortex the sample quickly at this point prior to the addition of the glycerol. For the addition of the glycerol it is the easiest to add this component by weight rather than trying to directly pipette

[4] Note that some tubes might be unsuitable for MS analysis as solvents such as methanol could lead to the dissolution of tube material. Thus, tubes should be checked regularly whether they are MS-compatible and ideally be washed before usage with the solvents that will be applied.

a fixed volume as this will most likely give results that are more reproducible. The mixture should then be vortexed until the matrix components appear to have dissolved. After vortexing, the solution should be sonicated for 15 min to ensure complete dissolution.[5] It is advisable to ensure that the temperature of the water in the sonicating water bath is at least at room temperature to avoid any greater precipitation due to a reduction in temperature if the solution is not completely dissolved. If the matrix components do precipitate, raising the water bath temperature to ~30 °C and sonicating the matrix solution for 30 min should be sufficient to dissolve the components. Once the components are fully dissolved they should remain so and this stock solution can be kept for several days if stored at 4 °C.

To generate the working matrix solution a small quantity ~10 μL of the stock solution should be diluted by a factor of 30 with the AP/MeOH solution. The reason for this dilution step is that it allows the matrix and analyte to concentrate to a much smaller volume when deposited on the target plate which will usually result in an increase in sensitivity as demonstrated by Cramer and Corless (2005). Whilst it is possible to dilute the matrix by a higher factor to further reduce the size of the matrix droplet and concentrate the analyte, additional gains in sensitivity are generally not seen and increases in the concentration of other undesirable components such as salts can lead to negative effects, including increased adduct ion formation and partial crystallization of the MALDI sample.

The AP in the liquid matrix system has an important role as it suppresses the formation of metal cation matrix clusters and the appearance of metal cation adduct formation with the analyte. This is a particular issue with liquid MALDI samples and an unfortunate by-product of the increased homogeneity, and in contrast to the crystalline MALDI samples where a fair portion of the metal cation content will not be incorporated into the analyte/matrix crystals due to partitioning effects. However, the liquid matrices are able to tolerate much higher amounts of some contamination and components such as buffer at levels which often prevent the crystallisation of the matrix in a 'dried–droplet' technique for solid MALDI.

For the use of AP as an additive it is important to note that at higher final concentrations the AP can also have a suppressive effect on the analyte ionization efficiency, resulting in an overall reduced analyte ion signal. This means that if the dilution factor is increased to try and improve the sensitivity, the concentration of AP in the dilution solution needs to be decreased so that the final concentration in the MALDI sample does not increase. Unfortunately, the effect of AP on metal cation suppression is linked to the ratio of AP to metal salt, which limits how much further the above system can be diluted and a gain in sensitivity can be observed.

To prepare the CHCA/3-AQ ILM the method is similar to the corresponding LSM with the obvious omission of the glycerol component, an adjustment of the AP concentration and the optional inclusion of octyl β-D-glucopyranoside (OGP) to aid

---

[5] In many LSM preparations it is common that two liquid phases occur. Thus, it is recommended to carefully check for the occurrence of two separate liquid phases and if they are formed to successfully disrupt their interface, e.g., with a clean pipette tip, and to thoroughly mix these, leaving just one homogenous phase.

with droplet formation as described in Towers et al. (2010). Briefly, CHCA, 3-AQ, and MeOH are combined in a ratio of 1:3:5 where 1 mg of solid is equal to 1 μL of solvent. This mixture is vortexed until the matrix components are dissolved and then sonicated for 15 min. A quantity of this stock solution can be diluted by a factor of 30 with aqueous 5 mM AP/50% MeOH/0.1% OGP to generate a working matrix solution.

The protocols for generating an ILM and LSM with ClCCA are quite similar to those for CHCA. However, as key difference there is a change in the ratio of ClCCA to 3-AQ compared to the ratio of CHCA to 3-AQ. For the ClCCA matrix the ratio is 1:5 instead of 1:3. In both cases, the respective value represents the first point, at which the matrix will dissolve and stay dissolved when under vacuum. Increasing the amount of 3-AQ will in both cases enhance the dissolution process. However, the increase of 3-AQ and its volume in relation to the final total sample volume might affect the overall desorption/ionization process as the concentration of the principle chromophore and ionization promoter decreases.

The glycerol-based ClCCA LSM matrix can be prepared in a similar fashion to the CHCA LSM, by combining ClCCA, 3-AQ, a solution of 10 mM AP 50% MeOH and glycerol in an MS-compatible tube in a ratio of 1:3:10:5 where 1 mg of solid is equal to 1 μL of solvent. To generate a working matrix solution an aliquot (e.g., 10 μL) of this stock solution should be diluted by a factor of 20 with the 10 mM AP/50% MeOH solution (Towers et al. 2010).

The ClCCA ILM can be prepared by dissolving ClCCA and 3-AQ in methanol in a ratio of 1:5:10 where 1 mg of solid is equal to 1 μL of solvent. The working matrix solution is then generated by diluting an aliquot of the stock solution by a factor of 20 with a 5 mM AP/50% MeOH/0.1% OGP solution.

Both ILMs and LSMs should be mixed 1:1 with the analyte solution prior to spotting on the target plate. With the ILM it is important to maintain this ratio when using aqueous analyte solutions as if the analyte solution is present in a higher ratio the MALDI sample droplets might not form correctly on the target.[6]

Depending on the composition of the liquid matrix the target plate surface has a considerable influence on the sample spot development. In general, better results will be achieved on smooth surfaces such as a polished stainless steel target plate as the shrinking droplet may adhere to any ridges or defects on a rougher target plate causing the droplet to split or collapse into a thin film, which is usually less ideal for long-term repeat measurements. For the glycerol-based LSM, Palmblad and Cramer (2007) showed how a conductive hydrophobic tape can be used as a sample plate surface to aid in droplet formation which allowed for an increase in analytical performance due to the superior droplet formation and potential decrease in analyte losses to the target surface. However, when using the hydrophobic tape or indeed a polished stainless steel target issues can be observed with regard to the precision of

[6] During the solvent evaporation phase the MeOH component evaporates first leaving a ring of ionic liquid with an aqueous core. As the aqueous component evaporates, the ionic liquid ring contracts presumably (re-)absorbing the analyte. If the aqueous component is too high the constraining ring does not form and the analyte is potentially lost to the target surface.

the final position of the droplet which can impede automated analysis due to the small size of the droplet. Structured target plates with a hydrophobic surface and hydrophillic anchors can also be highly useful as the anchors help to locate the droplet to a precise and repeatable position. However, the size of the anchor will determine the droplet width so it is necessary to choose an appropriate size for the level of dilution. Also, the incorporation of a small amount of nonionic detergent such as OGP into the sample droplet has been shown to aid in droplet formation (Towers et al. 2010). The amount of detergent, however, is a critical factor. The detergent aids in the droplet formation by lowering the adhesion to the surface of the target plate helping to maintain a compact droplet through the solvent evaporation process. However, if too much detergent is present, it will disrupt the surface tension of the droplet to a level that will cause the droplet to break apart.

Figure 1 shows the final liquid MALDI samples for the four sample preparations described above after the volatile solvents are evaporated. In Fig. 2 a comparison

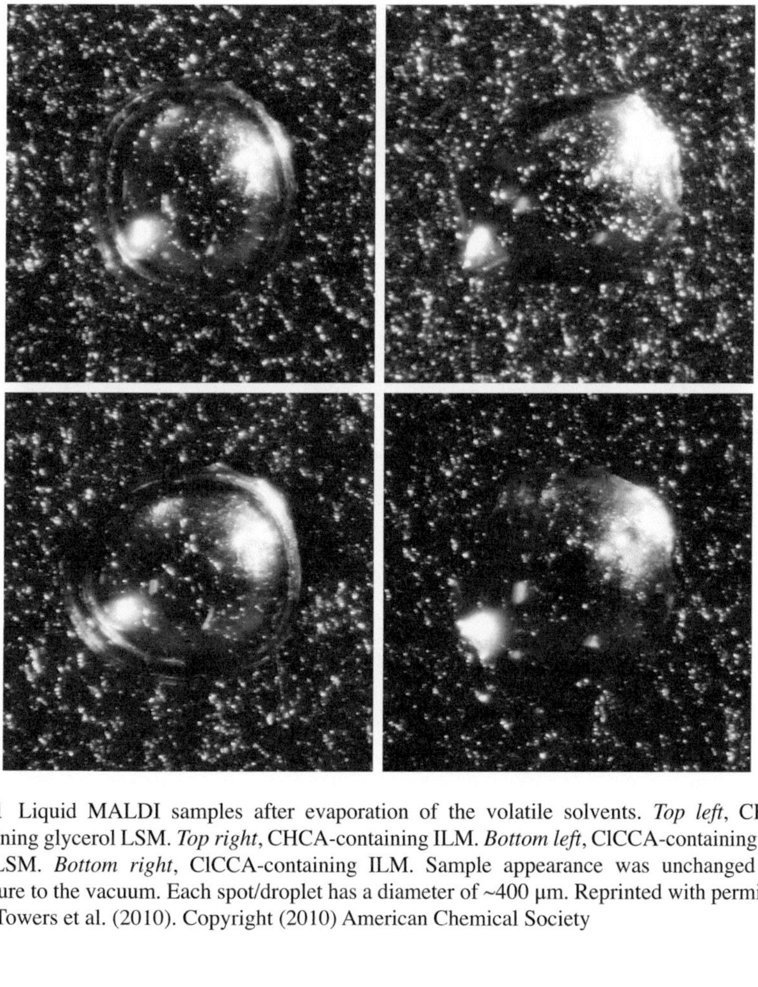

Fig. 1 Liquid MALDI samples after evaporation of the volatile solvents. *Top left*, CHCA-containing glycerol LSM. *Top right*, CHCA-containing ILM. *Bottom left*, ClCCA-containing glycerol LSM. *Bottom right*, ClCCA-containing ILM. Sample appearance was unchanged after exposure to the vacuum. Each spot/droplet has a diameter of ~400 μm. Reprinted with permission from Towers et al. (2010). Copyright (2010) American Chemical Society

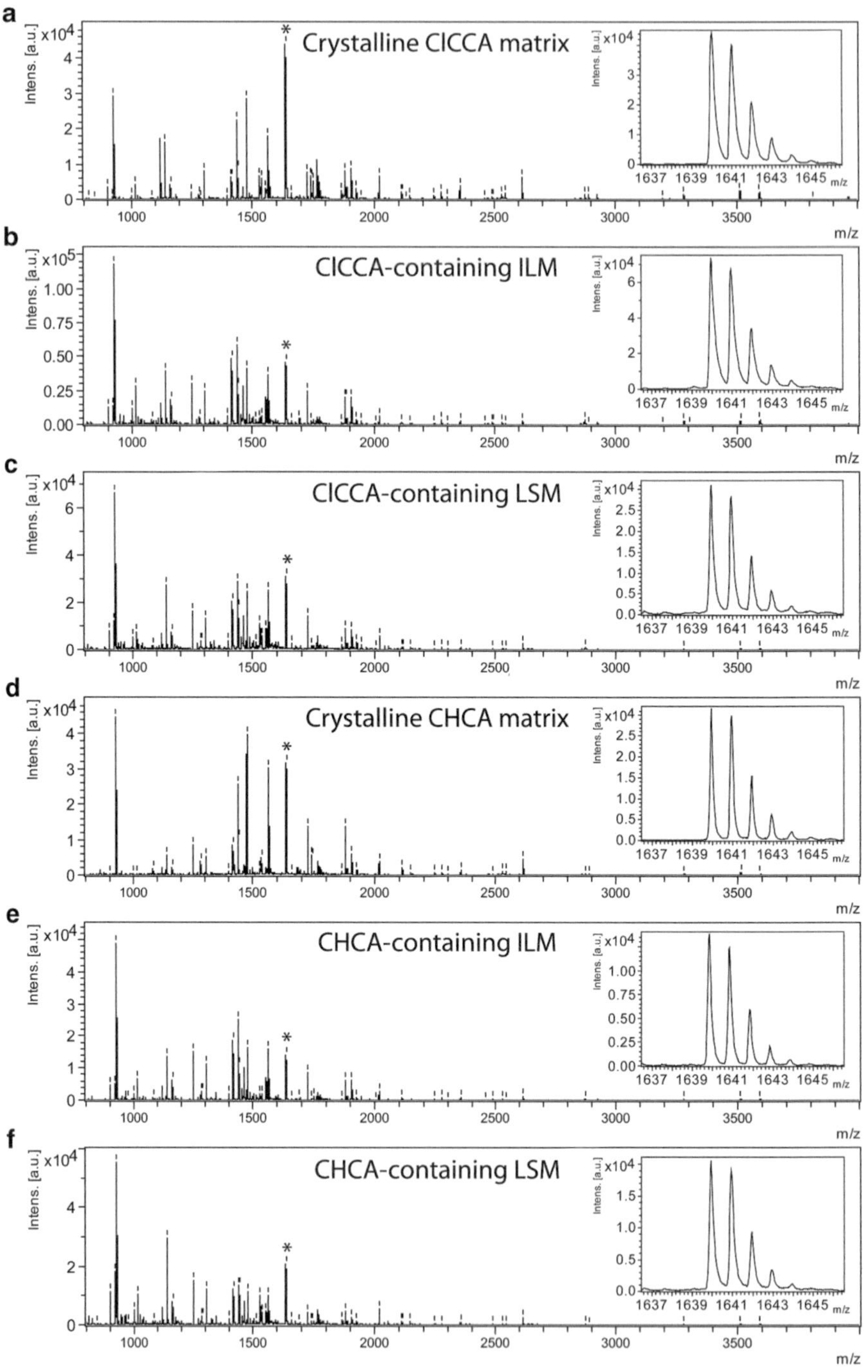

Fig. 2 Example mass spectra of an in-solution BSA digest with 100 fmol spotted on target for (**a**) crystalline ClCCA matrix, (**b**) ClCCA-containing ILM, (**c**) ClCCA-containing LSM, (**d**) crystalline CHCA matrix, (**e**) CHCA-containing ILM and (**f**) CHCA-containing LSM. A total of 500 single-shot spectra were acquired per spectrum. *Tick-marks* denote peaks matching theoretical BSA tryptic digest masses. *Asterisk* denotes expanded peak. Reprinted with permission from Towers et al. (2010). Copyright (2010) American Chemical Society

Table 1 Results of Mascot searches for the analysis of 100, 10, and 1 fmol on-target amounts of BSA in-solution digest (averaged over five replicates for each matrix, four replicates for 1 fmol of analyte)

Matrix	# of picked peaks	Sequence coverage (%)	# of matched peaks ± SD	Mascot score (sig. ≥70) ± SD
100 fmol				
ClCCA	142	79.4	58 ± 4	317.4 ± 32.8
ClCCA ILM	150	82.4	53 ± 1	269.4 ± 9.2
ClCCA LSM	150	75.4	49 ± 2	237.8 ± 14.8
CHCA	112	70.4	44 ± 5	241.2 ± 25.9
CHCA ILM	133	75.0	48 ± 5	256.4 ± 27.2
CHCA LSM	150	75.2	51 ± 2	248.0 ± 17.0
10 fmol				
ClCCA	105	74.8	49 ± 2	305.0 ± 15.9
ClCCA ILM	112	72.8	48 ± 3	288.6 ± 13.1
ClCCA LSM	87	54.8	35 ± 3	210.6 ± 26.3
CHCA	79	57.2	32 ± 3	181.0 ± 27.6
CHCA ILM	77	60.2	37 ± 5	251.6 ± 32.7
CHCA LSM	84	53.8	33 ± 2	182.8 ± 6.1
1 fmol				
ClCCA	70	41.3	24 ± 1	129.5 ± 21.0
ClCCA ILM	54	46.0	28 ± 2	199.0 ± 21.2
ClCCA LSM	28	26.3	13 ± 2	93.0 ± 21.2
CHCA	75	23.3	12 ± 3	32.5 ± 7.5
CHCA ILM	20	29.5	15 ± 5	148.3 ± 23.6
CHCA LSM	27	25.3	15 ± 1	115.0 ± 11.0

Peaks were picked between *m/z* 800–4000 with a signal-to-noise ratio ≥4 (≥2 for 1 fmol). Reprinted with permission from Towers et al. (2010). Copyright (2010) American Chemical Society

between liquid MALDI and solid MALDI mass spectra is shown using 100 fmol of a digest of bovine serum albumin (BSA) as analyte, while Table 1 summarizes the achievable performances in digest analyses using the example of a BSA digest and its dilutions for both liquid and solid MALDI, searching a protein database with the search engine Mascot (Matrix Science Ltd, London, UK).

The following concise step-by-step protocols summarize the essential steps of the above described liquid matrix preparations:

CHCA LSM

1. Add 10 mg of CHCA and 30 mg of 3-AQ to an MS-compatible microcentrifuge tube. Then add 50 μL of 10 mM AP/50% MeOH and vortex briefly.
2. By weight add 50 μL of glycerol (approximately 63 mg at RT [d=1.261 g/cm^3]) and vortex.
3. Sonicate for 15 min or until fully dissolved.
4. Take 10 μL of the stock solution and add 290 μL of 10 mM AP/50% MeOH.

CHCA ILM

1. Add 10 mg of CHCA and 30 mg of 3-AQ to an MS-compatible microcentrifuge tube. Then add 50 μL of MeOH and vortex.
2. Sonicate for 15 min or until fully dissolved.
3. Take 10 μL of the stock solution and add 290 μL of 5 mM AP/50% MeOH/0.1% OGP.

ClCCA LSM

1. Add 10 mg of ClCCA and 50 mg of 3-AQ to an MS-compatible microcentrifuge tube. Then add 100 μL of 10 mM AP/50% MeOH and vortex briefly.
2. By weight add 50 μL of glycerol (approximately 63 mg at RT [d=1.261 g/cm^3]) and vortex.
3. Sonicate for 15 min or until fully dissolved.
4. Take 10 μL of the stock solution and add 190 μL of 10 mM AP/50% MeOH.

ClCCA ILM

1. Add 10 mg of ClCCA and 50 mg of 3-AQ to an MS-compatible microcentrifuge tube. Then add 100 μL of MeOH and vortex.
2. Sonicate for 15 min or until fully dissolved.
3. Take 10 μL of the stock solution and add 190 μL of 5 mM AP/50% MeOH/0.1% OGP.

Last but not least, other ILMs (cf. Chapter 'Quantitative MALDI MS Using Ionic Liquid Matrices') as well as LSMs are under development. The Cramer research group has made some further progress in simplifying the preparation protocols for liquid matrices and using liquid MALDI for the generation of multiply charged MALDI ions (cf. Chapter 'Efficient Production of Multiply Charged MALDI Ions'). Some of these matrices also appear to be promising candidates for sensitive MALDI MS analysis in general. However, most of these tend to crystallize to some degree under high-vacuum conditions (as do most of the ILMs) but offer an interesting alternative to the cinnamic acid/base binary systems and their specific properties. One of these liquid matrices employs DHB and does not require any base. Its preparation is based on a simple two-step protocol. Briefly, (1) fully dissolve 10 mg of DHB in 100 μL of 50–70% organic solvent such as MeOH or acetonitrile, optionally including 10 mM AP, and (2) add 20–60 μL of glycerol and thoroughly mix. This liquid matrix should provide homogeneous and fully liquid MALDI samples if prepared in a ratio of 1:1 with the analyte solution and used under intermediate vacuum conditions (as found in Q-TOF MALDI ion sources) or under atmospheric pressure. Using MALDI samples prepared with this liquid matrix in high-vacuum ion sources as in MALDI-TOF MS instruments might result (although not necessarily) in partial crystallization. As with most MALDI sample preparations it is recommended to test a few different compositions in order to find suitable conditions for the specific experimental setup and analysis.

References

Armstrong DW, Zhang L-K, He L, Gross ML (2001) Ionic liquids as matrixes for matrix-assisted laser desorption/ionization mass spectrometry. Anal Chem 73(15):3679–3686. doi:10.1021/ac010259f

Beavis RC, Chaudhary T, Chait BT (1992) α-Cyano-4-hydroxycinnamic acid as a matrix for matrix-assisted laser desorption mass spectrometry. Org Mass Spectrom 27(2):156–158. doi:10.1002/oms.1210270217

Bungert D, Bastian S, Heckmann-Pohl D, Giffhorn F, Heinzle E, Tholey A (2004) Screening of sugar converting enzymes using quantitative MALDI-ToF mass spectrometry. Biotechnol Lett 26(13):1025–1030. doi:10.1023/B:BILE.0000032965.18721.62

Calvano CD, Carulli S, Palmisano F (2009) Aniline/α-cyano-4-hydroxycinnamic acid is a highly versatile ionic liquid for matrix-assisted laser desorption/ionization mass spectrometry. Rapid Commun Mass Spectrom 23(11):1659–1668. doi:10.1002/rcm.4053

Cramer R, Corless S (2005) Liquid ultraviolet matrix-assisted laser desorption/ionization—mass spectrometry for automated proteomic analysis. Proteomics 5(2):360–370. doi:10.1002/pmic.200400956

Fukuyama Y, Nakaya S, Yamazaki Y, Tanaka K (2008) Ionic liquid matrixes optimized for MALDI-MS of sulfated/sialylated/neutral oligosaccharides and glycopeptides. Anal Chem 80(6):2171–2179. doi:10.1021/ac7021986

Garaguso I, Borlak J (2008) Matrix layer sample preparation: an improved MALDI-MS peptide analysis method for proteomic studies. Proteomics 8(13):2583–2595. doi:10.1002/pmic.200701147

Glückmann M, Karas M (1999) The initial ion velocity and its dependence on matrix, analyte and preparation method in ultraviolet matrix-assisted laser desorption/ionization. J Mass Spectrom 34(5):467–477. doi:10.1002/(SICI)1096-9888(199905)34:5<467::AID-JMS809>3.0.CO;2-8

Gobom J, Schuerenberg M, Mueller M, Theiss D, Lehrach H, Nordhoff E (2001) α-Cyano-4-hydroxycinnamic acid affinity sample preparation. A protocol for MALDI-MS peptide analysis in proteomics. Anal Chem 73(3):434–438. doi:10.1021/ac001241s

Jaskolla TW, Lehmann W-D, Karas M (2008) 4-Chloro-α-cyanocinnamic acid is an advanced, rationally designed MALDI matrix. Proc Natl Acad Sci U S A 105(34):12200–12205. doi:10.1073/pnas.0803056105

Karas M, Ehring H, Nordhoff E, Stahl B, Strupat K, Hillenkamp F, Grehl M, Krebs B (1993) Matrix-assisted laser desorption/ionization mass spectrometry with additives to 2,5-dihydroxybenzoic acid. Org Mass Spectrom 28(12):1476–1481. doi:10.1002/oms.1210281219

Kjellström S, Jensen ON (2004) Phosphoric acid as a matrix additive for MALDI MS analysis of phosphopeptides and phosphoproteins. Anal Chem 76(17):5109–5117. doi:10.1021/ac0400257

Kolli VSK, Orlando R (1996) A new matrix for matrix-assisted laser desorption/ionization on magnetic sector instruments with point detectors. Rapid Commun Mass Spectrom 10(8):923–926. doi:10.1002/(SICI)1097-0231(19960610)10:8<923::AID-RCM610>3.0.CO;2-E

Kolli VSK, Orlando R (1997) A new strategy for MALDI on magnetic sector mass spectrometers with point detectors. Anal Chem 69(3):327–332. doi:10.1021/ac961033t

Laremore TN, Murugesan S, Park T-J, Avci FY, Zagorevski DV, Linhardt RJ (2006) Matrix-assisted laser desorption/ionization mass spectrometric analysis of uncomplexed highly sulfated oligosaccharides using ionic liquid matrices. Anal Chem 78(6):1774–1779. doi:10.1021/ac051121q

Laremore TN, Zhang F, Linhardt RJ (2007) Ionic liquid matrix for direct UV-MALDI-TOF-MS analysis of dermatan sulfate and chondroitin sulfate oligosaccharides. Anal Chem 79(4):1604–1610. doi:10.1021/ac061688m

Li YL, Gross ML (2004) Ionic-liquid matrices for quantitative analysis by MALDI-TOF mass spectrometry. J Am Soc Mass Spectrom 15(12):1833–1837. doi:10.1016/j.jasms.2004.08.011

Li YL, Gross ML, Hsu F-F (2005) Ionic-liquid matrices for improved analysis of phospholipids by MALDI-TOF mass spectrometry. J Am Soc Mass Spectrom 16(5):679–682. doi:10.1016/j.jasms.2005.01.017

Mank M, Stahl B, Boehm G (2004) 2,5-Dihydroxybenzoic acid butylamine and other ionic liquid matrixes for enhanced MALDI-MS analysis of biomolecules. Anal Chem 76(10):2938–2950. doi:10.1021/ac030354j

Palmblad M, Cramer R (2007) Liquid matrix deposition on conductive hydrophobic surfaces for tuning and quantitation in UV-MALDI mass spectrometry. J Am Soc Mass Spectrom 18(4):693–697. doi:10.1016/j.jasms.2006.11.013

Santos LS, Haddad R, Höehr NF, Pilli RA, Eberlin MN (2004) Fast screening of low molecular weight compounds by thin-layer chromatography and 'on-spot' MALDI-TOF mass spectrometry. Anal Chem 76(7):2144–2147. doi:10.1021/ac035387d

Strupat K, Karas M, Hillenkamp F (1991) 2,5-Dihydroxybenzoic acid: a new matrix for laser desorption—ionization mass spectrometry. Int J Mass Spectrom Ion Process 111:89–102. doi:10.1016/0168-1176(91)85050-V

Sze ETP, Chan TWD, Wang G (1998) Formulation of matrix solutions for use in matrix-assisted laser desorption/ionization of biomolecules. J Am Soc Mass Spectrom 9(2):166–174. doi:10.1016/S1044-0305(97)00237-7

Tanaka K, Waki H, Ido Y, Akita S, Yoshida Y, Yoshida T, Matsuo T (1988) Protein and polymer analyses up to m/z 100 000 by laser ionization time-of-flight mass spectrometry. Rapid Commun Mass Spectrom 2(8):151–153. doi:10.1002/rcm.1290020802

Tholey A (2006) Ionic liquid matrices with phosphoric acid as matrix additive for the facilitated analysis of phosphopeptides by matrix-assisted laser desorption/ionization mass spectrometry. Rapid Commun Mass Spectrom 20(11):1761–1768. doi:10.1002/rcm.2514

Tholey A, Zabet-Moghaddam M, Heinzle E (2006) Quantification of peptides for the monitoring of protease-catalyzed reactions by matrix-assisted laser desorption/ionization mass spectrometry using ionic liquid matrixes. Anal Chem 78(1):291–297. doi:10.1021/ac0514319

Towers MW, Cramer R (2007) Liquid matrices for analyses by UV-MALDI mass spectrometry. Spectroscopy 22(11):29–37

Towers MW, McKendrick JE, Cramer R (2010) Introduction of 4-chloro-α-cyanocinnamic acid liquid matrices for high sensitivity UV-MALDI MS. J Proteome Res 9(4):1931–1940. doi:10.1021/pr901089j

Wasserscheid P, Keim W (2000) Ionic liquids—new 'solutions' for transition metal catalysis. Angew Chem Int Ed 39(21):3772–3789. doi:10.1002/1521-3773(20001103)39:21<3772::AID-ANIE3772>3.0.CO;2-5

Zabet-Moghaddam M, Heinzle E, Tholey A (2004a) Qualitative and quantitative analysis of low molecular weight compounds by ultraviolet matrix-assisted laser desorption/ionization mass spectrometry using ionic liquid matrices. Rapid Commun Mass Spectrom 18(2):141–148. doi:10.1002/rcm.1293

Zabet-Moghaddam M, Krüger R, Heinzle E, Tholey A (2004b) Matrix-assisted laser desorption/ionization mass spectrometry for the characterization of ionic liquids and the analysis of amino acids, peptides and proteins in ionic liquids. J Mass Spectrom 39(12):1494–1505. doi:10.1002/jms.746

Zabet-Moghaddam M, Heinzle E, Lasaosa M, Tholey A (2006) Pyridinium-based ionic liquid matrices can improve the identification of proteins by peptide mass-fingerprint analysis with matrix-assisted laser desorption/ionization mass spectrometry. Anal Bioanal Chem 384(1):215–224. doi:10.1007/s00216-005-0130-6

Coupling Liquid MALDI MS to Liquid Chromatography

Kanjana Wiangnon and Rainer Cramer

Abstract Matrix-assisted laser desorption/ionization (MALDI) coupled with time-of-flight (TOF) mass spectrometry (MS) is a powerful tool for the analysis of biological samples, and nanoflow high-performance liquid chromatography (nanoHPLC) is a useful separation technique for the analysis of complex proteomics samples. The off-line combination of MALDI and nanoHPLC has been extensively investigated and straightforward techniques have been developed, focussing particularly on automated MALDI sample preparation that yields sensitive and reproducible spectra. Normally conventional solid MALDI matrices such as α-cyano-4-hydroxycinnamic acid (CHCA) are used for sample preparation. However, they have limited usefulness in quantitative measurements and automated data acquisition because of the formation of heterogeneous crystals, resulting in highly variable ion yields and desorption/ionization characteristics. Glycerol-based liquid support matrices (LSM) have been proposed as an alternative to the traditional solid matrices as they provide increased shot-to-shot reproducibility, leading to prolonged and stable ion signals and therefore better results. This chapter focuses on the integration of LSM MALDI matrices into the LC-MALDI MS/MS approach in identifying complex and large proteomes. The interface between LC and MALDI consists of a robotic spotter, which fractionates the eluent from the LC column into nanoliter volumes, and co-spots simultaneously the liquid matrix with the eluent fractions onto a MALDI target plate via sheath-flow. The efficiency of this method is demonstrated through the analysis of trypsin digests of both bovine serum albumin (BSA) and *Lactobacillus plantarum* WCFS1 proteins.

1 Introduction

Matrix-assisted laser desorption ionization (MALDI) mass spectrometry (MS) was introduced in the late 1980s by Hillenkamp and Karas (Karas and Hillenkamp 1988). MALDI is a very useful and effective technique for the MS analysis of (bio)molecules

K. Wiangnon • R. Cramer (✉)
Department of Chemistry, University of Reading, Whiteknights, Reading RG6 6AD, UK
e-mail: k.wiangnon@student.reading.ac.uk; r.k.cramer@reading.ac.uk

R. Cramer (ed.), *Advances in MALDI and Laser-Induced Soft Ionization Mass Spectrometry*, DOI 10.1007/978-3-319-04819-2_4

due to its inherently high sensitivity, simple sample preparation, ease of data interpretation (through the detection of mainly singly charged ions) and tolerance to contaminants such as salts, buffers, and other impurities (Cramer and Dreisewerd 2007; Nagra and Li 1995). It also produces mass spectra with little or no fragment ions due to the soft ionization process, so it is suitable for large and/or labile biomolecules such as proteins. MALDI sample preparation is typically achieved by mixing an analyte solution with a solution of a matrix such as 2,5-dihydroxybenzoic acid (DHB) or α-cyano-4-hydroxycinnamic acid (CHCA), resulting after drying in a solid (crystalline) sample with a high molar excess of the matrix in comparison to the analyte (>10,000:1). There are, however, disadvantages with solid MALDI samples, for instance the formation of inhomogeneous crystals which leads to 'hot spots'. In this situation the position of the laser needs to be changed frequently as the laser beam bores into the crystal with successive shots (Towers and Cramer 2007), thus desorbing (or better described as ablating) the MALDI sample from spots with different topology and morphology. These drawbacks are problematic for the successful automation of the analysis and quantitative MALDI MS measurement.

Liquid matrices were proposed as an alternative to solid matrices (Kolli and Orlando 1996; Sze et al. 1998). These early liquid matrices are based on the dissolution of an acidic solid matrix with a solubilizing helper compound, a basic organic reagent such as 3 aminoquinoline (3-AQ). The use of a viscous material with low-volatility such as glycerol in conjunction with this acid/base system supports the sample to remain liquid under vacuum conditions (Sze et al. 1998; Towers and Cramer 2007).[1] In contrast to solid matrices, liquid MALDI samples give greater sample homogeneity with a more consistent spot morphology and topology after irradiation due to their self-healing and renewing properties. As the spot remains liquid under vacuum conditions, a prolonged period of ion generation under these conditions is possible.

For the analysis of complex protein samples such as in proteomics, two-dimensional gel electrophoresis (2DE) is the classical method, separating proteins based on their molecular weight and charge (Beranova-Giorgianni 2003). After electrophoretic separation of the proteins, individual gel spots can be excised and digested with trypsin prior to protein identification analysis by MS. This methodology can provide high sensitivity and resolution but there are obvious drawbacks in terms of the time needed for the entire workflow, the difficulties in covering the whole proteome (particularly when some of the proteins have poor solubility such as membrane proteins), and the limited range of molecular weight (Wagner et al. 2003).

An alternative method to separate peptides in biological samples is liquid chromatography (LC) coupled either on- or off-line to a mass spectrometer. Generally, nanoLC is coupled on-line to electrospray ionization (ESI) instruments such as ion traps/Orbitrap mass analyzers. This combination allows for automated measurements with fast data acquisition and high-throughput analysis of complex proteolytic

---

[1] Please note that the liquid matrices in this chapter are glycerol-based liquid support matrices (LSMs) as described here and not ionic liquid matrices (ILMs), which lack the addition of glycerol. See also the Chapters 'Ionic Liquids and Other Liquid Matrices for Sensitive MALDI MS Analysis' and 'Quantitative MALDI-MS Using Ionic Liquid Matrices' in this book for more information on ionic liquid matrices.

digests, providing great sensitivity in the low-femtomole range, high resolution and good mass accuracy. However, in highly complex peptide mixtures, there may be co-eluting peptides from the LC column which would yield several different precursor ions over the MS/MS acquisition time. Due to the scan speed restrictions of the instrument, not all the peptides could be sequenced during MS/MS leading to a reduction in the proteome coverage (Mirgorodskaya et al. 2005; Yang et al. 2007). For this reason, the combination of the chromatographic separation and MALDI MS has been increasingly investigated as MALDI MS and MS/MS measurements are acquired 'off-line', i.e., independent on the chromatographic separation, allowing more time for data acquisition of all the peptides eluted from the LC column. Other benefits of LC-MALDI are that samples deposited on target can be stored for several days without degradation, and re-analysis with different data acquisition parameters is possible. Also, mainly singly charged peptide ions are detected in MALDI, whereas multiply charged ions are typically detected in ESI MS. This avoids the need for deconvolution and enables easy interpretation of the acquired MS spectrum, although it also limits fragmentation for MS/MS analysis and requires instruments with a higher *m/z* range, ideally >2000.[2]

The development of a suitable matrix-spotting method and also of a matrix that gives improved sample homogeneity is the key to achieve more reproducibility in automated MALDI MS measurements. The use of liquid matrices in such a spotting device would be beneficial due to the general properties of liquids, potentially also providing an automated on-line integration of liquid chromatography with MALDI as it is the case with ESI.

In this chapter, a suitable setup for automated spotting of liquid MALDI matrices is presented. In this setup, a nanoflow LC system for the separation of digested peptides is coupled with an automated sample spotting robot that simultaneously co-spots a liquid matrix aliquot with fractions of the LC eluent. The ability to use nanoLC-MALDI MS and MS/MS for identification of a bovine serum albumin (BSA) digest and Lactobacillus proteins resulted in successful identifications with high protein sequence coverages.

2 Applications

Recent progress in the field of LC-MALDI MS/MS have included fractionation of the LC eluent and the development of an automated MALDI spotter and software that allows the rapid acquisition, high-speed processing and combining of tens of thousands of peptide spectra as well as searching the processed data set against a protein sequence database (Mirgorodskaya et al. 2005).

The design of the LC-MALDI interface, the method of transferring the LC eluent and matrix solution and ultimately their on-target deposition, is an important aspect.

---

[2] See also the discussion/introduction to multiply charged MALDI ions in the Chapter 'Efficient Production of Multiply Charged MALDI Ions' in this book.

In earlier studies, the matrix was either pre-spotted (Mirgorodskaya et al. 2005) or mixed with the LC eluent using a T-junction. In the latter case the fused silica tubing carrying the eluted sample is connected to a T piece where the matrix solution is introduced. The two solutions are then co-spotted onto the MALDI target (Melchior et al. 2010). Other studies have demonstrated the use of nanoLC coupled to a robotic spotting device, in which the MALDI spots were mixed using a liquid sheath-flow (PROTEINEER fc, Bruker). These studies also made use of the WARP-LC and BioTools software from Bruker. The associated workflow proved to be successful for large-scale proteome analysis (Maccarrone et al. 2010). A few researchers have built a spotter device of their own. For example, Hioki et al. have made a column probe, incorporating a depositing probe onto the nanoLC column (Hioki et al. 2014). It directly spots the eluted peptides onto the MALDI target plate where the matrix is pre-deposited. The advantage of this combination is a reduction in post-column diffusion. Hioki et al. demonstrated the successful detection of a trace element spiked into a complex sample (Hioki et al. 2014). Finally, Pereira and co-workers have revealed the use of an oil stream for producing a droplet (Pereira et al. 2013). Here the LC fraction and MALDI matrix were mixed at a T-junction and then delivered into an oil stream. The resulting sample spots were homogenous and thus more reproducible compared to the solid MALDI samples (Pereira et al. 2013). In proteomics, most analyses have focused on applying LC-ESI MS for peptide sequencing. The peptide mixtures from a proteolytic digest are typically separated by HPLC coupled to a tandem mass spectrometer, in which the resultant peptides are fragmented by collision-induced dissociation (CID). Each MS/MS acquisition is searched against a protein sequence database for matching peptides. In contrast, following digestion with trypsin, MALDI-TOF analysis typically employs peptide mass fingerprinting, searching individual protein digests without MS/MS analysis (Trauger et al. 2002). Gel electrophoresis is often used as a means of reducing sample complexity prior to MALDI-TOF MS analysis. However, for this, proteins need to be separated first if a complex protein sample is to be analyzed.

A few papers have now shown the advantages in interfacing off-line LC with MALDI for the analysis of a complex peptide mixture. The successful identification of both bovine serum albumin (BSA) digests at the 1-fmol level and complex proteins from *Escherichia coli* have been achieved by nanoLC-MALDI MS/MS with high reproducibility (Mirgorodskaya et al. 2005). Due to the differences between MALDI and ESI, they each have specific advantages over the other. In a study by Yang et al., the LC-MALDI methodology gave almost 50% more protein identifications compared to the LC-ESI MS/MS analysis of the same highly complex sample (Yang et al. 2007). The unique proteins obtained from LC-MALDI were larger and slightly more hydrophilic than the unique proteins identified by ESI. These results also showed that the LC-MALDI approach has an advantage over ESI when the sample contains ion suppression agents or contaminants (Yang et al. 2007). Therefore, combining the results from both LC-ESI MS/MS and LC-MALDI MS/MS have improved the number of proteins identified and the protein coverage of complex protein mixtures (Bodnar et al. 2003; Hattan et al. 2005; Yang et al. 2007). Moreover, the analysis of heterogeneous compounds such as glycoproteins is far more suited to the MALDI approach because heterogeneous compounds produce highly complex spectra, requiring greater sample fractionation/separation if ESI is used (Trauger et al. 2002).

Quantitation is also a challenging area in proteomics research. However, the main problem with MALDI is the poor reproducibility because of the nature of the solid MALDI sample. One potential solution to overcome the problem of poor reproducibility would be the use of an internal standard. The incorporation of a stable isotope-labeled standard allows for greater precision and high reproducibility in MS. A variety of isotope labeling methods have been developed, such as isotope-coded affinity tags (ICAT), isobaric tags for relative and absolute quantitation (iTRAQ), stable isotope labeling with amino acids in cell culture (SILAC), and absolute quantitation of abundance (AQUA) (Szajli et al. 2008). Yang et al. also used iTRAQ-labeling with LC-MALDI in their research (Yang et al. 2007). Generally, label-free relative quantitation has been restricted to LC-ESI due to the greater ion signal fluctuation and differential ion suppression in MALDI with solid (crystalline) samples. However, there has been one study which has demonstrated the use of label-free LC-MALDI in secretome analysis. For this purpose the statistical tools and software were specifically designed. The researchers were successful in quantifying proteins within 1.5–20 fold changes from the control with satisfactory statistics but fold changes of <0.05 were hardly measurable and had high coefficient-of-variation and relative-error values (Riffault et al. 2015).

Due to their great homogeneity and signal stability, liquid MALDI samples have been successfully used for peptide quantitation with high accuracy and precision (Li and Gross 2004; Palmblad and Cramer 2007), but not in an automated LC-MALDI workflow. The potential benefits of using liquid matrices also with LC-MALDI for quantitative applications in large-scale proteomics are therefore highly promising.

3 Materials and Protocols

A major reason for using liquid MALDI matrices in an LC-MALDI workflow is their potential to provide automated and quantitative measurements with great reproducibility. In proteomics, this is ideally achieved without sacrificing the protein and proteome sequence coverage that is achievable with solid MALDI samples. As a starting point a less complex sample such as a single protein like bovine serum albumin (BSA) is recommended. The example of a more complex protein sample, the lysis of *Lactobacillus plantarum* WCFS 1, is also presented. *L. plantarum* WCFS 1 is an important bacterium for the food industry and only a few proteomic studies have been carried out so far.

3.1 Materials

BSA, 3-aminoquinoline (3-AQ) and all solvents were purchased from Sigma-Aldrich (Poole, UK). Peptide calibration standard II (#222570) and CHCA were purchased from Bruker UK Ltd. (Coventry, UK). All chemicals, solvents, and peptides were used without further purification. *L. plantarum* WCFS 1 was originally isolated from human saliva (National Collection of Industrial and Marine Bacteria, Aberdeen, UK). It was kindly donated by Mr. Pornpoj Srisukchayakul, Food and Nutritional Sciences, University of Reading, Reading, UK.

3.2 Preparation of Samples

For *L. plantarum* WCFS 1, the workflow starts with growing the bacterial cells in broth media followed by mechanical lysis. The resulting protein solution is subsequently digested with trypsin. Finally, the sample is cleaned and concentrated using solid phase extraction (SPE) and a speed vacuum concentrator. Figure 1 provides the details of this workflow.

For BSA, in-solution digests are performed using a similar protocol as for *L. plantarum* WCFS 1.

3.3 Automated LC-MALDI MS Analysis of Protein Digests

3.3.1 Liquid MALDI Sample Preparation

The glycerol-based liquid support matrix (LSM) is prepared by dissolving CHCA and 3-AQ in 50% methanol/50% 10 mM ammonium phosphate (solution X) and glycerol in the ratio of 1:3:5:5 (w:w:v:v/[mg:mg:μL:μL]). The mixture is vortexed and sonicated for 15 min until all solutes are dissolved to give a clear yellow solution. This stock matrix solution is further diluted 20-fold in solution X prior to the analysis. The matrix solution is mixed with the peptide standard solution in a ratio of 1:1 (v/v), and 1 μL of matrix-analyte solution is spotted onto a 600 μm AnchorChip target plate (Bruker UK Ltd.) and allowed to dry at ambient temperature.

3.3.2 NanoHPLC and Automated MALDI Sample Spotting

In our setup, an Ultimate nanoLC system (Dionex, Hemel Hempstead, UK) is coupled directly to a PROTEINEER fc liquid handler (Bruker UK Ltd.). Both systems can be controlled by the Hystar software (Bruker UK Ltd.). The LC system has two pumps, of which the loading pump (which is a Switchos system) switches between the trap column (Acclaim PepMap 300 μm i.d. × 1 cm, 3 μm C_{18} particles) and the analytical column (Acclaim PepMap 75 μm i.d. × 25 cm, 3 μm C_{18} particles). The second pump is the Ultimate LC pump with a NAN-75 cartridge for splitting the flow rate (from 187 μL/min to 300 nL/min). An autosampler is used for the injection and loading of the sample onto the trap column. The loading solvent is the mobile phase solution A, comprising of 2% ACN/0.01% trifluoroacetic acid (TFA). A total of 1 pmol of the tryptic BSA digest or 1 μg of the bacterial protein digest is

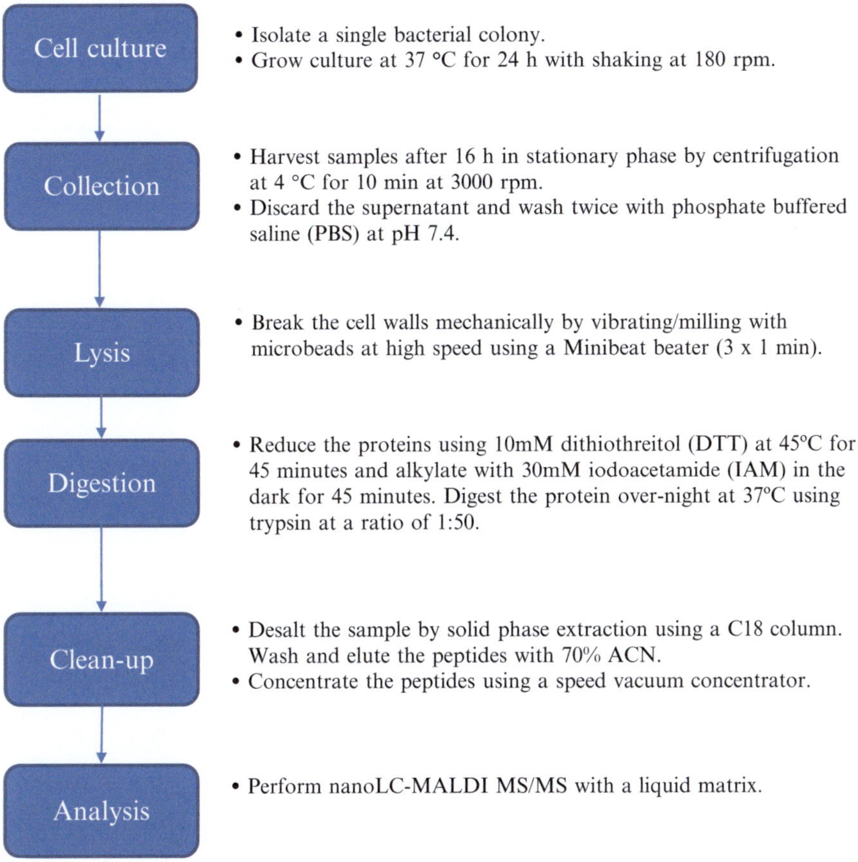

Fig. 1 Overall workflow for the preparation of protein digests from bacterial cells

injected at a flow rate of 20 μL/min onto the trap column. After 5 min of loading, the trap column is switched to an in-line position with the analytical column at a flow rate of 300 nL/min. The solvent used for the LC system is made up of the mobile phase solution A (2% ACN/0.01% TFA; see above) and the mobile phase solution B (90% ACN/0.01% TFA). A gradient of increasing mobile phase solution B is used, from 3 to 55% in 125 min. The UV detector is set to measure at 214 nm. The 3-nL UV flow cell is joined to the spotter by fused silica tubing (20 μm i.d., 360 μm o.d. × 100 cm).[3] The peptides run through the flow cell and fractions are spotted (30-s depositions)[4] onto the MALDI target plate from 20 to 120 min, as shown in Fig. 2.

---

[3] The length of the fused-silica tubing between the UV flow cell and the automated spotter should be kept as short as possible to minimize band spreading of the eluted sample. The internal diameter of the fused-silica tubing should be 20 μm for nanoHPLC and 50 μm for capillary HPLC.

[4] For highly complex samples, the fractionation time interval should be no more than 10–15 s but can be increased up to 30 s to obtain higher sensitivity.

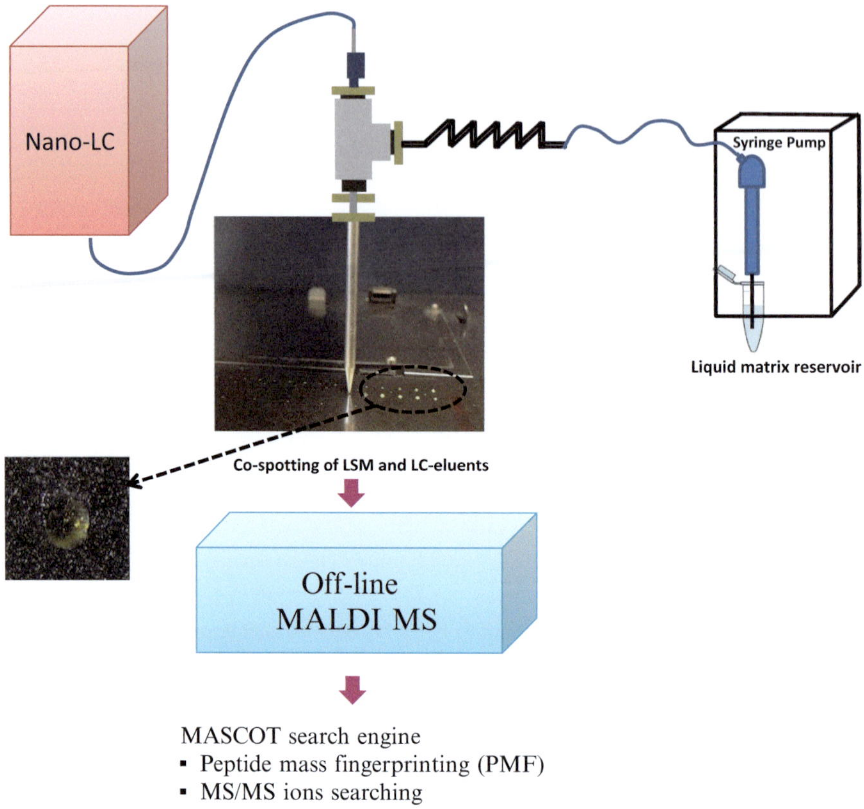

Fig. 2 Schematic summary of the nanoHPLC-MALDI MS spotting system, co-spotting the liquid support matrix (LSM) with the LC eluents

The liquid matrix is simultaneously deposited (0.5 μL/spot) by a probe which is connected to a syringe pump to aspirate and dispense the matrix solution.[5] Prior to sample and matrix deposition, the PROTEINEER fc robot was taught the *x*–*y*–*z* coordinates of the 600-μm AnchorChip target plate[6] to ensure the tip of the spotting capillary[7] closely approaches but does not touch the surface of the target (i.e., <0.5 mm above the surface of the target). The probe needle must be washed before and after each run with a solvent that dissolves the matrix.

[5] Where continuous flow is selected the matrix flow rate should be set higher than normally required because of the viscosity of the liquid matrix. For discontinuous flow, the volume should be higher than normally required for the same reason.

[6] Note that other MALDI-MS instruments and thus target plates can be used with the appropriate adaptations/teaching.

[7] The fused-silica capillary should protrude no more than 0.1 mm from the needle probe.

3.3.3 Mass Spectrometer Settings

Mass spectrometry measurements are performed on an Ultraflex TOF/TOF (Bruker UK Ltd.) instrument, employing the positive ion and reflectron mode. The instrument is calibrated using a standard peptide mixture mixed with the liquid matrix. MS data are collected with the *ion source 1* set to 25.00 kV and *ion source 2* set to 21.75 kV. The *lens voltage* is set to 8.50 kV and *reflector 1* is set to 26.30 kV with *reflector 2* at 13.75 kV. Matrix suppression below 500 Da is used. The MS/MS measurement on an Ultraflex instrument is performed using the LIFT technology with its default calibration, which is typically carried out by Bruker as part of the Preventative Maintenance Contract.[8]

3.3.4 MS Data Acquisition and Analysis[9]

Analysis of the fractionated samples is performed automatically using the AutoXecute function in flexControl (Bruker UK Ltd.). The data is collected from the center of the spot without changing the desorption position, with a laser repetition rate of 25 Hz. Spectra of 500 laser shots are summed at each desorption position. After MS data acquisition, MS spectral peaks are picked using flexAnalysis software version 3.0 (Bruker UK Ltd.). Peak picking is performed automatically using the Sophisticated Numerical Annotation Procedure (SNAP) algorithm with the criterion that a maximum of 300 peaks are picked with an S/N of at least three and a resolution higher than 500. The WARP-LC software (Bruker UK Ltd.) is used to establish a peak list of the entire LC experiment after peak picking. WARP-LC is then used for the automatic selection of peptide signals and their subsequent MS/MS analysis. MS and MS/MS spectra are sent by WARP-LC to the BioTools software (Bruker UK Ltd.) as combined peak lists for database searching (see Sect. 3.4).

3.3.5 MS/MS Data Acquisition and Processing

Automated MS/MS data acquisition is performed by using the LIFT technology of the Ultraflex TOF/TOF instrument. For this, LIFT spectra are obtained from the accumulation of 300 parent ion spectra and 500 single-shot fragment ion spectra for each parent/precursor ion. A maximum of ten precursor ion signals per fraction are analyzed starting preferentially in the range of *m/z* 800–3500. The LIFT spectral peaks are picked using flexAnalysis after baseline subtraction (Median with a 0.8 flatness) and smoothing (Savitzky-Golay, 4 cycles with a width of *m/z* 0.15). Peak picking is set to a default maximum of 200 peaks with an S/N of at least three using the centroid algorithm.

---

[8] LIFT calibration is not straightforward and is best left to an experienced service engineer.

[9] In our lab we use the following software versions for this step: 3.0 (flexAnalysis), 1.1 (WARP-LC), 3.1 (BioTools).

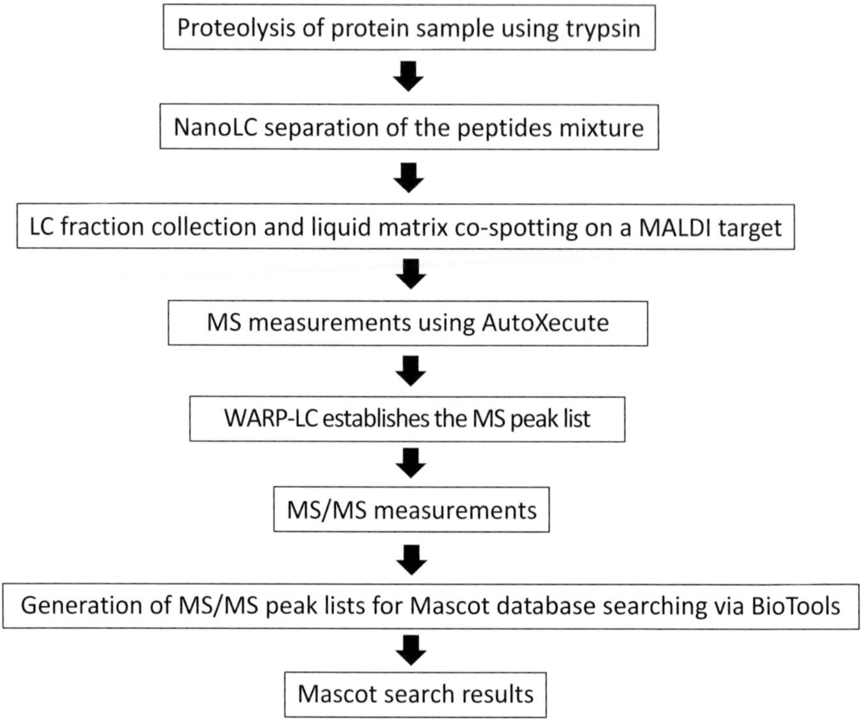

Fig. 3 LC-MALDI TOF/TOF MS/MS analysis workflow for the analysis of complex protein digests

3.4 Database Searching

In our lab, the peak lists are automatically searched against an in-house copy of the Swiss-Prot protein database using Mascot Server 2.4.1 (Matrix Science, London, UK). Peptide mass fingerprinting (PMF) is performed with a mass tolerance of 50 ppm, allowing for up to one missed cleavage, with a fixed modification of cysteines by carbamidomethylation and variable modifications of acetylation at the *N*-terminus and methionine oxidation. The MS/MS ions searching is usually done with the peptide mass tolerance set at 50 ppm and the MS/MS ion tolerance at 0.8 Da. The modifications of fixed carbamidomethylation of cysteines and variable oxidation of methionines are selected. A maximum of one missed cleavage is allowed. The entire workflow for nanoLC-MALDI MS and MS/MS data acquisition and analysis is shown in Fig. 3. An example liquid MALDI MS spectrum from a nanoHPLC fraction of an *L. plantarum* WCFS1 digest and the annotated MS/MS spectrum from the precursor ion signal at *m/z* 1613.935 identifying the elongation factor Tu protein are shown in Fig. 4. Details of the 20 highest scoring proteins from the entire analysis of the *L. plantarum* WCFS1 digest used for Fig. 4 are summarized in Table 1.

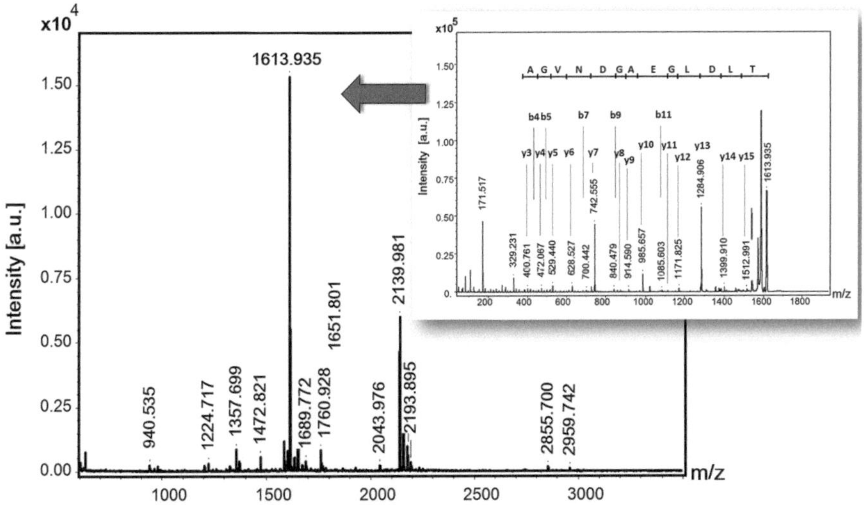

Fig. 4 Liquid MALDI MS and MS/MS spectrum obtained from a nanoHPLC fraction of a *Lactobacillus plantarum* WCFS1 digest at a retention time of 54 min. The MS/MS spectrum (*inset*) shows the fragment ions obtained for the peak at *m/z* 1613.935. Based on Mascot MS/MS ions searching, the peptide was identified as TLDLGEAGDNVGALLR of the elongation factor Tu protein

Table 1 The top ten highest scoring *L. plantarum* WCFS1 proteins obtained by whole proteome digest analysis using nanoHPLC coupled to liquid MALDI MS and MS/MS[a]

Protein accession number	Protein description	MW (kDa)	Mascot score
sp|Q88VE0|EFTU	Elongation factor Tu	43.4	582
tr|F9UM10|F9UM10	Glyceraldehyde 3-phosphate dehydrogenase	36.6	433
sp|Q88YM5|CH60	60 kDa chaperonin	57.4	358
sp|Q88VJ2|LDHD	D-lactate dehydrogenase	37.2	309
tr|F9UTT2|F9UTT2	Fructose-bisphosphate aldolase	31.0	218
sp|Q88YH5|PGK	Phosphoglycerate kinase	42.8	206
sp|Q88YH3|ENO1	Enolase 1	48.1	189
tr|F9UPM3|F9UPM3	Pyruvate kinase	62.9	145
tr|F9UPL0|F9UPL0	30S ribosomal protein S1	47.1	131
sp|Q88YH4|TPIS	Triosephosphate isomerase	27.1	127

[a]For database searching, Mascot MS/MS ions searching was employed, searching against the *Lactobacillus plantarum* WCFS1 database with 3088 entries, which was downloaded on March 31, 2014, from HAMAP (High-quality Automated and Manual Annotation of microbial Proteomes; http://hamap.expasy.org/proteomes/LACPL.html)

References

Beranova-Giorgianni S (2003) Proteome analysis by two-dimensional gel electrophoresis and mass spectrometry: strengths and limitations. TrAC Trends Anal Chem 22(5):273–281

Bodnar WM, Blackburn R, Krise J, Moseley MA (2003) Exploiting the complementary nature of LC/MALDI/MS/MS and LC/ESI/MS/MS for increased proteome coverage. J Am Soc Mass Spectrom 14(9):971–979

Cramer R, Dreisewerd K (2007) UV matrix-assisted laser desorption ionization: principles, instrumentation, and applications. In: Gross ML, Caprioli RM (eds) The encyclopedia of mass spectrometry, vol 6. Elsevier, Amsterdam, pp 646–661

Hattan SJ, Marchese J, Khainovski N, Martin S, Juhasz P (2005) Comparative study of [Three] LC-MALDI workflows for the analysis of complex proteomic samples. J Proteome Res 4(6):1931–1941

Hioki Y, Tanimura R, Iwamoto S, Tanaka K (2014) Nano-LC/MALDI-MS using a column-integrated spotting probe for analysis of complex biomolecule samples. Anal Chem 86(5):2549–2558

Karas M, Hillenkamp F (1988) Laser desorption ionization of proteins with molecular masses exceeding 10,000 daltons. Anal Chem 60(20):2299–2301

Kolli VSK, Orlando R (1996) A new matrix for matrix-assisted laser desorption/ionization on magnetic sector instruments with point detectors. Rapid Commun Mass Spectrom 10(8):923–926

Li YL, Gross ML (2004) Ionic-liquid matrices for quantitative analysis by MALDI-TOF mass spectrometry. J Am Soc Mass Spectrom 15(12):1833–1837

Maccarrone G, Turck C, Martins-de-Souza D (2010) Shotgun mass spectrometry workflow combining IEF and LC-MALDI-TOF/TOF. Protein J 29(2):99–102

Melchior K, Tholey A, Heisel S, Keller A, Lenhof H, Meese E, Huber CG (2010) Protein- versus peptide fractionation in the first dimension of two-dimensional high-performance liquid chromatography-matrix-assisted laser desorption/ionization tandem mass spectrometry for qualitative proteome analysis of tissue samples. J Chromatogr A 1217(40):6159–6168

Mirgorodskaya E, Braeuer C, Fucini P, Lehrach H, Gobom J (2005) Nanoflow liquid chromatography coupled to matrix-assisted laser desorption/ionization mass spectrometry: sample preparation, data analysis, and application to the analysis of complex peptide mixtures. Proteomics 5(2):399–408

Nagra DS, Li L (1995) Liquid chromatography-time-of-flight mass spectrometry with continuous-flow matrix-assisted laser desorption ionization. J Chromatogr A 711(2):235–245

Palmblad M, Cramer R (2007) Liquid matrix deposition on conductive hydrophobic surfaces for tuning and quantitation in UV-MALDI mass spectrometry. J Am Soc Mass Spectrom 18(4):693–697. doi:10.1016/j.jasms.2006.11.013

Pereira F, Niu X, deMello AJ (2013) A nano LC-MALDI mass spectrometry droplet interface for the analysis of complex protein samples. PLoS One 8(5):1–10

Riffault M, Moulin D, Grossin L, Mainard D, Magdalou J, Vincourt J (2015) Label-free relative quantification applied to LC-MALDI acquisition for rapid analysis of chondrocyte secretion modulation. J Proteomics 114:263–273

Szajli E, Feher T, Medzihradszky KF (2008) Investigating the quantitative nature of MALDI-TOF MS. Mol Cell Proteomics 7(12):2410–2418

Sze ETP, Chan TWD, Wang G (1998) Formulation of matrix solutions for use in matrix-assisted laser desorption/ionization of biomolecules. J Am Soc Mass Spectrom 9(2):166–174

Towers M, Cramer R (2007) Liquid matrices for analyses by UV-MALDI mass spectrometry. Spectroscopy 22:24–31

Trauger SA, Webb W, Siuzdak G (2002) Peptide and protein analysis with mass spectrometry. Spectroscopy 16(1):15–28

Wagner Y, Sickmann A, Meyer HE, Daum G (2003) Multidimensional nano-HPLC for analysis of protein complexes. J Am Soc Mass Spectrom 14(9):1003–1011

Yang Y, Zhang S, Howe K, Wilson DB, Moser F, Irwin D, Thannhauser TW (2007) A comparison of nLC-ESI-MS/MS and nLC-MALDI-MS/MS for geLC-based protein identification and iTRAQ-based shotgun quantitative proteomics. J Biomol Tech 18(4):226–237

Quantitative MALDI MS Using Ionic Liquid Matrices

Joanna Tucher, Prasath Somasundaram, and Andreas Tholey

Abstract Since the introduction of ionic liquid matrices (ILM) for matrix-assisted laser desorption/ionization (MALDI) mass spectrometry, manifold applications for a range of different substance classes (e.g., amino acids, peptides or proteins) were described. ILM are composed of an equimolar mixture of classically used acidic MALDI matrices and organic bases, which allows for an almost infinite number of combinations with different properties. A major advantage offered by many ILM compared to classical crystalline matrices is a highly homogeneous sample distribution, which is especially important in regard to quantification.

In general the concentration of an analyte can be determined based on the intensity of its corresponding ion in the mass spectrum (MS^1 quantification). For this purpose the use of an internal standard with a high similarity to the analyte of interest is usually recommended. However, in certain cases it is possible to quantify the analyte without the use of an internal standard. Here we describe two exemplary applications of MALDI MS using ILM for the determination of enzyme activities.

1 Introduction

Matrix-assisted laser desorption/ionization (MALDI) mass spectrometry (MS) has become a major analytical technique in various sectors of scientific research, industrial applications and pharmaceutical or food quality control. Low molecular weight compounds (Cohen and Gusev 2002), e.g., amino acids, lipids, sugars, or vitamins, can be analyzed as well as molecules of higher molecular weight such as peptides, proteins, or oligonucleotides (Karas and Hillenkamp 1988). Key features of this technology are its relatively simple and fast sample preparation, a comparatively high tolerance towards impurities (e.g., salts or detergents), low sample consumption as well as straightforward interpretable spectra. The latter is mainly caused by the preferred formation of singly charged ions.

J. Tucher • P. Somasundaram • A. Tholey (✉)
AG Systematic Proteome Research & Bioanalytics, Institute for Experimental Medicine, Christian-Albrechts-Universität zu Kiel, Kiel, Germany
e-mail: j.tucher@iem.uni-kiel.de; p.somasundaram@iem.uni-kiel.de; a.tholey@iem.uni-kiel.de

R. Cramer (ed.), *Advances in MALDI and Laser-Induced Soft Ionization Mass Spectrometry*, DOI 10.1007/978-3-319-04819-2_5

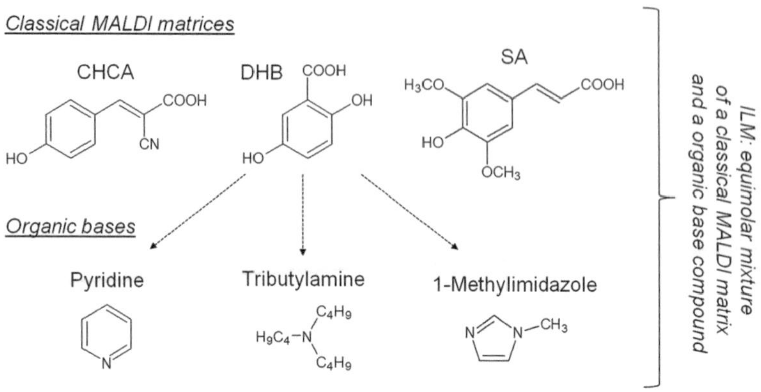

Fig. 1 Structures of exemplary classical MALDI matrices and organic bases commonly used for ILM preparations

The careful choice of a suitable matrix for a certain application is of fundamental importance for the success of an experiment as the matrix can influence the ionization/desorption behavior, trigger the formation of adducts or may affect the stability of the analyte of interest. This fact concerns qualitative as well as quantitative applications of MALDI MS. In general MALDI matrices must fulfill several criteria: they have to be stable under high vacuum conditions (low vapor pressure), and absorb the laser light at the emitted wavelength. Additionally, matrices should be inert towards the analyte and either co-crystallize (solid matrices) with the analyte or solubilize (liquid matrices) the latter (Fig. 1).

Whereas qualitative measurements are applied routinely, quantification by MALDI MS is hampered by several challenges. First, signal intensities are not only dependent on the amount of analyte present in the sample but also of its chemical composition. The ionization behavior of peptides, for example, is influenced by the basicity of the single amino acids, certain amino acid modifications and the sequence itself. Internal standards with high chemical similarity to the analyte, at best stable isotope-labeled counterparts of the analyte, are required for quantification. Second, complex samples or impurities can favor ion suppression effects. Accordingly, analytes have to be purified prior to MALDI MS measurement or the sample complexity has to be reduced, e.g., by offline-LC separation and subsequent spotting onto the target. Last, sample spots using classical crystalline MALDI matrices such as α-cyano-4-hydroxycinnamic acid (CHCA), 2,5-dihydroxybenzoic acid (DHB) or sinapinic acid (SA) are indicated by rather inhomogeneous distribution of the analyte over the spot. These so-called 'hot spots' or 'sweet spots' trigger a poor spot-to-spot and shot-to-shot reproducibility of the measurement, substantially hampering quantification. Consequently, protocols using conventional MALDI matrices have to average a representative amount of spectra over the whole spot area by automated laser movement to achieve reliable results. In addition the homogeneity of the matrix-analyte co-crystal has to be optimized, e.g., by fast evaporation of the matrix solvent (Nicola et al. 1995) or the addition of co-matrices (Gusev et al. 1996; Distler and

Allison 2001) such as fucose. Another possibility for improved sample homogeneity is the application of liquid matrices. Glycerol-based liquid matrices, for example, are composed of crystalline matrices dissolved in a highly viscous and vacuum stable liquid, such as glycerol, by the help of a solubilizing reagent (Sze et al. 1998). A second approach, which will be focused on in the following chapter, is the direct application of liquid components, e.g., ionic liquids, as MALDI matrices.

Ionic liquids are defined as salts with melting points below 100 °C. Due to their outstanding properties, these compounds have gained widespread applications in almost all disciplines of analytical chemistry in recent years (Sun and Armstrong 2010). In 2001, the use of ionic liquids as MALDI matrices was reported for the first time by Armstrong and coworkers (Armstrong et al. 2001). However, it turned out, that so-called room-temperature ionic liquids (RTIL) were not generally suitable as MALDI matrices; one reason was the lack of suitable UV-absorption. Therefore, the authors introduced a novel class of ionic liquids, which they synthesized by the equimolar combination of classical acidic MALDI matrices (e.g., CHCA, DHB, SA) with different organic bases (e.g., pyridine, tributylamine, 1-methylimidazole). These novel matrices, which showed comparable or in some applications even improved MALDI matrix properties, were later called ionic liquid matrices (ILMs) (Zabet-Moghaddam et al. 2004; Tholey and Heinzle 2006). As several of these ILMs crystallize at room temperature, they are termed ion matrices as well (Lemaire et al. 2006). Two methods for the preparation of ILMs will be described in the practical part of this chapter.

The option of choosing different acidic matrix compounds and organic bases for the preparation of an ILM allows for an almost infinite number of combinations (Berthod et al. 2009; Crank and Armstrong 2009; Towers et al. 2010; Gabriel et al. 2014) with unique properties, giving the opportunity of creating tailor-made matrices. However, unlike demonstrated for crystalline matrices (Jaskolla et al. 2008), up to now no prediction of final properties of a particular acid–base combination is possible. Thus, the test of properties is still a matter of trial-and-error experiments. Despite this, ILMs provide a number of interesting features. Even after salt formation the suitability of the parent acid compound for a distinct class of analytes persists; e.g., CHCA, which is one of the best suited matrices for the analysis of peptides, retains this property after conversion to an ILM. A second feature of ILM is the strong reduction of matrix signals in the low *m/z* region (Vaidyanathan et al. 2006). In contrast to the crystalline acidic parent matrices, only weak or even no signals of the ILM itself are usually present in the mass spectrum; only the base compound of the ILM forms a strong signal. This creates an inherent advantage for the analysis of low molecular weight compounds by MALDI MS.

ILMs are not superior over their classical crystalline counterparts in every respect. The formation of sodium or potassium adducts in many ILMs is increased compared to the use of solid matrices. This feature is in particular useful for the analysis of carbohydrates (Harvey 1999). However, in case of peptides, alkali adducts hamper data interpretation as spectra become more complex. The simultaneous occurrence of different species of the analyte (e.g., $[M+H]^+$, $[M+Na]^+$, $[M+K]^+$) causes sensitivity loss and consequently increases the limit of detection. It has to be noted, that the introduction of the so-called substoichiometric ILM, composed by a 2:1

ratio of CHCA and pyridine, can lead to improved spectra qualities in the analysis of peptide mass fingerprint spectra (Zabet-Moghaddam et al. 2006). Furthermore the addition of ammonium salts such as ammonium phosphate, as employed in glycerol-based liquid matrices (Cramer and Corless 2005) and previously introduced for the reduction of alkali adduct ion formation in nucleotide analysis, also enhances spectra quality. In case of intact proteins, the formation of alkali adducts leads to significant peak broadening, thus making ILMs less suitable for this application.

Sample spots using ILMs appear like a viscous film on the target after sample preparation. Within this film the analyte is very homogenously distributed (Zabet-Moghaddam et al. 2004; Mank et al. 2004; Tholey et al. 2006; Tholey 2006). Thus, the formation of 'hot spots' is widely suppressed. Even in acid–base combinations which appear to be crystalline at room temperature the sample distribution is improved compared to conventional crystalline MALDI matrices. This feature renders ILMs useful especially in regard to quantification based on MS^1 signal intensities (Zabet-Moghaddam et al. 2004; Li and Gross 2004). Principally quantification can be performed on the level of MS or MS/MS (aka MS^2 or tandem MS) spectra. In this chapter we will only focus on the quantification based on MS^1.

In the following sections we describe the use of ILMs for the quantification of low molecular weight biomolecules (carbohydrates) and of biopolymers (peptides). We present two applications for the determination of enzyme activities by quantitative MALDI MS. However, the principles elucidated in the two examples are easily transferrable to a number of other applications.

2 Applications

Enzyme-catalyzed reactions play a central role in various fields of biotechnology and scientific research. Novel enzyme variants, either from natural sources or created by means of directed evolution (e.g., site-directed mutagenesis) of existing biocatalysts, are constantly requested. This creates a high demand for powerful screening technologies for the most active candidates within a selection of enzymes (enzyme screening) and accordingly for the determination of the product and/or substrate concentration of the enzyme-catalyzed reaction. Typically, optical methods relying on UV/Vis absorption or fluorescence are the method of choice for this purpose. However, these approaches are restricted to the monitoring of reactions in which either the substrates or products of the reactions show alterations in terms of absorption/fluorescence. As substrates of interest often do not contain the latter, artificial substrates are applied, which can falsify results by affecting the enzymatic reaction. In addition optical methods are sensitive towards impurities, such as buffers or cofactors. Other methods applied in enzyme screening are HPLC, GC/MS, or NMR. Mass spectrometric methods gained increasing importance for biocatalyst screening in the last years (Tholey and Heinzle 2002; Reetz 2003; Greis 2007; de Rond et al. 2015). Mass spectrometry can provide not only quantitative information, but is also suitable for unambiguous identification and structural analysis of the investigated molecules, which is an advantageous key feature of this technology compared to optical screening methods. This property of mass spectrometry is especially beneficial for the understanding of unknown conversion pathways by new enzymes.

Quantitative MALDI MS has been demonstrated to be a powerful technology for the determination of enzyme activities, as it provides easy and straightforward sample preparation and at least a moderate tolerance against salts, which allows for a direct monitoring of reaction mixtures without prior purification steps (Kang et al. 2000; Wittmann and Heinzle 2001).

In many cases substrates and products of enzyme-catalyzed reactions are compounds with molecular weights below 500 Da. Prerequisites for qualitative and quantitative analysis of low molecular weight compounds by MALDI MS in general are (1) a clear distinction of matrix and analyte signals in the low *m/z* region of the spectrum, and (2) the use of a molar matrix-to-analyte ratio (M/A ratio) of approximately 10:1 to 100:1 (Kang et al. 2000). Furthermore (3) suitable internal standards have to be applied to prevent errors deriving from signal suppression effects. These standards should either exhibit physicochemical similarity to the analyte of interest, as in case of peptides achieved by the exchange of a single amino acid for example, or at best should be its isotopically labeled counterpart (e.g., ^{2}H, ^{13}C, ^{15}N, ^{18}O). Last, (4) poor reproducibility caused by 'hot spot' formation should be minimized by automated measurement protocols and by increasing the sample homogeneity as far as possible. As ILMs are characterized by a strong reduction of matrix signals in the lower mass region and the formation of highly homogeneous sample preparations, they offer reasonable properties for the quantitative analysis of low molecular weight compounds, e.g., the substrates and products of enzyme-catalyzed reactions.

Quantification using internal standards—enzyme screening for the identification of the most active biocatalyst.

In 2004, Bungert and coworkers presented a method to identify the most active variant of the enzyme pyranose oxidase (POx) using ILMs and quantitative MALDI MS (Bungert et al. 2004). This enzyme catalyzes the conversion of glucose into glucosone (Fig. 2a). During this reaction, hydrogen peroxide is liberated as a by-product, which can be degraded into water and oxygen by catalase (Giffhorn 2000). For the screening, ten different variants of POx, overexpressed in *Peniophora* sp., were incubated with the substrate glucose in 96-well microtiter plates. To prevent an inactivation of POx by increasing concentrations of hydrogen peroxide, catalase was added to the mixture as well.

MALDI MS-based quantification of the substrate glucose was achieved by using fully ^{13}C-labeled glucose as an internal standard. An ILM composed of equimolar amounts of DHB and pyridine was used as matrix. Prior to determination of enzyme activities, a calibration curve was measured, monitoring the relative intensities of analyte to internal standard in dependence of the molar ratio of these two compounds. This step is the most critical and time consuming in the overall screening process. Major goals are to elucidate the range of linear correlation and to fit the analyte amounts with the needs of the enzymatic reaction. For the screening of the enzymatic

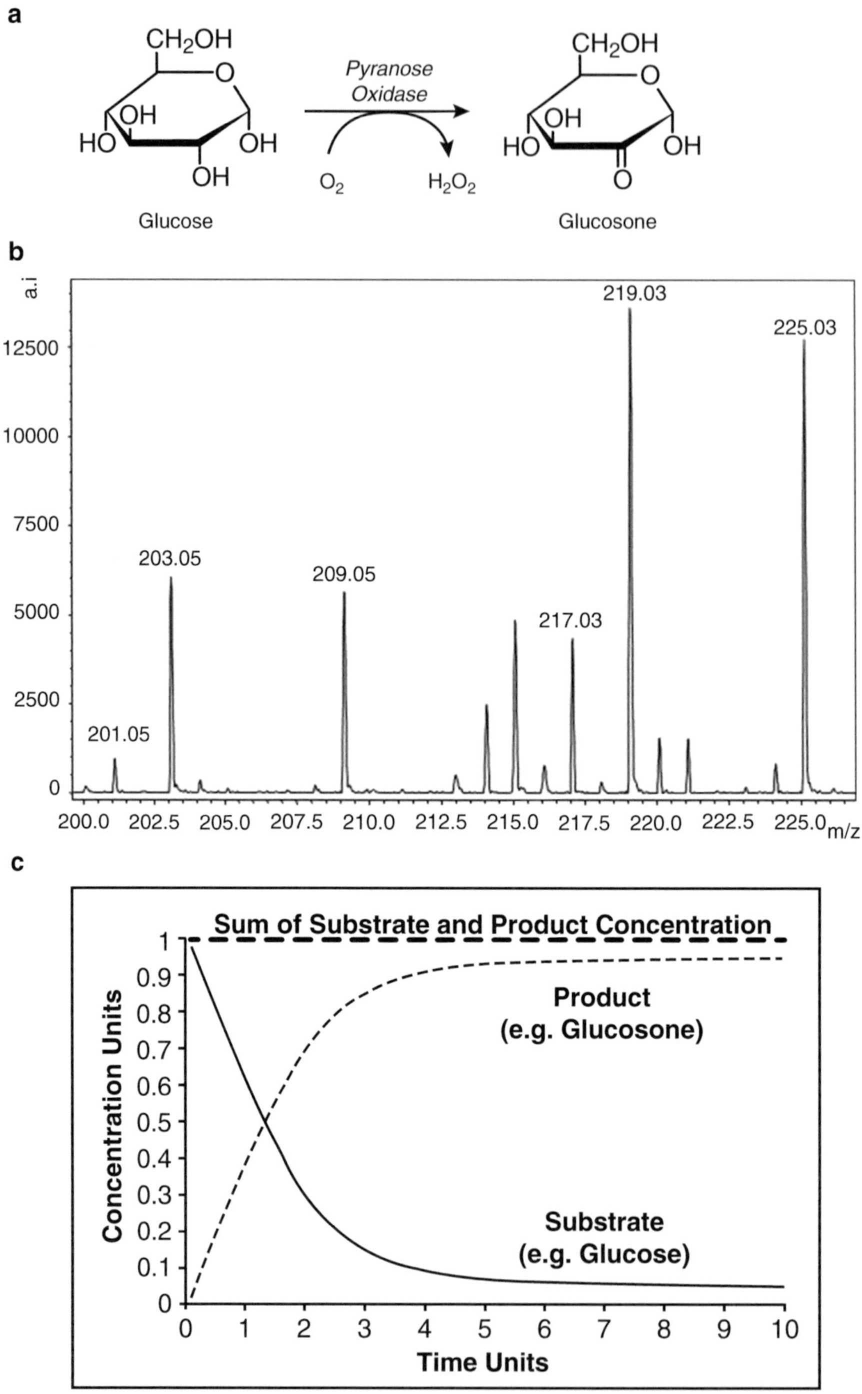

Fig. 2 (**a**) Reaction scheme of the pyranose oxidase-catalyzed conversion of D-glucose into 2-keto-D-glucose (glucosone). (**b**) MALDI MS spectrum of low molecular weight substrates and products of the pyranose oxidase-catalyzed reaction. Sodium ($[M+Na]^+$) (*m/z* 203.05) and potassium

activity, at certain time points, aliquots were taken and the enzymatic reaction was stopped by the addition of trifluoroacetic acid (TFA). Afterwards the solution was mixed with the internal standard. After sample preparation using the dried-droplet technique, samples were spotted each five times on the target for MALDI MS measurement.

Spectra of the experiments showed alkali adducts of the substrate and product (Fig. 2b). As the $[M+K]^+$ signals were more intense in comparison to the $[M+Na]^+$ signals, potassium adducts were chosen for quantification. Concentrations of glucose were assigned by comparing the signal intensity of the POx substrate to the intensity of the added internal standard ^{13}C-labeled glucose. Based on the calibration curve, the concentration of the substrate glucose could be determined easily based on the signal intensity ratio of the analyte and internal standard. The product concentration was calculated based on ^{13}C-labeled glucose as well, because isotopically labeled glucosone was not commercially available. As the ion response of glucosone in comparison to glucose was unknown, a correction factor to transform glucosone signals into the scale of glucose signals was determined. This was possible under the assumption that the sum of substrate and product amount is constant throughout the enzymatic reaction (Fig. 2c).

Monitoring substrate degradation and subsequent product generation enabled the identification of the most active variant of POx easily. The determination of glucose and glucosone by HPLC proved the reliability of the MALDI MS-based method. Compared to HPLC runs, which took 22.0 min for each sample, the fast measurement by MALDI MS took only 3.5 min for five replicates of each sample (42 s per spot), using a 20 Hz laser. Contemporary MALDI mass spectrometers can even manage the same workload with significantly reduced measurement times.

In summary, enzyme screening using MALDI MS exhibits various different advantages. Besides the benefits already mentioned in the introduction, such as simple sample preparation or low sample consumption, samples can be stored for re-measurement for a certain period of time.

Standard-free quantification—substrate screening in monitoring trypsin-catalyzed peptide degradation.

While for the quantification of low molecular weight compounds isotopically labeled standards are available, for more complex biomolecules, e.g., peptides and polynucleotides, this can be a limiting factor, in particular in large-scale screening processes. Although in most cases the application of structural homologous standards can deliver satisfying quantitative results, peak suppressing effects and potential peak overlapping should always be taken into account (Gusev et al. 1996).

Fig. 2 (continued) adducts ($[M+K]^+$) (*m/z* 219.03) of glucose and its fully ^{13}C-labeled analogue ($[U\text{-}^{13}C]$) (*m/z* 209.05 and 225.03) as well as of the product glucosone (*m/z* 201.05 ($[M+Na]^+$), 217.03 ($[M+K]^+$)) were detected in positive mode. All other peaks not assigned here belong to matrix components. Figure with permission from (Bungert et al. 2004). (**c**) Schematic time-dependent change of substrate and product concentrations during an enzymatic reaction. The sum of substrate and product concentrations at a certain time-point is constant all over the curve the proteolytic reaction were determined. Results obtained in parallel by relative quantification using a homologous internal standard (Trp^{11}-neurotensin) and HPLC-UV (at 214 nm) as an orthogonal method were in good agreement to those achieved by absolute quantification. This confirms the application of standard-free analysis by MALDI MS as a compatible quantitative method.

Considering the previously mentioned ion suppressive effects, if more than one substrate is analyzed simultaneously, this method can still be used for monitoring multisubstrate conversions in a rather semi-quantitative way. Therefore the authors investigated the evolution of tryptic digestion of an equimolar mixture of five peptides (angiotensin II, substance P, neurotensin, ACTH (1–17) and ACTH (18–39)). This more complex analyte environment impeded the direct determination of kinetic parameters for substrates with more than one cleavage site as different cleavage sites with different surrounding amino acid sequences result in various kinetic parameters. Once products are generated from the first proteolysis event they can serve as substrates for the following reactions and therefore can be consumed during the reaction. Additionally, the same problems of peak suppression (Knochenmuss and Zenobi 2003) hampered the overall accuracy. Therefore the authors chose the signal intensities rather than calculated concentrations to monitor the reaction, which means kinetic parameters, e.g., v_{max} and K_M, could not be determined in this experiment. However, the evolution of signal intensities was considered to be sufficient enough for a semi-quantitative method capable of screening several substrates for an enzyme.

In conclusion, ILMs applied with elevated M/A ratios allow for the direct quantification of peptides without the need of an internal standard. As for the understanding of enzyme activities the initial conversion velocity is the most relevant parameter, the limitation of the dynamic range to one order of magnitude also does not hamper the substrate screening. The described method is not limited to monitor proteolytic reactions but also applicable for most enzyme-catalyzed reactions as well. The key prerequisite is obviously a detectable difference in mass for substrate and product molecules.

3 Materials and Protocols

3.1 Materials

3.1.1 Laboratory and Technical Equipment

- Pipettes and pipette tips, reaction tubes
- Sonication bath
- Flash evaporator/rotary evaporator
- Vacuum oven/vacuum centrifuge
- Stainless steel MALDI MS target
- MALDI-TOF mass spectrometer

3.1.2 Chemicals

- MALDI matrix compound (e.g., CHCA, DHB, or SA)
- Organic base compound (e.g., pyridine, tributylamine or 1-methylimidazole)
- Methanol, ethanol, or acetonitrile
- Deionized water (18.2 MΩ/cm)
- Trifluoroacetic acid (TFA)
- Internal standard
- Mass calibration standard

3.2 Preparation of ILMs

In general, two different protocols are available for the preparation of ILMs. The major difference between both protocols is the performance of a solvent evaporation step as suggested in the original protocol (Armstrong et al. 2001). In the second protocol the acidic and basic component are simply premixed in appropriate solvents and are immediately used for further sample preparation (Zabet-Moghaddam et al. 2004). The original method is in particular suited for longer storage of the prepared ILM. No significant differences in the MALDI behavior have been observed for the two protocols up to now.

3.2.1 Preparation with Solvent Evaporation (Armstrong et al. 2001)

- Prepare the ILM by dissolving the crystalline matrix (~0.5 g) in methanol (15 mL) and add an equimolar amount of organic base.
- Sonicate the mixture for 5 min, filtrate and remove the solvent by evaporation.
- Dry the product in a vacuum oven at room temperature in order to remove residual solvent traces. The resulting organic salt can be stored at 4 °C in the fridge for a few days and can be dissolved freshly in the required solvents prior to measurement.

3.2.2 Fast Preparation (Zabet-Moghaddam et al. 2004)

- Alternatively, prepare the ILM by dissolving the crystalline matrix directly in the solvent mixture, which is used for sample preparation, e.g., 70% acetonitrile, and add the equimolar amount of pure organic base.
- Sonicate the mixture for 5 min.
- ILMs prepared in this way should be made freshly on each day of measurement as degradation occurs faster in dissolved than in solid form of the matrix.

3.3 First Step Towards Quantification

- Analyze the ion response of your analyte of interest by MALDI MS and identify the amount on target required to get a reasonable signal-to-noise ratio in the mass spectrum.
- Assign the linear range for the analyte by plotting the signal intensities for a broad variety of concentrations around the amount on target determined in the previous step. The linear range of an analytical method is defined as the range in which the signal intensity increases directly proportional with the analyte concentration.
- For highly contaminated samples it is recommended to purify the samples prior to MALDI MS measurement. Due to sample losses during purification the internal standard should be added beforehand.

3.4 Setting up a Calibration Curve

3.4.1 General Rules

- In general, a suitable calibration curve can either span the whole linear range or can be limited to the region of interest in case your experimental setup allows you to estimate the approximate concentration of analyte a priori.
- As a general rule, the more data-points you collect to set up the curve, the more accurate the determination of the unknown concentration of the analyte. The final concentration of the analyte should be determined in the central part of the curve.

3.4.2 Setting up a Calibration Curve Using an Internal Standard

- Prepare dilution series of the analyte covering the concentration range of interest. Use the same buffer solutions as in your final experiment before mixing the analyte solution with matrix as this influences the ion response. Finally, spike the calibration solutions with a constant amount of internal standard, dry (e.g., 5 min under air stream or under vacuum) and measure the calibration solutions (each concentration at least three times) by MALDI MS.
- Generate the calibration curve by plotting analyte concentrations on the *x*-axis and signal intensity ratios of analyte and internal standard on the *y*-axis of the diagram.

3.4.3 Setting up a Calibration Curve Without an Internal Standard

- Proceed as described above with the preparation of the analyte dilution series but without the addition of an internal standard.
- Generate the calibration curve by plotting analyte concentrations on the *x*-axis and signal intensities of the analyte on the *y*-axis of the diagram.

3.5 *Sample Preparation for MALDI MS Measurement and Data Analysis*

- Mix the sample (with internal standard) and the ILM of choice at a suitable molar M/A ratio. Spot at least three replicates of the sample on the target, dry the sample spots as described in Sect. 3.4.2, and acquire the MALDI MS data.
- Calculate the signal intensity ratio of analyte and internal standard or, in case of standard-free quantification, simply determine the signal intensity of the analyte from the mass spectrum. Use the calibration curve to determine the unknown concentration of the analyte.

3.6 *General Remarks*

- Note that certain ILM preparations lead to the formation of small crystals after ejecting the target out of the high vacuum. This effect does usually not hamper the quality of a potential re-analysis of the sample preparations.
- Addition of small amounts of acids can help to promote ionization of different analytes. For example, the analysis of phosphopeptides can be facilitated by the addition of 1% H_3PO_4 to the ILM (Tholey 2006).

References

Armstrong DW, Zhang LK, He LF, Gross ML (2001) Ionic liquids as matrixes for matrix-assisted laser desorption/ionization mass spectrometry. Anal Chem 73(15):3679–3686

Berthod A, Crank JA, Rundlett KL, Armstrong DW (2009) A second-generation ionic liquid matrix-assisted laser desorption/ionization matrix for effective mass spectrometric analysis of biodegradable polymers. Rapid Commun Mass Spectrom 23(21):3409–3422

Bungert D, Bastian S, Heckmann-Pohl DM, Giffhorn F, Heinzle E, Tholey A (2004) Screening of sugar converting enzymes using quantitative MALDI-ToF mass spectrometry. Biotechnol Lett 26(13):1025–1030

Cohen LH, Gusev AI (2002) Small molecule analysis by MALDI mass spectrometry. Anal Bioanal Chem 373(7):571–586

Cramer R, Corless S (2005) Liquid ultraviolet matrix-assisted laser desorption/ionization—mass spectrometry for automated proteomic analysis. Proteomics 5(2):360–370

Crank JA, Armstrong DW (2009) Towards a second generation of ionic liquid matrices (ILMs) for MALDI-MS of peptides, proteins, and carbohydrates. J Am Soc Mass Spectrom 20(10):1790–1800

de Rond T, Danielewicz M, Northen T (2015) High throughput screening of enzyme activity with mass spectrometry imaging. Curr Opin Biotechnol 31:1–9

Distler AM, Allison J (2001) Improved MALDI-MS analysis of oligonucleotides through the use of fucose as a matrix additive. Anal Chem 73(20):5000–5003

Dreisewerd K (2003) The desorption process in MALDI. Chem Rev 103(2):395–426

Dreisewerd K (2014) Recent methodological advances in MALDI mass spectrometry. Anal Bioanal Chem 406(9–10):2261–2278

Gabriel SJ, Pfeifer D, Schwarzinger C, Panne U, Weidner SM (2014) Matrix-assisted laser desorption/ionization time-of-flight mass spectrometric imaging of synthetic polymer sample spots prepared using ionic liquid matrices. Rapid Commun Mass Spectrom 28(5):489–498

Giffhorn F (2000) Fungal pyranose oxidases: occurrence, properties and biotechnical applications in carbohydrate chemistry. Appl Microbiol Biotechnol 54(6):727–740

Greis KD (2007) Mass spectrometry for enzyme assays and inhibitor screening: an emerging application in pharmaceutical research. Mass Spectrom Rev 26(3):324–339

Gusev AI, Wilkinson WR, Proctor A, Hercules DM (1996) Direct quantitative analysis of peptides using matrix assisted laser desorption ionization. Anal Bioanal Chem 354(4):455–463

Harvey DJ (1999) Matrix-assisted laser desorption/ionization mass spectrometry of carbohydrates. Mass Spectrom Rev 18(6):349–450

Hillenkamp F, Karas M, Beavis RC, Chait BT (1991) Matrix-assisted laser desorption/ionization mass spectrometry of biopolymers. Anal Chem 63(24):1193A–1203A

Jaskolla TW, Lehmann WD, Karas M (2008) 4-Chloro-alpha-cyanocinnamic acid is an advanced, rationally designed MALDI matrix. Proc Natl Acad Sci U S A 105(34):12200–12205

Kang MJ, Tholey A, Heinzle E (2000) Quantitation of low molecular mass substrates and products of enzyme catalyzed reactions using matrix-assisted laser desorption/ionization time-of-flight mass spectrometry. Rapid Commun Mass Spectrom 14(21):1972–1978

Karas M, Hillenkamp F (1988) Laser desorption ionization of proteins with molecular masses exceeding 10,000 daltons. Anal Chem 60(20):2299–2301

Knochenmuss R, Zenobi R (2003) MALDI ionization: the role of in-plume processes. Chem Rev 103(2):441–452

Lemaire R, Tabet JC, Ducoroy P, Hendra JB, Salzet M, Fournier I (2006) Solid ionic matrixes for direct tissue analysis and MALDI imaging. Anal Chem 78(3):809–819

Li YL, Gross ML (2004) Ionic-liquid matrices for quantitative analysis by MALDI-TOF mass spectrometry. J Am Soc Mass Spectrom 15(12):1833–1837

Mank M, Stahl B, Boehm G (2004) 2,5-Dihydroxybenzoic acid butylamine and other ionic liquid matrixes for enhanced MALDI-MS analysis of biomolecules. Anal Chem 76(10):2938–2950

Nicola AJ, Gusev AI, Proctor A, Jackson EK, Hercules DM (1995) Application of the fast-evaporation sample preparation method for improving quantification of angiotensin II by matrix-assisted laser desorption/ionization. Rapid Commun Mass Spectrom 9(12):1164–1171

Palmblad M, Cramer R (2007) Liquid matrix deposition on conductive hydrophobic surfaces for tuning and quantitation in UV-MALDI mass spectrometry. J Am Soc Mass Spectrom 18(4):693–697

Reetz MT (2003) An overview of high-throughput screening systems for enantioselective enzymatic transformations. Methods Mol Biol 230:259–282

Sun P, Armstrong DW (2010) Ionic liquids in analytical chemistry. Anal Chim Acta 661(1):1–16

Sze ET, Chan TW, Wang G (1998) Formulation of matrix solutions for use in matrix-assisted laser desorption/ionization of biomolecules. J Am Soc Mass Spectrom 9(2):166–174

Tholey A (2006) Ionic liquid matrices with phosphoric acid as matrix additive for the facilitated analysis of phosphopeptides by matrix-assisted laser desorption/ionization mass spectrometry. Rapid Commun Mass Spectrom 20(11):1761–1768

Tholey A, Heinzle E (2002) Methods for biocatalyst screening. Adv Biochem Eng Biotechnol 74:1–19

Tholey A, Heinzle E (2006) Ionic (liquid) matrices for matrix-assisted laser desorption/ionization mass spectrometry-applications and perspectives. Anal Bioanal Chem 386(1):24–37

Tholey A, Zabet-Moghaddam M, Heinzle E (2006) Quantification of peptides for the monitoring of protease-catalyzed reactions by matrix-assisted laser desorption/ionization mass spectrometry using ionic liquid matrixes. Anal Chem 78(1):291–297

Towers MW, McKendrick JE, Cramer R (2010) Introduction of 4-chloro-alpha-cyanocinnamic acid liquid matrices for high sensitivity UV-MALDI MS. J Proteome Res 9(4):1931–1940

Vaidyanathan S, Gaskell S, Goodacre R (2006) Matrix-suppressed laser desorption/ionisation mass spectrometry and its suitability for metabolome analyses. Rapid Commun Mass Spectrom 20(8):1192–1198

Wittmann C, Heinzle E (2001) MALDI-TOF MS for quantification of substrates and products in cultivations of Corynebacterium glutamicum. Biotechnol Bioeng 72(6):642–647

Zabet-Moghaddam M, Heinzle E, Tholey A (2004) Qualitative and quantitative analysis of low molecular weight compounds by ultraviolet matrix-assisted laser desorption/ionization mass spectrometry using ionic liquid matrices. Rapid Commun Mass Spectrom 18(2):141–148

Zabet-Moghaddam M, Heinzle E, Lasaosa M, Tholey A (2006) Pyridinium-based ionic liquid matrices can improve the identification of proteins by peptide mass-fingerprint analysis with matrix-assisted laser desorption/ionization mass spectrometry. Anal Bioanal Chem 384(1):215–224

Part III
Advances in MALDI Mass Spectrometry Imaging

Despite all achievements made by using relative quantification methods, standard-free and absolute quantification, e.g., for the analysis of enzyme activities as well as for proteomics-related applications, is still of high interest.

However, absolute quantification in MALDI MS is hampered by the dependence of the ion formation on the applied laser energy/fluence (Dreisewerd 2003), the appearance of spatially isolated 'hot spots' in the sample due to the inhomogeneous incorporation of analytes into the matrix crystal lattice and a typically nonlinear increase of signal intensities; the latter can usually be observed over a broad range of matrix-to-analyte (M/A) ratios (Dreisewerd 2003). Furthermore a reduction of sample complexity, e.g., by solid phase extraction or chromatographic fractionation, might be necessary to reduce ion suppression effects (Dreisewerd 2014). Replacement of crystalline matrices by ILMs can circumvent the mentioned drawbacks to some extent. It was shown that the homogeneous sample preparations achievable by the use of ILMs are also suitable for quantitative analysis using MALDI MS and allow the direct quantification of peptides (Tholey et al. 2006). The correct adjustment of molar M/A ratios was identified as major critical parameter in this respect.

Typical M/A ratios used for the analysis of peptides are in the range of 10^3–10^5 (Hillenkamp et al. 1991). However, for the absolute quantification of peptides the M/A ratio has to be noticeably increased, which was shown by the analysis of neurotensin measured with an ILM consisting of an equimolar ratio of CHCA and 3-(dimethylamino)-1-propylamine (DMAPA). Using M/A ratios of 35,700–500,000 a linear correlation between the amount of peptide on target and signal intensity was observed (Tholey et al. 2006). The dynamic range of linearity was determined to one order of magnitude. This range is slightly lower compared to those achievable with relative quantification using an internal standard (Li and Gross 2004). Using glycerol-based liquid matrices, Palmblad and Cramer have shown a linear dynamic range of more than two orders of magnitude without the use of an internal standard (Palmblad and Cramer 2007). When using an internal standard, the ratio between the analyte and the internal standard is the critical parameter (Kang et al. 2000). In addition several factors influence the dynamic range, e.g., (1) the applied laser energy/fluence, (2) the limitation of applicable matrix amount onto the target and (3) the limit of detection as well as (4) the ionization efficiency of the analyte itself. However, in case of unknown analyte concentrations the elucidation of the best M/A ratio is a prerequisite task; but if the sample complexity hampers this first step, relative quantification with an internal standard is the method of choice. For the screening of enzyme reactions, where the initial substrate concentration is a known parameter, the absolute quantification without an internal standard is possible as long as the number of analytes to be detected simultaneously is low enough to minimize peak suppressing effects.

Nevertheless, this method enabled to monitor enzyme-catalyzed reactions, e.g., the time-dependent degradation of peptides and peptide mixtures by trypsin (Tholey et al. 2006). This model situation represents a method for enzyme screening, e.g., when a set of different isoforms of an enzyme is screened to determine the most active one by conversion of the same substrate. By acquisition of a calibration curve, variation of the initial substrate concentration of neurotensin and the application of a classical enzyme kinetic method (Lineweaver-Burk plot) the K_M- and v_{max}-value of

Techniques for Fingermark Analysis Using MALDI MS: A Practical Overview

Simona Francese

Abstract Matrix-assisted laser desorption/ionization (MALDI), mass spectrometry profiling (MSP), and mass spectrometry imaging (MSI) are continuously pushing the boundaries of many applications in which they can be employed. In the field of forensic science, MALDI MS has proven to be a useful technique but reports were sporadic and only in an academic context until 2009. However, in the past 6 years it has been demonstrated that MALDI MSP and MSI can be practically implemented as an analytical forensic tool in the context of fingermark/fingerprint analysis. Previously always regarded 'simply' as physical evidence, fingermarks are an invaluable source of chemical information that can be exploited to provide intelligence on an individual's lifestyle and activities taking place prior to the crime. A solid proof of principle of the versatility and robustness of MALDI MSP and MSI has gained MALDI a place in the newly launched *Fingermark Visualisation Manual* edited by the Home Office of the United Kingdom (UK), whose guidelines are followed in the UK and are highly regarded worldwide. Protocols are currently being developed to be used operationally within primarily UK law enforcement. In this chapter, a practical overview is given for informed method development. Protocols that have been successful in answering specific research questions will also be outlined along with the most significant and interesting applications of MALDI MSP and MSI in fingermark/fingerprint analysis.

1 Introduction

There is little doubt that MALDI-based techniques are very versatile and can assist in the elucidation of a number of different research questions in the life sciences. The year 2009 saw the birth of yet another application of MALDI, namely the chemical mapping of fingermarks.

S. Francese (✉)
Biomedical Research Centre, Sheffield Hallam University, Sheffield, UK
e-mail: s.francese@shu.ac.uk

R. Cramer (ed.), *Advances in MALDI and Laser-Induced Soft Ionization Mass Spectrometry*, DOI 10.1007/978-3-319-04819-2_6

For over a hundred years, fingermarks/fingerprints[1] have been seen 'only' as a source of physical evidence; the fingermark/fingerprint ridge[2] pattern, including the ridge flow (level 1 details) and local characteristics of the ridge flow called *minutiae* (level 2 details) have been exploited for suspect identification on the basis that each individual has unique fingerprints. Physical comparison of level 1 and level 2 details between a ridge impression found at a crime scene and fingerprints stored in databases is the process through which fingerprint experts may identify a suspect, following strict country-specific evaluation and cross-examination rules for declaring a match.

Despite the advent of DNA analysis, 'fingerprinting' is still the most used and successful form of biometric identification, leading the way in the UK for suspect identification and being responsible for two thirds of all identifications (N. Denison, Head of Identification for the Humber and Yorkshire Region UK, personal communication).

Although hugely successful, 'fingerprinting'[3] has its pitfalls and does not succeed when the quality of the visualized crime scene ridge impression is inadequate; this could be due to its original nature (smudged, partial, empty, overlapped marks) or the inadequacy of the forensic enhancement workflow, the choice of which is heavily dependent on the level of expertise/competence of the CSI,[4] the surface of deposition, the environmental conditions to which the ridge impression was exposed, and the age of the fingermark. Some of these factors, such as environmental conditions and age, are either very difficult or impossible to correctly evaluate at the crime scene, making the visualization and development of the fingermark not trivial, despite world-leading guidelines, published in the *Fingermark Visualization Manual* edited by the Home Office, UK (Bandey et al. 2014). Additionally, should the suspect not have been previously convicted, their fingerprints will not be present in the National Database, making the quality of the crime scene ridge impression unimportant for database search, comparison, and match.

However, ridge impressions are not 'simply' physical evidence but indeed a very valuable source of chemical intelligence. They originate from the transfer of sweat to a surface from an individual's fingertip upon contact. If this contact is intentional, such as control ridge impressions taken at police stations or airports, we should use the term 'fingerprints' as opposed to 'fingermarks,' which instead result from the accidental and involuntary contact with surfaces. Sweat carries a plethora of biological substances and consequently latent fingermarks[5] are a complex biological matrix featuring the presence of inorganic and organic constituents such as alkali metals/salts, amino acids, lipids, and proteins (Knowles 1978; Ramotowski 2001),

---

[1] Fingerprints are obtained intentionally, for instance at police stations or airports, while fingermarks are a result of accidental and involuntary contact with surfaces (more details below).

[2] In fingerprints, ridges are the 'lines' generated by the raised parts of the skin on fingertips.

[3] In a forensic context 'fingerprinting' refers to the process of identifying an individual through their fingerprints. This is different from 'fingerprinting' used in proteomics where this term refers to the identification of a protein usually through a bottom-up approach.

[4] Crime Scene Investigator.

[5] The most used definition of latent fingermark is 'a fingermark invisible to the naked eye.'

which we can define as *endogenous* as they come from within the body. In addition, any substance our fingertips come in contact with can be transferred to a surface and subsequently detected in a fingermark; we can commonly refer to these as *exogenous* substances. Examples of these substances as obvious sources of forensic intelligence may include drugs of abuse, explosives, condom lubricants, make-up, paint, and blood.[6] Our group also refers to a third category of species as *semi-exogenous* when they are not naturally present in our body but they are, in different ways, introduced and excreted (intact or metabolized) through sweat, such as substances (metabolites) derived from prescription medications or drugs (Bradshaw et al. 2015), including smoke (Benton et al. 2010a, b) and even consumed food and drinks (Bradshaw et al. 2012).

These three categories of chemical species, endogenous, exogenous, and semi-exogenous are an invaluable source of intelligence. For example, lipids and proteins (endogenous species) can often act as biomarkers, a term used in the biomedical field for (biomolecular) species indicating a physiological state of the individual, being pathological (diseases), pharmacological (response induced by medications), or biological (molecular make-up placing the individual into a biological category such as diabetic). It is therefore entirely conceivable to think of the detection of these substances as a forensic opportunity to provide additional intelligence on the suspect, as will be discussed in Sect. 2.

Given the huge potential of fingermarks as a source of 'enhanced' intelligence and the profound implications and impact this could have on a more efficient Criminal Justice System (with attached savings to the public purse), it is not surprising then that the analytical science community, with a keen interest in forensic science, has made tremendous efforts in the last few years to develop, either existing or new, analytical tools specifically for the detection, and in some cases, mapping of fingermark residuals. A plethora of different and possibly suitable analytical techniques have been reported in the literature for the analysis of fingermarks but only a number of the authors have shown dedication to this area of research with follow-ups and clear developments. These techniques can be grouped into spectroscopic and mass spectrometric, with the latter category being the strongest in terms of diversity of the ionization techniques used (DESI MS, SALDI MS, DART MS, SIMS, and MALDI MS) and 'presence' in the literature. Figure 1 shows a representation of the above analytical efforts in the form of number of publications per technique, starting from 2008 onwards.

Although DESI was the first mass spectrometric technique showing its potential for fingermark profiling and imaging (Ifa et al. 2008), only MALDI MS and SIMS are presently featured in the recently re-edited Home Office *Fingermark Visualization Manual*.[7] Compared to the other techniques reported in Fig. 1, and despite inevitable

[6] Blood may be present on the surface prior to the deposition of the fingermark (mark in blood) or landing some time after the fingermark deposition (coincidental association), or coming from the fingertip of the owner of the mark (blood mark).

[7] MALDI is reported as a new process having high potential to be implemented in the forensic practice (Cat C) but still under development to reach the required maturity (TLR 3–4).

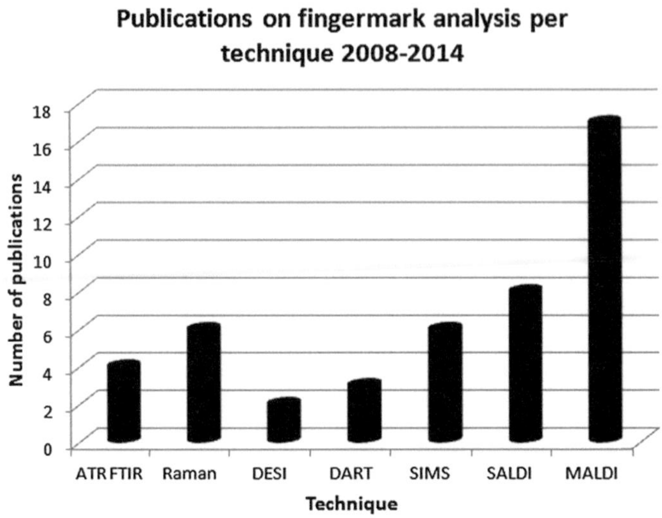

Fig. 1 Histogram showing the number of publications featuring the use of either ATR FTIR, Raman (in its various forms including SERS and CARS), DESI, DART, SIMS, SALDI, or MALDI (in profiling and/or imaging mode) for the analysis of fingermarks. Figures reported refer to peer-reviewed papers in 89% of the cases while the remaining 11% originate from online magazines, technical articles, or application notes. This information has been obtained by searching both Pubmed and ISI Web of Science databases using the keywords fingerprints, fingermarks, MALDI, ATR FTIR, Raman, DESI, DART, SIMS, SALDI, and MALDI

limitations as with any analytical technique, MALDI MS has clearly demonstrated its potential, particularly with respect to versatility, robustness and compatibility with current fingermark enhancement techniques (FET) in a number of different scenarios, in both profiling and imaging mode (Bradshaw et al. 2013a), and having the highest number of publications to date. In terms of feasibility, a leap forward was made with the invention of the *dry–wet* method (Ferguson et al. 2011). This 2-step method enables visualization of fingermarks on surfaces by: 1) brushing the surface with α-cyano-4-hydroxycinnamic acid acid (CHCA), a common MALDI matrix, and recovering the evidence through tape-lifting, and 2 homogeneously and rapidly spraying the mark with a solvent mixture, in which both the matrix and the analytes of interest are soluble, making the mark ready for MALDI MSI analysis. Together with its wide compatibility with current FET, this method probably initiated the superiority of MALDI over those analytical techniques that can only be used in academic settings. Therefore, for MALDI, the *dry–wet* method 'bridges the gap' by actually making the MS analysis of 'real' fingermark evidence possible. Applications and protocols for this method are discussed in Sects. 2 and 3, respectively.

Though the option is now available to locate, recover, photograph, and analyze the mark with the *dry–wet* method (albeit further development is in progress for wider applicability), it is not expected that this method entirely replaces a century of forensic practice to visualize marks on surfaces. Therefore, in order to integrate MALDI in the armory of techniques for fingermark analysis, the application of

MALDI has been demonstrated to be compatible with the pre-application of existing chemical and physical developers currently used to localize marks at crime scenes. FET compatibility is strongly desired by the UK Home Office as it would allow new and emerging technologies to provide additional forensic intelligence to both inform investigations and provide court room evidence. With regard to mass spectrometry, two UK groups have demonstrated a certain degree of compatibility of the techniques SIMS and SALDI MS with cyanoacrylate fuming, with SIMS showing compatibility also with powder suspension (Bailey et al. 2013; Sundar and Rowell et al. 2014). Our group has shown much wider compatibility of MALDI (for MSP, MSI or both) as it can be used in tandem with some blood enhancement techniques (BET) (Bradshaw et al. 2014), enhancement powders, vacuum metal deposition, 1,8-diazafluoren-9-one, and ninhydrin (Bradshaw et al. 2013a) on a number of different surfaces and under different environmental conditions in laboratory settings and, in some cases, in pseudo-operational trials in collaboration with West Yorkshire Police (Bradshaw et al. 2014). For operational deployment, these findings are very encouraging and justify further research into the development of sample preparation and analytical protocols for MALDI MSP and MSI of fingermarks in forensics.

Indeed, since the pioneering start in 2008, the rapid evolution in the applicability of MALDI MS-based techniques for fingermark analysis required extensive methodological developments, taking into account the complex and varying chemical nature of fingermarks, the most appropriate corresponding options for their preparation as well as their correct storage. In this chapter, the most notable protocols of MALDI MS in the field of fingermark analysis will be discussed together with the limitations of the technique. 'Tips' will be given to assist both keen beginners and experienced MALDI users approaching fingermark analysis for the first time, thus enabling their contribution towards the development of operationally deployable protocols.

2 Applications

Valuable intelligence from latent fingermarks can be obtained using MALDI in profiling and/or imaging mode. In profiling mode (using MS and MS/MS), MALDI can be used for the quick detection or confirmation (in minutes) of the presence of forensically interesting substances. Speed in the acquisition of the intelligence is very desirable and the profiling mode can be employed on fingermarks previously visualized using a FET, which already yielded a suitable image of the ridge impression for suspect identification purposes. Like some other analytical techniques, MALDI can also be used for chemical imaging; this feature enables multiple images of the same mark to be obtained in under 90 min from sample preparation to the end of acquisition (depending on the mass spectrometer and laser frequency), potentially succeeding where FET failed or further 'improving' the quality of the mark image. Multiple images may be stitched together (Fig. 2a) or superimposed (Fig. 2b) to improve ridge continuity, to increase completeness of the mark image, or even to separate

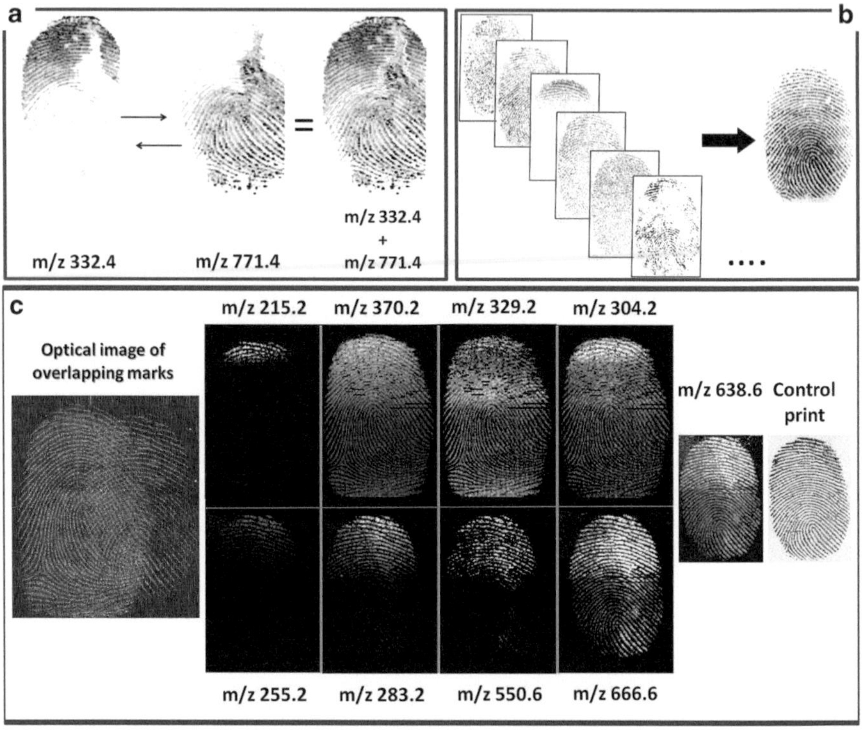

Fig. 2 Improving the quality of fingermark images by MALDI MSI: (**a**) stitching *m/z* images together and (**b**) superimposing multiple *m/z* images of the same mark to improve ridge continuity; (**c**) separation of overlapping marks through interrogation of multiple *m/z* ions (Francese et al. 2013b, Analyst, **138**, 4215-4228, adapted and reproduced by permission of The Royal Society of Chemistry)

overlapping ridge impressions as previously shown by our group and illustrated in Fig. 2c (Francese et al. 2013b; Bradshaw et al. 2012). This in turn improves the quality of the database search and increases the chance of suspect identification.

When providing intelligence to inform investigations or court cases, mapping forensically interesting chemicals in fingermarks is also very important to make robust associative evidence. The use of MSI shows whether the chemical is, in fact, present on the ridges, rather than on the surface of the deposition where the mark lies. In the latter case, the detection of this chemical is just a contamination and this changes the forensic situation significantly.

For example, in a hypothetical rape case where the perpetrator had worn a condom, the ability to determine whether the condom lubricants are present on the ridges (i.e., condom lubricants-contaminated fingertip, indicating that the accused had handled the condom) or on both the ridges and the furrows[8] (i.e., mark made in

[8] In fingerprints/fingermarks, furrows, otherwise termed valleys, are the 'voids' in between ridges and are generated by the lower parts of the skin on fingertips.

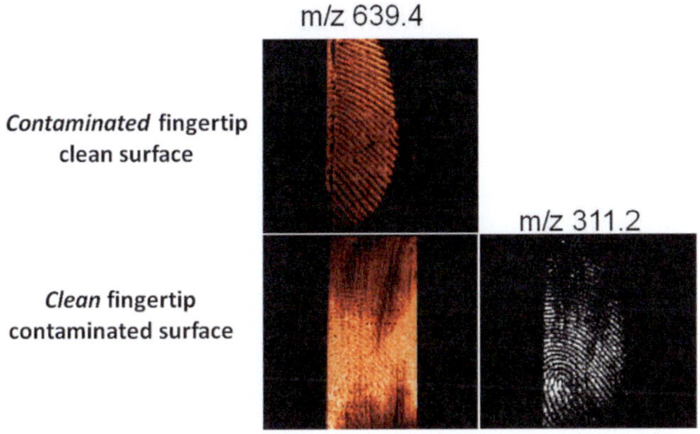

Fig. 3 MALDI MSI of *Trojan Enz*-contaminated fingermarks. The ion at *m/z* 639.5 (9-mer-nonoxynol 9) is solely distributed on the ridges of a mark if the fingertip came in contact with the condom lubricant prior to touching a surface. The same ion is distributed both on the ridges and the furrows if the mark was generated by a clean fingertip upon contact with a contaminated surface and conclusions can be corroborated by imaging another ion such as an endogenous fatty acid (*m/z* 311.2), the sole distribution of which on the ridges proves that the nonoxynol-9 image is not an artifact (Bradshaw et al. 2013b, Analyst, **138**, 2546-2557, adapted and reproduced by permission of The Royal Society of Chemistry)

condom lubricants, pointing at sexual activity prior to the alleged crime) might make the difference between an acquittal or a conviction. The opportunity to discriminate between these two scenarios has been illustrated by our group (Bradshaw et al. 2013b) and is shown in Fig. 3.

Indeed, the detection of condom lubricants may provide important associative evidence, which becomes probative when the lubricants profile detected in a mark matches that of an unopened condom packet found in possession of the alleged offender or within the victim's vaginal swab (if this is taken soon after the crime). It has also been demonstrated that 'condom discrimination' at a brand level (and in some cases at a type level) is achievable (Bradshaw et al. 2013b).

In the above work, a MALDI MSP sample preparation protocol was also suggested to obtain comprehensive information on the condom lubricant composition; polydimethylsiloxane (PDMS) was never detected using MALDI MSP and MSI in fresh contaminated marks. However, a set of experiments performed with the purpose of evaluating robustness of the methodology for aged condom-contaminated marks revealed that it was only possible to detect PDMS in aged marks, that is, marks kept at 25 °C and 60% relative humidity for a month. It can be speculated that the ageing treatment caused a decrease in the presence of species that are either more volatile or more prone to degradation than PDMS (under those conditions) and that these species may act as PDMS suppressants in fresh (non-aged) fingermarks. Detection of PDMS and ethoxylate-based polymers will be briefly discussed in Sect. 3.5.

In addition to lubricant detection and mapping by MALDI MSI, a protocol was devised to make this information more robust and retrievable if this technology was used in conjunction with other spectroscopic ('non-destructive') techniques such as Raman and ATR-FTIR spectroscopy (Sect. 3.5). This protocol provided complementarity of information on the condom lubricant composition together with the detection and mapping of the fingermarks' endogenous species as illustrated in Fig. 4.

Although much has been done to allow the detection of exogenous substances, valuable intelligence can also originate from the exploitation of the endogenous content of fingermarks. Lipids are certainly the easiest class of molecules to ionize as demonstrated by the first publication in the literature showing MALDI MSI analysis of endogenous lipids in ungroomed[9] marks (Wolstenholme et al. 2009). They can be readily detected using CHCA as matrix in positive ionization mode. Although we found that this matrix was the most efficient overall, certain classes of fatty acids (nonpolar such as squalene) cannot be efficiently ionized, and doping the matrix with cations such as Li^+ is certainly an option. Matrix-spotting and spraying techniques as well as the patented *dry–wet* method (see below in this section) work well to ionize lipids. In terms of spraying, our group has resorted to the SunCollect autosprayer (SunChrom, Friedrichsdorf, Germany) employing pneumatic atomization and being capable to spray an entire fingermark within 30 min for the matrix solution and within 10 min just for the solvent.

Recently, another automated matrix sprayer was trialed, the TM sprayer (HTX Technologies, NC, USA), and a very efficient protocol, allowing clear and detailed ridge visualization in under 1 min, was devised for the interrogation of lipids, which will be illustrated in Sect. 3.2. Lipids can be used for improving the quality of the ridge detail (Fig. 2) but they can also be an indicator of pathological states (biomarkers). Therefore, their detection and mapping may have both forensic and biomedical (prognostic) applications.

Another interesting class of biomolecules in fingermarks are peptides and proteins. Our group has exploited the detection of these species from the fingermarks of 80 donors to 'biologically categorize' and distinguish between fingermarks belonging to men and those belonging to women (Fig. 5), with an 85% prediction accuracy to date (Francese et al. 2011; Ferguson et al. 2012). This advance has unleashed our imagination and that of other researchers with respect to the achievable information on the individual's lifestyle and physiological state, using what could be termed 'offender chemical profiling.' Further work is in progress in order to increase the accuracy of prediction and robustness of the method by using a larger number of donors with no participation exclusion criteria, which were applied to the previous study.

Fingermark proteins originating from either the victim or the offender can also be detected and mapped. This has been the case when applying MALDI MSP and MSI to the specific detection of blood in fingermarks (or in bloodstains). A reliable

[9] Ungroomed fingermarks are 'lipid-depleted' marks from fingers that have been washed and have subsequently not touched any body parts or contaminants (to test the sensitivity of the technology). For a more comprehensive description, see Sect. 3.1.

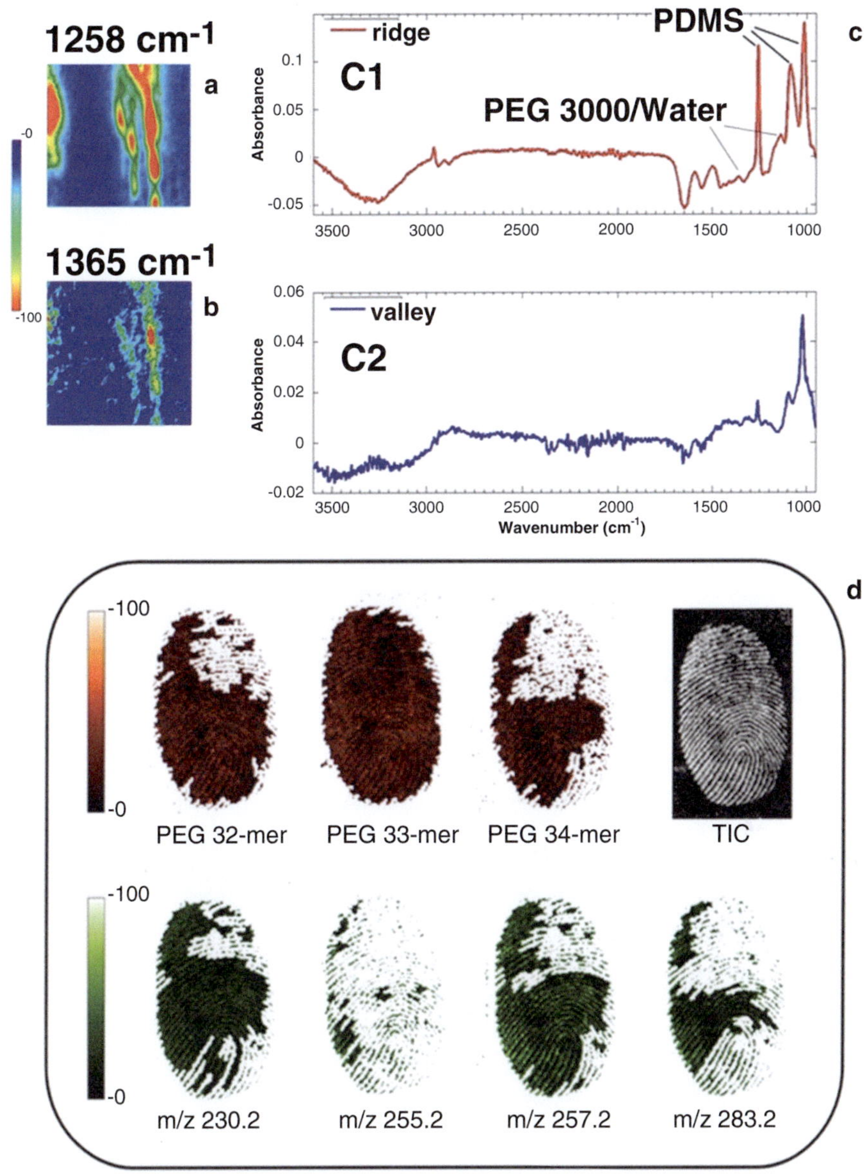

Fig. 4 Combined ATR-FTIR imaging/MALDI MSI analysis of a *Condomi Max Love*-contaminated fingermark. Panels A and B show the ATR-FTIR image of PDMS at 1258 cm^{-1} and the ATR-FTIR image of PEG at 1365 cm^{-1} present on the same ridges of a *Condomi Max Love*-contaminated fingermark selected region. Panel C shows two ATR-FTIR spectra from the ridge (C1) and the valley (C2) (high and low concentration of the two polymers respectively). Panel D displays the MALDI MS images of the PEG 32-mer, 33-mer, and 34-mer ion signals, the complete ridge pattern provided by the image of the total ion current (TIC), as well as a small sample set of the many fatty acids detected (Bradshaw et al. 2013b, Analyst, **138**, 2546-2557, adapted and reproduced by permission of The Royal Society of Chemistry)

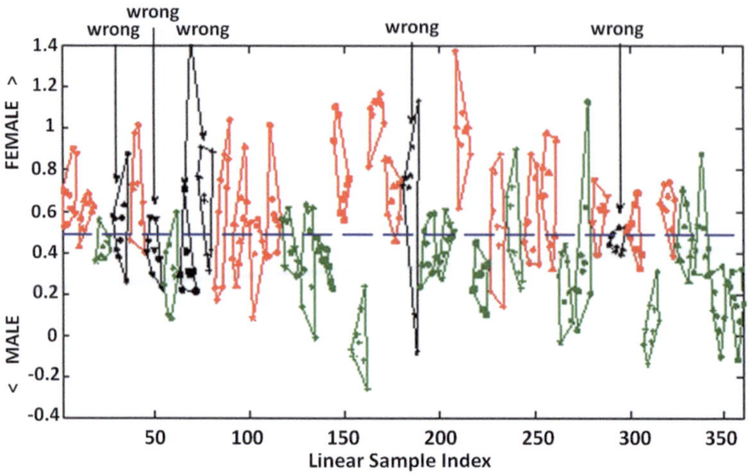

Fig. 5 Sex classification results and validation using an independent test set within partial least square discriminant analysis (PLSDA). Multiple spectra from each donor are grouped together and can be found within the area defined by a line connecting the peripheral data points. Prediction above the *blue threshold line* results in a female and below the threshold in a male sex assignment. Donors plotted in *red* are correctly predicted as female, *green* represents correct male prediction, and donors plotted in *black* and labeled with 'wrong' have more than half of their spectra assigned to the opposite sex (Ferguson et al. 2012, Analyst, **137**, 4686-4692, adapted and reproduced by permission of The Royal Society of Chemistry)

identification of this fluid at the scene of a violent crime bears considerable forensic value. A few forensic tests are currently applied to the detection of blood in both bloodstains and fingermarks. However, these tests are presumptive as they can only indicate the presence of blood rather than confirming it (Bradshaw et al. 2014). This is due to the lack of specificity as the molecular triggers of the color change may come from a different source than blood. It is common knowledge that MALDI MS is a versatile technique in terms of the *m/z* range of the species that can be detected and is a particularly suitable technique for large molecules. Reports already showed the ability to detect both hemoglobin and heme in different fields of application such as microbiology, plant biology, and biomedicine (Yang et al. 2013; McComb et al. 1998). Encouraged by these preliminary reports, our group developed a method to detect both heme and hemoglobin from fingermarks and bloodstains. When devising a methodology with operational deployment in mind, one should try to minimize the number of protocols to apply. Therefore, after testing a range of conditions, the optimization of one CHCA matrix system was achieved (Sect. 3). Although MALDI MS has a suitable dynamic range for detecting both heme and hemoglobin simultaneously, the calibration of the MALDI instrument that was employed (Voyager DE-STR) did not cope with the large mass range, and the mass accuracy for the heme species was severely compromised. However, the method developed, using one matrix system for both molecules, allowed imaging analysis to map heme and a subsequent profiling analysis to detect hemoglobin in a 7-day-old mark

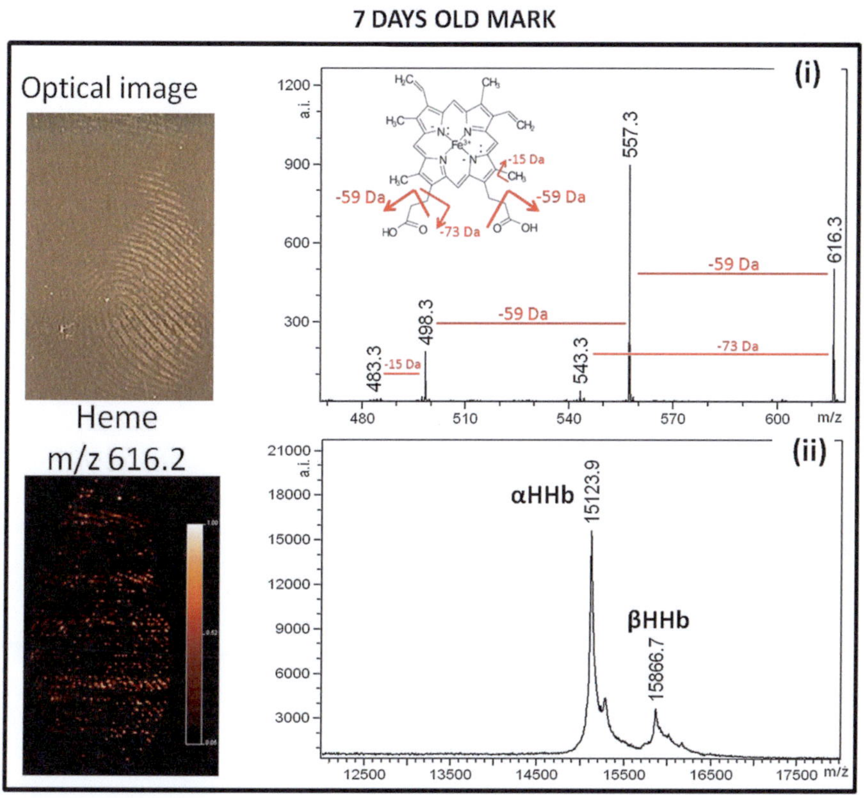

Fig. 6 MALDI MSI of a 7-day-aged blood mark. Heme (*m/z* 616.2) was successfully imaged and confirmed by the product ions at *m/z* 557.3 and 498.3 in the MS/MS spectrum of the parent ion at *m/z* 616.3 taken from a small area of the mark (**i**). Presence of human hemoglobin was ascertained by MALDI MSP analysis on the same mark in the higher mass range (**ii**) in a subsequent MALDI MS acquisition. Reprinted from Bradshaw et al. 2014, Direct detection of blood in fingermarks by MALDI MS profiling and imaging, Sci. Justice, **54**, 110-117, with permission from Elsevier

(Fig. 6), using two different calibration mixtures, one more appropriate to a narrower and lower mass range (phosphorous red for heme analysis) and the other more suitable for higher molecular weight species (protein standards mixture of ubiquitin, cytochrome C, hemoglobin, and apomyoglobin for hemoglobin analysis). This strategy also allowed us to confirm the presence of heme by MS/MS.

The devised protocol proved to be applicable to real crime scene samples as tested during a site visit alongside officers from the West Yorkshire CSI, UK. On that occasion a bloodstain, recovered and testing positive for blood by using the conventional Kastle Meyer Test, was also analyzed by MALDI MS, leading to the detection of the hemoglobin in addition to the heme ion signal (Bradshaw et al. 2014).

A significant part of the fingermark analysis work using MALDI is also dedicated to make this application compatible with BET. In this context, MALDI MSI can be compatibly used with amino acid reagents (ninhydrin) and Acid Black 1 (protein

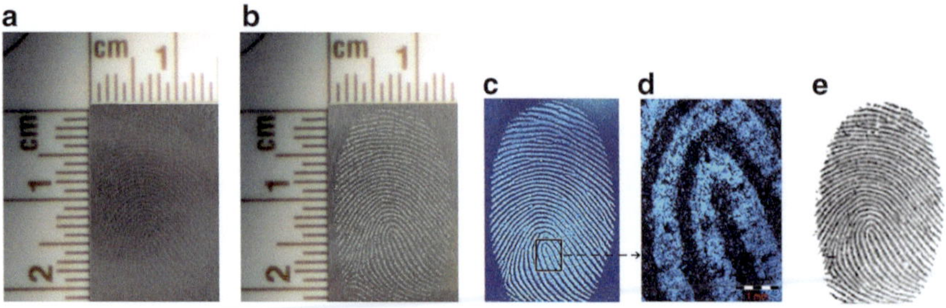

Fig. 7 Potential forensic fingermark analysis workflow using the *dry–wet* method. The latent fingermark (**a**) can be enhanced by brushing a MALDI matrix over the mark and photographed (**b**). The dusted fingermark image can be captured using UV light at 365 nm for a more accurate database comparison (**c**). The dusted fingermark can also be subsequently inspected by using fluorescence microscopy for accurate visualization of the minutiae: (**d**) illustrates a high magnification image of the fingermark loop highlighted by the box in Panel (**c**). The fingermark can finally undergo MALDI MSI analysis for obtaining chemical information: (**e**) shows the mass image of the *m/z* species at *m/z* 304 previously identified as dimethylbenzylammonium ion. Reprinted with permission from Ferguson et al. 2011, Anal. Chem., **83**, 5585–5591. Copyright 2011 American Chemical Society

staining) in the case of fresh marks. However, a general decrease in the ion signal intensity for heme and hemoglobin is observed, and quantification of the loss of these signals compared to undeveloped marks remains a challenge. Though blood has been successfully and reliably detected in 7-day-old marks, work is in progress to determine the robustness of the methodology with older marks and for pre-developed marks (using other BET). A proteomic approach is currently being applied to make the techniques robust against possible hemoglobin degradation and loss of amino acid residues due to time and environmental conditions.

In many of the applications referenced above, the *dry–wet* method was successfully employed. As explained earlier, this novel method is considered a significant step forward towards implementing MALDI into forensic casework. In our workflows, marks are visualized by brushing the surfaces with the MALDI matrix (Fig. 7a), the visualized mark can then be photographed (Fig. 7b) or directly removed through tape-lifting. As the matrix is an UV-absorbing substance, UV image capture can also be utilized (Fig. 7c) as well as fluorescent microscopy (Fig. 7d) to reveal fine details of the ridge pattern, including pores. Photos or digital scans can be retained as evidence and, provided that the chain of custody is preserved, the same mark can be subjected to MALDI MSI (if additional images are needed) or MALDI MSP for additional intelligence as shown in Fig. 7e.

This workflow has been successfully trialed on a large range of deposition surfaces (including glass, different types of metal, wood, plastic, leather, cardboard, and ceramic tiles) of which Fig. 8 shows some examples.

Normally, the matrix of choice for MALDI MS fingermark analysis is CHCA as it has so far proven to be the most efficient and versatile. However, a novel matrix has been recently reported by our group to work in some cases more effectively than

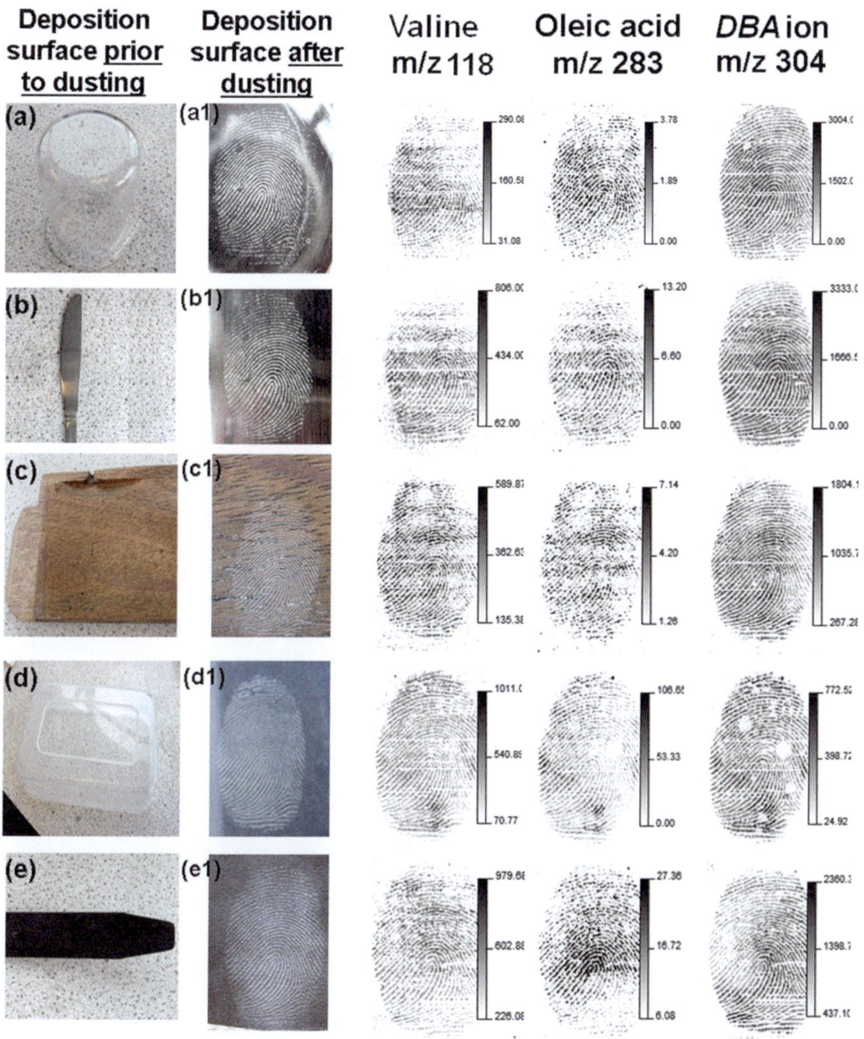

Fig. 8 Recovery and MALDI MSI analysis of ungroomed fingermarks lifted from different surfaces (panels **a**–**e**) using the *dry–wet* method. Representative MS images of an endogenous amino acid, an endogenous fatty acid, and an exogenous compound (dimethylbenzylammonium ion, DBA, *m/z* 304) are shown for each deposition surface. Reprinted with permission from Ferguson et al. 2011, Anal. Chem., **83**, 5585–5591. Copyright 2011 American Chemical Society

CHCA (Francese et al. 2013a) as shown in Fig. 9. This novel matrix compound is curcumin, which is a polyphenolic compound naturally present in the *Curcuma longa* plant and also known as tumeric, displaying an extended structure of conjugated double bonds and making it a very efficient UV-absorbing molecule. Curcumin has proven to be versatile as it assists with the detection of pharmaceuti-

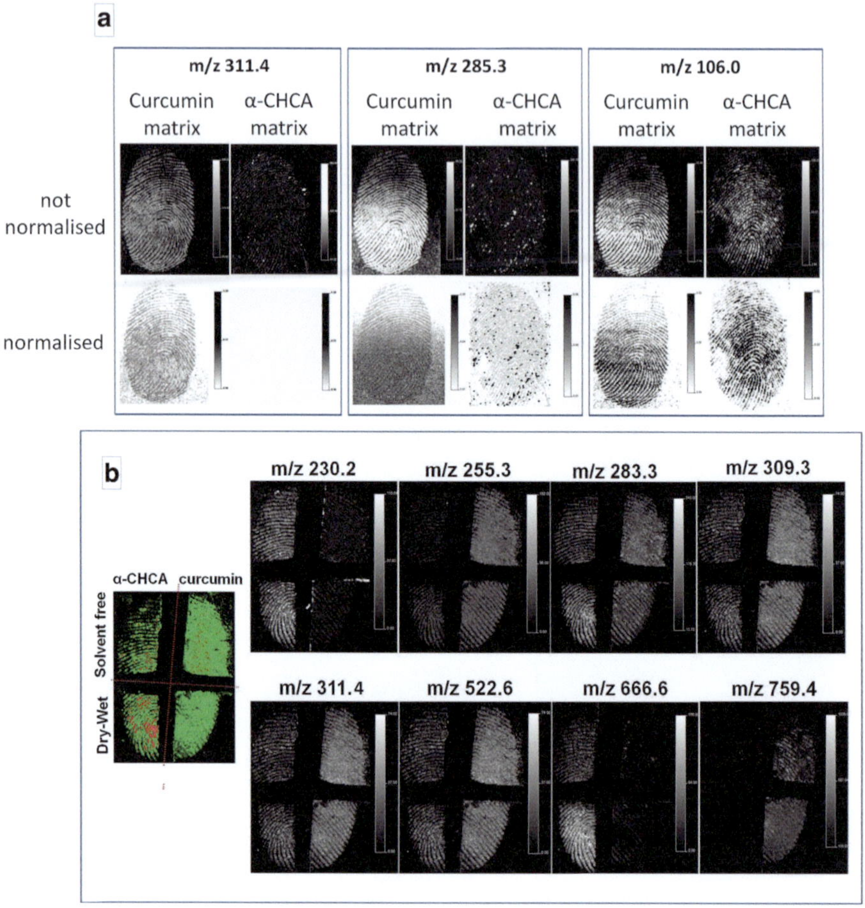

Fig. 9 Comparative MALDI MSI of ungroomed fingermarks using either synthetic curcumin or CHCA as MALDI matrix. Panel A shows the molecular maps of three endogenous species imaged either by using curcumin or CHCA and the *dry–wet* method. In a separate experiment (Panel B), an ungroomed mark was split in quarters which were each treated with one of the following methods: (*upper left*) solvent-free using CHCA, (*lower left*) *dry–wet* method using CHCA, (*upper right*) solvent-free using curcumin, and (*lower right*) *dry–wet* method using curcumin. Lipid molecular maps were reported for a comparative evaluation of the four methods. Reprinted with permission from Francese et al. 2013a, Anal. Chem., **85**, 5240–5248. Copyright 2013 American Chemical Society

cals, lipids, peptides, and proteins. It is fluorescent and therefore fingermark images can be captured using commercially available crime lights. In pseudo-operational trials with the West Yorkshire Police, it was shown that curcumin can be used on a vast range of colored surfaces and that only a very small amount of curcumin is needed to visualize marks as it has a strong adherence to surfaces. The latter is not surprising as one of its uses is in the coloring industry. However, this characteristic could pose some limitation to the operational application of curcumin as it has proven to be extremely difficult to wash it off the surface with possible liability in terms of damage to the property.

All of the referenced applications have benefited from an extremely careful approach to sample preparation and the development of optimized protocols. Some of these protocols as well as suggestions derived from our experience in fingermark preparation are shared in the following sections.

3 Materials and Protocols

3.1 How to Prepare Fingermarks for Method Development

For all samples and specimens to be subjected to MALDI MSP and MSI, preparation is crucial and is a strong determinant for the success of such analyses. Fingermarks are no exception and care must be taken in the selection of the sample preparation conditions. There are fundamentally four types of fingermarks that can be prepared and investigated:

(a) *Eccrine*: produced by (1) thoroughly washing hands[10]; (2) drying them with tissue paper; (3) resting them in a closed plastic bag for 10 min to produce excess sweat; (4) the deposition of the mark on the surface of interest.
(b) *Groomed*: produced by (1) thoroughly washing hands; (2) drying them with tissue paper; (3) rubbing the fingertips against the base of the neck or forehead[11]; (4) the deposition of the mark on the surface of interest.
(c) *Ungroomed*: produced by (1) thoroughly washing hands; (2) drying them with tissue paper; (3) carrying out 15 min of activities such as typing on a keyboard[12] whilst avoiding touching any part of the body or coming into contact with other contaminants; (4) the deposition of the mark on the surface of interest.
(d) *Natural*: produced by depositing a mark on the surface of interest without any prior treatment.

Knowledge of the (bio)chemistry associated with the different types of fingermarks is paramount for designing the experiment and understanding its outcome.

Eccrine marks directly relate to the production of eccrine sweat. In general terms, this is generated by the two to four million eccrine sweat glands (located in the dermis) present throughout the body, with the highest density on the palms of the hands and the soles of the feet. Eccrine sweat is largely constituted of water (98%) though many

[10] Common protocols include the use of soap and plenty of water or a 50/50 solution of ethanol/water.

[11] Grooming fingertips by touching the face is also viable, though one needs to be aware of possible contamination deriving from make-up or moisturizers, which may cause ion suppression.

[12] Previously cleaned with a 50/50 solution of ethanol/water.

inorganic and organic species are present (Ramotowski 2001). These are the types of marks containing the least abundance of lipids. Amino acids are the most abundant species of the organic content with their amounts being reported to be several times higher in sweat than in plasma (Miklaszewska 1968). Peptides and proteins are also present in eccrine sweat, although the epidermis is another source of these species as a result of the desquamation process (Drapel et al. 2009; Girod et al. 2012) as well as of the defense against microorganisms (antimicrobial peptides and proteins). Inorganic constituents include ammonia, chloride, bromide, fluoride, phosphate, sulfate, sodium, and potassium amongst others. These species can exert considerable ion suppression in MALDI. Therefore, sample preparation methods are needed to counteract the effect of their presence and enable ionization of the species of interest.

Because of the way they are generated, groomed marks are rich in sebum originating from the contact with parts of the body containing sebaceous glands. Sebaceous glands are the second major class of secretory glands and are present throughout the body except for the palms and the dorsum of the feet (Strauss et al. 1991) with the highest prevalence around the face and scalp. Together with traces of organic materials including aldehydes, amines, alcohols, phospholipids, and sulfides amongst others, sebaceous sweat mainly contains triglycerides, free fatty acids,[13] wax esters, squalene, cholesterol, and cholesterol esters.

Groomed marks are valuable for studying lipids but may need additional sample preparation when investigating other substances of both endogenous and exogenous nature as lipids will in fact exert ion suppression. Also, due to the diversity and abundance of lipids, they may be isobaric with forensically interesting substances (e.g., drugs, medications, explosives), thus masking their presence. A third type of gland, named apocrine, is located predominantly in the genital and axillary regions of the dermis, and therefore it is believed that fingermarks mainly comprise the species present in the eccrine and sebaceous glands.

Ungroomed marks contain species predominantly arising from the epidermis material and the eccrine glands, and therefore all of the above described inorganic and organic constituents can be found in addition to lipids, which are still detectable (Wolstenholme et al. 2009). Here, the potential ion suppression exerted by the salts is not observed, likely due to their much lower abundance. Because of the way they are prepared, these marks are considered 'depleted specimens,' i.e., containing the minimum amount of biological material, though this has not been quantified. This type of marks should only contain endogenous or semi-exogenous material and no external contaminants. However, this is not always the case as the presence of exogenous species depends on their persistence.

Though eccrine, groomed and ungroomed marks do not have a fixed chemical composition, as this varies according to the individual and their physiological state and age, *natural* marks have the most variable chemistry as fingertips may have been in contact with any body part as well as with surfaces and external contaminants. Therefore the resulting fingermarks may contain a wide range of endogenous, semi-endogenous, and exogenous substances.

[13] Saturated, mono- and polyunsaturated.

Compared to other specimens, fingermarks probably require extra sample preparation planning as the kind of answers being sought dictate the type of fingermark samples that needs preparing. In order to test new methodologies (or modified methods) and techniques for fingermark development or analysis, recommendations are in place with regard to what kind of fingermarks should be used as test samples. The International Fingermark Research Group, of which the author of this chapter is a member, has provided guidelines for developing a methodology through application and implementation in casework (IFRG 2014). In particular, these guidelines say that eccrine and groomed marks are not suitable for method/technique evaluation as they are 'unrealistic' samples, though may be used for initial pilot studies. Generally speaking, ungroomed fingermarks should be employed in method development because prior substance depletion makes them useful to probe sensitivity and feasibility of the methodology, whereas for validation purposes, natural fingermarks should be used to test the robustness of the method/technique being applied. Although these guidelines were more specifically written for techniques used by crime scene investigators to visualize the marks (enhancement or detection), in the view of the author of this chapter, they are also relevant for analytical techniques exploiting the fingermark's chemical composition to provide additional information including multiple images of the same mark for identification purposes. However, the case is different for controlled fingermark/fingertip smear[14] collection, for example in hospitals, work environments, or drug rehabilitation clinics. Here, methods can be specifically developed for the type of fingermark/fingertip smear that yields the highest abundance of the substances of interest. Depending on the fingermark type and molecular target, protocols have been devised for the detection of lipids, peptides, proteins, metabolites, and exogenous contaminants by MALDI MSP and MSI. The majority of these protocols will be summarized in the next sections together with lessons learnt in the detection and imaging of the above species.

3.2 *Define Your Strategic Approach: Profiling or Imaging?*

In the classic approach to fingermark preparation for MALDI MS analyses, the matrix solution is either spotted on the mark (manually or robotically) to quickly gain a molecular profile or homogeneously sprayed on the mark (mainly robotically) for imaging purposes (Fig. 10).

At the beginning of any MALDI MSI method development, it is recommended to start with the design of profiling experiments via matrix-spotting and their optimization for two main reasons:

---

[14] Fingertip smears are specimens that do not allow a linkage to the biometric data of the individual. Like fingermarks, it is possible to obtain eccrine, groomed, ungroomed, and natural smears by preparing the fingertips as previously described for fingermarks and pressing the fingertip on a surface sliding it to the side.

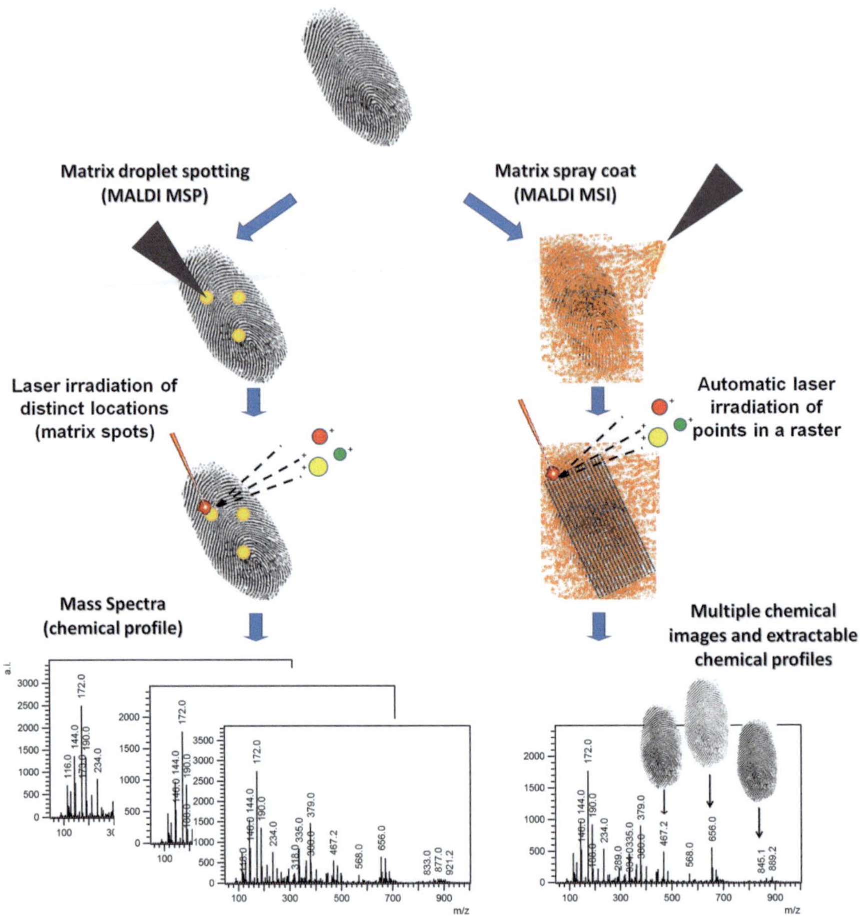

Fig. 10 Classic workflows for MALDI MSP (profiling, *left*) and MALDI MSI (imaging, *right*) applied to fingermark analysis. In profiling experiments, the MALDI matrix is usually spotted and mass spectral profiles from specific fingermark locations are obtained. In imaging experiments, the matrix is usually sprayed and spectra are automatically acquired at each *x*,*y* fingermark location in a raster fashion (Francese et al. 2013b, Analyst, **138**, 4215-4228, adapted and reproduced by permission of The Royal Society of Chemistry)

1. Analyte extraction, desorption, and ionization are generally much more efficient with spotting than with spray-coating techniques.
2. If the molecular target of interest has low or no ionization efficiency, imaging experiments are an expensive and time-consuming way to find this out.

Where possible, method development should start with the analysis of the matched standard using conventional MALDI sample preparations such as the dried-droplet method (especially true for exogenous and semi-exogenous substances). Sample preparation conditions (matrix, matrix concentration, solvent mixture) and instrumental parameters for both MS and MS/MS analyses can be optimized allowing a rapid discovery of the molecular targets that can be ionized and the conditions leading to maximum ionization efficiency. Using an iterative exclusion process, a set of optimized conditions can then be found and applied to the reference molecular target/s spotted on the mark at different concentrations. This set of experiments will allow to gain insights into the ion suppression effects exerted by the fingermark molecular make-up as well as the determination of the limit of detection (LOD), which is typically lower than in imaging mode (e.g., when the matrix is applied through spray-coating).

In such experiments, to reduce response variability (e.g., due to the well-known 'sweet spots' in MALDI) across the marks tested with different concentrations, the generation of reproducible fingermarks (e.g., through the same pressure, contact time and contact angle) is recommended. Within the author's group a custom-built device is used for this purpose, termed the Reed-Stanton press rig (Reed et al. 2015 *in preparation*), though another device, named fingerprint sampler, with different features has already been patented and published (Fieldhouse 2011) and is commercially available. Alternatively, the mark can be split into the number of regions required in order to analyze different analyte concentrations (and different matrix deposition methods), thus maintaining the same chemical environment for comparative evaluation. Note that also in this case, replicates are still necessary to reach statistically significant conclusions.

After the matrix spray-coating has been optimized and assessed by using MALDI MSP,[15] the method development for MALDI MSI should be straightforward, provided that the laser pulse repetition rate does not need adjusting to obtain the right compromise between speed and sensitivity.

When MALDI MSP and MSI are fully optimized, the methodology can then be applied to real samples. A workflow summarizing the recommendations for MALDI MSI method development is depicted in Fig. 11.

- **TIP**: *Whether or not a fingermark sampler is used, it is recommended to rub the fingertips against each other and use different fingers for the generation of the replicate marks and NOT the same finger as otherwise biological material depletion will affect the validity of the experiment conclusions.*

3.3 *Matrix Deposition: Spray-Coating and the* Dry–Wet *Method*

Initially, for many of the applications reported by our group, the classic matrix spray-coating technique was employed for imaging purposes. As described earlier, the SunCollect autosprayer was the device employed for all of the sample preparation method developments. This sprayer is simple to operate and is overall

[15] This includes optimization of the instrumental settings.

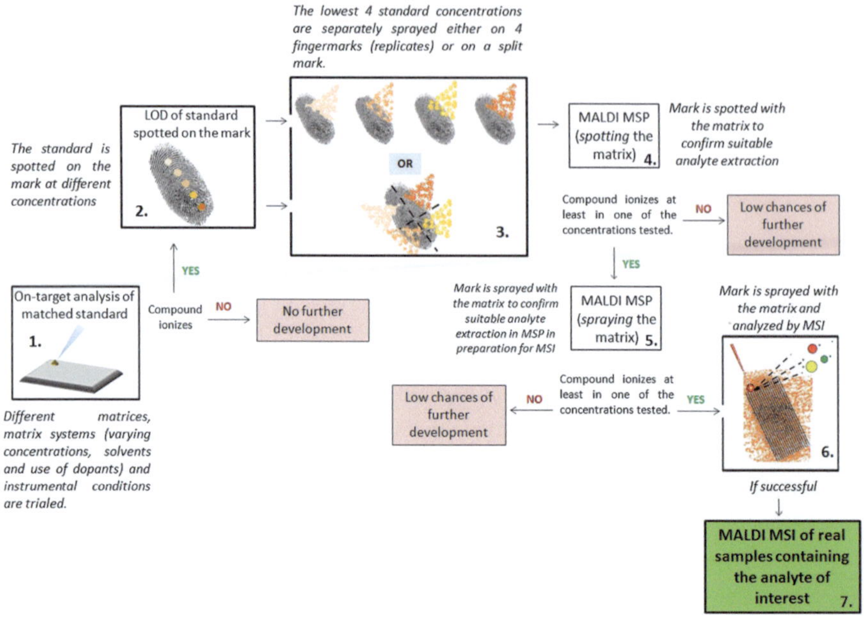

Fig. 11 Recommended workflow to develop a MALDI MSP and a MALDI MSI method for chemical species analysis in fingermarks. (**1**) Different matrices, matrix systems (including varying concentrations, solvents and use of dopants), and instrumental conditions are trialed; (**2**) the standard is spotted on a fingermark at different concentrations with the optimal MALDI sample preparation from step 1 and analyzed by MALDI MS; (**3**) the lowest four standard concentrations are separately sprayed on either four fingermarks (replicates) or four sections of a split mark; (**4**) matrix is spotted on the standard-containing mark(s) to confirm suitable analyte desorption/ionization conditions. Here instrumental conditions can already be optimized; (**5**) matrix is now sprayed onto fingermarks with various amounts of spray-coated standard and the mark is analyzed by MSP in preparation for imaging experiments; (**6**) the standard spray-coated mark is sprayed with the matrix and analyzed by MALDI MSI to confirm suitable analyte extraction and mapping in imaging mode. Here matrix deposition parameters are optimized for the lowest detectable concentration of the spray-coated standard; (**7**) if step 6 is successful, the method can be transferred to a real fingermark sample naturally containing the analyte

reproducible in the amount of matrix deposited. However, geometrical and physical parameters such as nozzle-to-surface distance, N_2 pressure, number of layers, and flow rate per layer require optimization. A few protocols are illustrated in this section for various applications.

In collaboration with HTX Technologies LLC (Carrboro, NC, USA), our group developed a matrix spray-coating method on a conceptually different automatic depositor, namely the TM sprayer (Bradshaw et al. 2013c). In order to achieve homogenous matrix deposition and maximum analyte extraction with minimal to no delocalization, this device employs a heated and pressurized matrix solution. Different variables such as N_2 pressure, nozzle temperature, number of passes, pump flow rate, velocity, and the track spacing were systematically investigated.

Table 1 Protocol parameters for the spray-coating method using the TM Sprayer (HTX Technologies) and CHCA as matrix[a]

Preparation of ungroomed fingermarks using the TM Sprayer (HTX Technologies, NC, USA)						
Fingermark quarter	N_2 pressure (psi)	Nozzle T (°C)	Number of passes	Pump flow rate (mL/min)	Velocity (mm/min)	Track spacing (mm)
1	10	75	1	0.08	1300	1.5

[a]In the author's lab, this set of parameters provides the best results overall in terms of fingermark ridge detail and speed of preparation

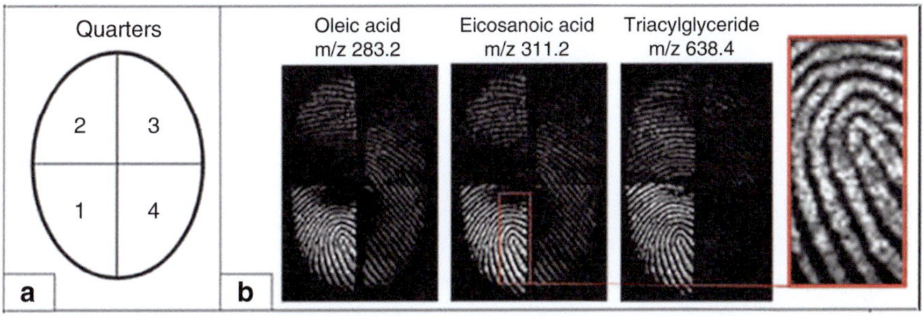

Fig. 12 Optimization of matrix deposition using a TM Sprayer (HTX Technologies, LLC (USA). (**a**) Fingermark divided in quarters and numbered according to the number of passes performed. (**b**) Corresponding MALDI molecular images of three endogenous species; oleic acid (*m/z* 283.2), eicosanoic acid (*m/z* 311.2), and a triacylglyceride (*m/z* 638.4), with a magnified region for eicosanoic acid. Keeping all the other parameters constant, one layer of matrix generated the most satisfying result (see also Bradshaw et al. 2013)

Eventually, all of these parameters were optimized and then kept constant, apart from the number of spray-coating passes which was changed from 1 to 4 and individually applied to quarters of a fingermark. The best performing method used just one layer of matrix (Table 1) and allowed the highest possible quality of ridge detail images of lipids (Fig. 12) (analyzed in positive ion mode) in less than 1 min in total (*one-layer* method).

However efficient, both the classic dried-droplet and spray-coating methods cannot be applied to crime scene marks, if these are not somehow previously enhanced and removed from surfaces (unless, in the latter case, the mark is present on a flat and suitably thin surface). As previously described, the *dry–wet* method represents a significant step here for the implementation of MALDI MSI into the operational fingerprint analysis workflows.

While our group has been devising protocols to enable MALDI MSP and MSI analysis on FET-enhanced marks, the *dry–wet* method is also under further development to establish exactly on which surfaces and under which environmental circumstances, the MALDI matrix can efficiently act as a dual agent, i.e., in addition to its conventional MALDI matrix role also as a fingermark enhancer on surfaces.

During method development studies, our group has found that there are mainly three factors affecting the efficiency of the *dry–wet* method, namely the type of brush, matrix powder excess, and matrix particle size. There are different types of brushes that CSIs use according to the powder to be deployed and the surface. They mainly differ in the chemical nature of the material, the cross-section of the heads, and stiffness of the hairs.

- **TIP**: *Amongst the types of brushes we have tried, the zephyr brushes are the most efficient to deploy the MALDI matrix. It is important that the shaft of the brush is rotated between thumb and fingers while dusting for marks to avoid smudging, merging, or removal of the ridges.*

The second and the third factor should not be surprising for experienced MALDI users and with regard to the existing literature (Sugiura et al. 2006; Puolitaival et al. 2008). Large matrix excess may lead to suboptimal or completely out of range matrix:analyte ratios (often reported to be ideal around a 10,000:1 ratio), leading to the absence of signal (or very low signal intensity).

- **TIP**: *To remove excess matrix, use a quick blast of compressed air* (e.g., *from a can when at crime scenes) using a distance of about 10 cm.*

As the matrix is dispensed 'dry' in the first step of the method, the particle size affects the size of the matrix-analyte co-crystals upon evaporation of the solvent that is sprayed on the matrix/fingermark as the second step of the process. Therefore, it is very important to use a matrix particle size as small as possible. In recent work, a remarkable difference in the ion intensity and ridge continuity was shown with the quality of the results gradually improving from using commercially available unground matrix, manually ground matrix, mechanically ground matrix sieved through a 38 μm sieve to ground matrix with particle sizes of <10 μm (Fig. 13) (Ferguson et al. 2013b). Once dusted with the matrix, the mark is typically lifted by forensic tape and mounted face up on a MALDI target plate using double-sided conductive tape for subsequent solvent spraying. A working protocol that our group has optimized for the *dry–wet* method preparation using milled matrix is illustrated in Table 2 together with the instrumental parameters for a modified API Q-Star Pulsar *i* hybrid quadrupole time-of-flight (qTOF) instrument (Applied Biosystems), the instrument on which the bulk of our 6 years of research on latent fingermark analysis has been undertaken.

As discussed in the introduction, the novel matrix curcumin was trialed with the *dry–wet* method and the ionization efficiency was compared with that obtained using both the conventional spray-coating method and by leaving the curcumin-dusted fingermark unsprayed. The optimized protocol for the application of cur-

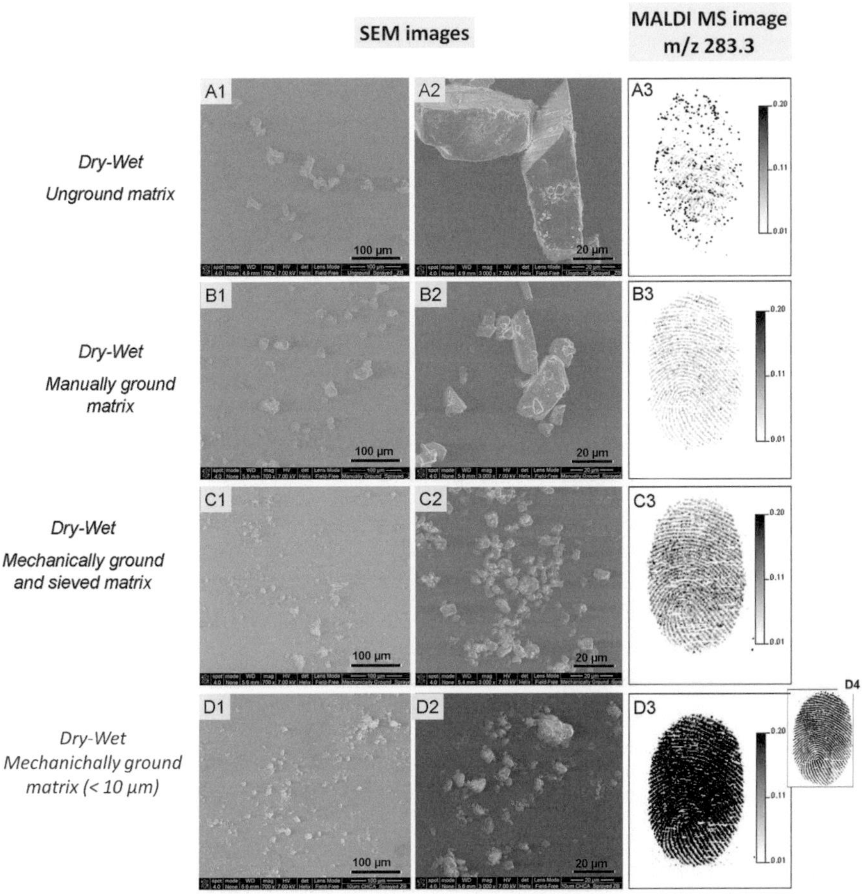

Fig. 13 Comparative analysis of ungroomed fingermark image quality prepared with the *dry–wet* method. Panels A1–D1 show a 700× magnification SEM image of the dusted fingermarks using different matrix particle sizes. Panels A2–D2 show a 3000× magnification. Panels A3–D3 show the corresponding MALDI MS images of the ion signal at *m/z* 283.3. Panel D4 displays the normalized image of the same mark shown in panel D3, which has not undergone normalization (Ferguson et al. 2013b, J. Mass Spectrom. 2013, **48**, 677–684, adapted and reproduced with permission from John Wiley and Sons)

cumin onto fingermarks for the *dry–wet* method was the same as for CHCA with respect to solvent spraying and instrumental parameters as shown in Table 2 (except for the declustering potential 2 and focusing potential set at 10 and 20 a.u., respectively), and permitted the analysis of lipids.[16]

[16]The *dry–wet* method employing curcumin has not been trialed on other classes of molecules.

Table 2 Protocol parameters for the application of fine CHCA matrix particles using the *dry–wet* method with a zephyr brush (cf. Ferguson et al. 2011)

Preparation of fingermarks using the *dry–wet* method for MALDI MSI analysis			
Dry–wet method application		Mass spectrometer and instrumental parameters	
Matrix	α-cyano-4-hydroxycinnamic acid (CHCA)	Mass spectrometer	Q-Star Pulsar *i* (Applied Biosystems)
Milling instrument	Bench top, single grinding station PM100 Planetary Ball Mill (Retsch)	Mode/laser repetition rate	Continuous raster imaging in positive ion mode/5 KHz
Milling conditions/ final particle size	450 rpm for 30 min in agate jar with 15 × 20 mm grinding balls/< 10 μm	Declustering potential 2/Focusing potential	15/10
Autosprayer for solvent spraying	SunCollect (SunChrom)	Accumulation time	0.117 s
Solvent spray conditions	5 layers of 70:30 of an ACN/0.5% TFA solution at a rate of 5 μL/min	Spatial resolution	100 μm × 150 μm

Having very similar protocols is advantageous for method implementation in the forensic practice as, generally speaking, the lower the number of protocols (or variability in the protocols) the quicker they are accepted and integrated in practice.

3.4 Detection and Mapping of Peptides and Proteins

The detection of fingermark peptides and proteins may have profound implications for forensic practice as well as in the biomedical field. In the former case, our group has demonstrated the ability to detect the sex of individuals through their fingermarks by exploiting the peptide/protein profiles (Ferguson et al. 2012).

- **TIP 1**: Use a *dedicated* zephyr brush to dust curcumin, thus avoiding cross-contamination and 'unexpected' results.
- **TIP 2**: Use only a small amount as the 'staining' and the adherence to fingermarks are greater with this matrix than with CHCA.
- **TIP 3**: Use synthesized curcumin as opposed to plant-derived commercially available curcumin for greater ionization efficiency and less interfering background signals.
- **TIP 4**: If you do not use the *dry–wet* method but the classic spray-coating method on the SunCollect autosprayer, use a dedicated capillary for each matrix to avoid cross-contamination as matrices take a very long time to clean off the capillary line.

This forensic analysis provides additional intelligence, especially important if DNA cannot be recovered. Implications in the wider biomedical context relate to the exploitation of peptides/proteins for noninvasive screening of disease biomarkers. It is therefore important to devise protocols for their optimal detection. However, to date, the detection of peptides and proteins from fingermarks has been by far the most challenging piece of research and further method development is still in progress. The challenge is mainly due to their low abundance (the total protein concentration in sweat has been reported to be between 15 and 25 mg/dL) and their main origin from the eccrine glands (Ramotowski 2001).[17]

The challenges in peptide/protein detection are often fingermark-specific. Eccrine marks should contain most of the proteins but the electrolyte concentration here is also the highest amongst the different types of marks listed in Sect. 3.1. As a result, there is a strong ion suppression phenomenon preventing the detection of peptides and proteins (Ferguson et al. 2012). Groomed marks exhibit the presence of peptides/proteins but the high presence of lipids (as a consequence of grooming) also exerts ion suppression, greatly decreasing the peptide/protein ion population as well as the corresponding MS signal intensity. Ungroomed marks are somewhere in between these two types though proteins are less abundant than in eccrine marks. Natural marks may be problematic as their variable composition (and possible presence of external contaminants) may affect peptide/protein ionization through similar suppression phenomena. We have previously shown that, similarly to the treatment of other specimens subjected to MALDI MSP and MSI, washing steps can greatly improve the detection of peptides/proteins. For example, groomed fingermarks yield a higher peptide/protein ion population and signal intensity when washings are performed to selectively eliminate lipids. Denatured ethanol was the most efficient solvent out of the three trialed which also included chloroform and acetone (Ferguson et al. 2012). These washings were performed by pipetting 750 μL of the solvent over the fingermark deposited on a suitable support which was held at an angle on a flat surface.

If marks are collected for biomarker discovery studies or in anticipation of their use for medical diagnostics, eccrine marks may be good specimens to use as they can be produced in a controlled way and contain peptides/proteins in higher concentration. However, the ion suppression phenomenon needs to be counteracted. As the challenge here is the presence of electrolytes, we have taken up one of the MALDI MSI sample clean-up strategies applied to tissue samples, employing low concentration of high MALDI-tolerant salts, such as ammonium acetate, to preliminarily treat the mark (Wang et al. 2011). In this preparation, the ammonium and acetate ions will advantageously engage the endogenous chloride and the sodium/potassium ions, respectively. The type of application of ammonium acetate (submersion or buffer pipetting), the volume, the concentration, and the contact time with this buffer all play a role. Our current (so far unpublished) protocol is depicted in Table 3.

[17] Protein-producing apocrine glands are rarely a source of proteins in fingermarks.

Table 3 Parameters for the preparation and analysis of peptides and proteins in eccrine marks

Preparation of the eccrine mark and analysis			
Hands are washed with 50% $EtOH_{aq}$ and kept for 15 min in sealed bag; fingertips are then rubbed together to spread sweat evenly			
Buffer used	Ammonium acetate (AA)	Mass spectrometer	Voyager DE-STR
Concentration	150 mM	Accelerating voltage	25 kV
Application mode	Soak in AA for 5 min	Grid voltage	93%
Buffer volume	500 mL	Delay time	150 ns

Benefits of this method are shown in Fig. 14, where the MALDI spectrum of an unwashed eccrine mark is compared with that of a washed one using the protocol of Table 3.

Sometimes, even successfully optimized MALDI MSP protocols using the dried-droplet method cannot be successfully developed in imaging methods employing matrix spray-coating. In our laboratories, we have trialed a number of different sample preparation protocols to image proteins with samples being analyzed on a Voyager DE-STR MALDI-TOF instrument (Applied Biosystems) equipped with a solid state laser operating at 60 Hz. However, results as reported by Ferguson (2013) in her Ph.D. thesis were disappointing. Generally, with all the sample preparation methods trialed, the intensity of peptide/protein ion signals in imaging mode (with matrix spray-coating) was either too low or signals were not detected at all. Actual presence of peptides and proteins was verified using the classic dried-droplet method, thus indicating that the protein abundance and extraction efficiency in MALDI MSI were insufficient. However, since detection proved to be satisfactory when the dried-droplet method of matrix deposition was applied, the lack of peptide/protein ion signals using the automatic acoustic ejector spotter for sample preparation is somewhat surprising. Amongst the methods trialed, only the one spraying the matrix at 5 mg/mL with 25:25:50 acetonitrile/ethanol/0.5% TFA by the SunCollect autospraying system yielded a couple of signals previously detected with the dried-droplet method (Fig. 15). It was also possible to obtain an image for the ion signal at *m/z* 4819.5, which was putatively identified as dermicidin (Ferguson et al. 2012) but highly speckled and bearing no hint of ridge detail (Fig. 15c). Though the lack of peptide and protein detection may be a limitation of MALDI MSI for the analysis of latent fingermarks, these experiments are not yet conclusive. Many more conditions can be trialed and our experiments (unpublished data) suggest that fingermark/fingerprint peptides/proteins are more hydrophobic than previously thought. Furthermore, their low abundance is most likely another decisive factor.

Experiments performed on an Ultraflex III MALDI-TOF/TOF mass spectrometer (Bruker Daltonik, Germany) equipped with a SMART beam laser operating at 200 Hz gave more encouraging results. Here, the *dry–wet* method was applied where an ungroomed mark was dusted with CHCA and sprayed with 3 layers of

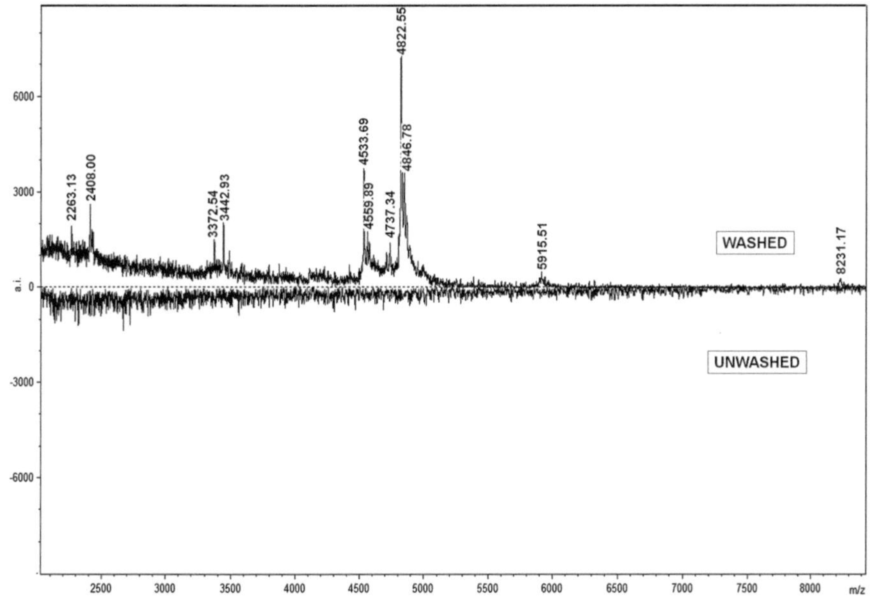

Fig. 14 MALDI MS spectra of peptides and small proteins in eccrine fingermarks. *Top panel*: eccrine fingermark washed using the method reported in Table 3; *bottom panel*: unwashed fingermark

70:30 ACN/0.5% TFA (at a rate of 2 μL/min and a capillary distance of 31.75 mm) prior to the analysis by MALDI MSI at a spatial resolution of 200 μm × 200 μm. As Fig. 16 shows, despite the mark being only partial and highly speckled, there is a 'hint' of ridge detail.

These initial results indicate that perhaps using a different mass spectrometer with higher sensitivity and spatial resolution in conjunction with the most promising sample preparation methods trialed (and further tweaking of the preparation conditions) may, in a not too distant future, yield protein/peptide images. This work is in progress in our laboratory.

Abundant proteins from an external source (external to the fingermarks) have not been such a challenge. The specific detection and mapping of blood through detecting the hemoglobin protein (α and β chains), in addition to the heme group, was achieved through the application of a standard MALDI MS protein analysis protocol. Although sinapinic acid (SA) was trialed as one of the best matrices for protein analysis, CHCA was the most efficient matrix to detect hemoglobin both in simulated laboratory conditions and for crime scene evidence, though the evidence recovered and analyzed from the latter were bloodstains rather than blood marks. The possibility to use the same matrix for molecules so different in size enables both

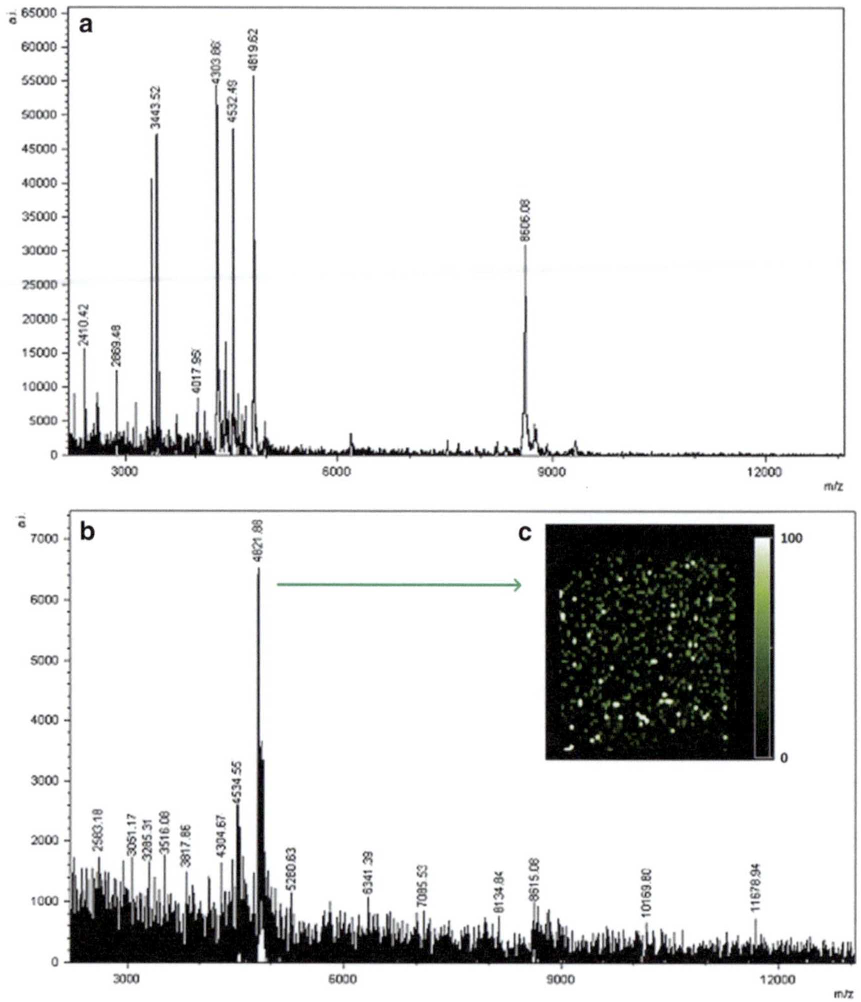

Fig. 15 MALDI MSP and MSI of a split groomed mark. One half was prepared using the dried-droplet method spotting 5 mg/mL CHCA in 25:25:50 acetonitrile/ethanol/0.5% TFA (**a**) while the other half was prepared using the same matrix solution but sprayed using the SunCollect auto-spraying system (**b**). Fingermark of the MS spectrum shown in (**b**) was then imaged and panel (**c**) shows the image of the peak at *m/z* 4821.9, corresponding to the putatively identified dermicidin (the recorded and expected *m/z* values differ by 2 units). No ridge details could be observed. Analyses were performed on a Voyager DE-STR MALDI-TOF mass spectrometer (Ferguson 2013; permission granted by the author)

the analysis of heme and that of hemoglobin on the same mark in two subsequent acquisitions (instrumental parameters are different in the two cases), without washing off and changing the matrix. The detection of two blood-specific molecules makes blood analysis much more reliable. The protocol adopted for both hemoglobin

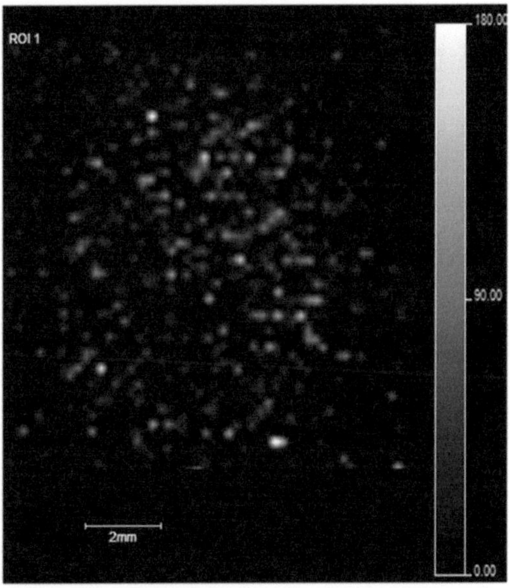

Fig. 16 MALDI MSI of an ungroomed fingermark at a spatial resolution of 200 μm × 200 μm. The mark was prepared using the *dry–wet* method by dusting with CHCA and spraying a 70:30 ACN/0.5% TFA solution using a SunCollect autosprayer. MALDI MSI was performed on an Ultraflex III MALDI-TOF/TOF mass spectrometer

profiling and imaging is depicted in Table 4.[18] This protocol is also compatible with the prior application of the currently used ninhydrin and Acid Black 1 enhancement protocols for the presumptive detection of blood. Despite the fact that the instrument adopted for these analyses is relatively old and does not have a high sensitivity, preliminary experiments on the hemoglobin standard have shown that we can detect this protein at a concentration 1000 times lower than the physiological concentration (~13–18 g/dL for healthy adult males and ~11.5–16.5 g/dL for healthy adult females).

3.5 *MALDI MSI of Condom Lubricants in Fingermarks*

Two years of research were spent on devising and optimizing protocols for imaging condom-contaminated fingermarks with the ultimate aim to provide information for suspect identification (ridge pattern images) as well as the identification of the condom lubricant chemicals (associative/probative evidence). In some cases, condom brand/type identification was achieved. Condom-contaminated fingermarks/fingerprints can

[18] The protocol (instrumental conditions) for the detection, the mapping, and the MS/MS of heme can be retrieved in (Bradshaw et al. 2014).

Table 4 Parameters for the preparation and MSP/MSI analysis of blood fingermarks

Preparation of blood marks for MALDI MSP and MSI analysis			
MALDI MSP analysis			
Matrix (spotted)	20 mg/mL CHCA in 70:30 ACN: 0.5% TFA_{aq} (0.5 µL per spot)	Mass spectrometer	Voyager DE-STR
Application mode	Spotted	Accelerating voltage	25 kV
		Grid voltage	90%
		Delay time	300 ns
		Number of shots	150
MALDI MSI analysis			
Matrix (sprayed)	5 mg/mL CHCA in 70:30 ACN: 0.5% TFA_{aq}	Mass Spectrometer	Voyager DE-STR
Application mode	Sprayed	Accelerating voltage	25 kV
Spraying conditions	SunCollect (SunChrom) 5 layers at a rate of 2 µL/min	Spatial resolution	150 µm × 150 µm (spot-to-spot classic method)
		Grid voltage	90%
		Delay time	300 ns
		Number of shots	100

be easily prepared by washing hands with a 50:50 ethanol:water solution, rubbing fingertips against the condom for a few seconds and subsequently touching the deposition surface of choice.

- **TIP 1**: If you want to preserve the ridge details and avoid smudging due to the oily nature of the lubricants, lightly touch the surface three times in three different areas and analyze the third mark.
- **TIP 2**: Condom lubricants (polymers) are very persistent substances. Therefore, to avoid carry-over effects, use different fingers to deposit replicate marks or marks containing different condom lubricants. Wash hands thoroughly and ensure that no polymer is left, which could interfere with other analyses when still using your fingertips to generate other fingermarks/fingerprints.
- **TIP 3**: Make sure that the sample is completely dry as otherwise pumping down the MALDI mass spectrometer could become challenging.

In general, as well as in condom lubricant analyses, 'hot spots' are very much a problem when the matrix is spotted rather than sprayed, which is in agreement with the findings and observations by others (Hensel et al. 1997; Hanton 2001). 'Hot spots' are mainly due to the differential incorporation of the analyte(s) into the matrix crystals, depending on the analyte, matrix and crystallization conditions, and thus

Table 5 Parameters for the preparation and MALDI MSI analysis of condom-contaminated marks

Preparation of condom-contaminated fingermarks for MALDI MSI analysis			
Matrix and standards application		Instrumental parameters	
Matrix	10 mg/mL dithranol in 45:40:15 acetonitrile: dichloromethane: tetrahydrofuran	Mass spectrometer	Q-star Pulsar *i* (Applied Biosystems)
Application mode	Spray-coating using manual sprayer Eclipse HP-CS (Iwata Medea Inc.)	Mode/laser repetition rate	Continuous raster imaging/positive ion mode
Total volume delivered	20 mL	Declustering potential 2/focusing potential	10/30 a.u.
		Accumulation time	0.117 s
		Spatial resolution	150 μm × 150 μm

leading to significant variability in the ion signal intensity and mass shifts from spot to spot. For polymers, this has been previously described as 'polymer segregation' (Gruendling et al. 2010), preventing homogenous polymer-matrix co-crystallization.

In our studies, employing MALDI MSI and a matrix spray-coating technique, a dithranol-based matrix application was optimized, enabling the detection and mapping of ethoxylate-based polymers and polydimethylsiloxanes (PDMS). The main method used to perform MALDI MSI of condom-contaminated marks is detailed in Table 5.

Here the SunCollect autosprayer could not be used as the solvent mixture can damage the capillary and the tubing of the device. A manual sprayer (airbrush) was therefore used instead, which is an adequate alternative in the hands of an experienced user. With this device, the distance between the nozzle and the surface as well as the number of passes and the time interval between each pass are all very important parameters to prevent analyte delocalization whilst achieving maximum analyte extraction and optimal co-crystallization with the matrix. Although the protocol shown in Table 5 enables the detection and mapping of ethoxylate-based polymers as well as PDMS, the latter was only accidentally detected after condom-contaminated marks aged at 25 °C and 60% relative humidity for a month (cf. Sect. 1 Introduction). To speed up sample preparation and to achieve completeness of information about the condom lubricant profile, an alternative protocol step was devised, whereby the contaminated mark was kept at 37 °C for a minimum of 10 min. Whether this will be required operationally, given that the majority of sexual assaults are reported weeks, sometimes months after the crime, remains to be seen.

To exemplify the integration of this technology in a workflow where less destructive techniques are used first, MALDI MSI was employed in tandem with ATR-FTIR analysis (Bradshaw et al. 2013b). The development of this workflow required careful planning on the application of these two techniques knowing that ATR FTIR is to be applied first. The feasibility and efficiency of this new type of workflow is somewhat limited by both the materials used and the high-vacuum environment of conventional MALDI sources. In particular, the gelatin lifters employed to remove the mark from the surface to be subsequently submitted to ATR FTIR are an unsuitable sample support for MALDI MS due to the difficulty in achieving the required high vacuum of the MS instrument. This constraint prevents the same mark to be analyzed sequentially by ATR FTIR and MALDI MSI and dictates a first fingermark lift using gelatin tape to be analyzed by ATR FTIR and a second lift of the remaining fingermark material with conventional tape to be submitted to MALDI MSI analysis. However, as shown in Fig. 4, MALDI MSI is sensitive enough to detect the relevant chemicals (polymers and endogenous substances) even in a secondary lift, albeit with lower ion intensity and ridge coverage.

3.6 Further Strategy to Improve Information Reliability of MALDI MSI When Imaging Small Molecules

Whilst molecular detection is achieved in the majority of the cases through optimized protocols, reliable molecular identity is not trivial. Depending on the sample, the presence of isobaric species, including matrix clusters, may often lead to questionable molecular identification if solely based on the *m/z* value. Therefore, the application of tandem mass spectrometry (MS/MS) is often a requirement. However, even the application of MS/MS may incur the same problem with no reliable identification. For instance, this is the case if the selection mass window of the precursor ion is not narrow enough, thus allowing ions with similar *m/z* to be fragmented, generating an overpopulated and very complicated MS/MS spectrum. Certainly both high resolution mass spectrometry (HRMS) and Ion Mobility MS (IMS) are two excellent ways to overcome the problem of isobaric ions, although they do come with significant financial implications.

The inclusion of standards combined with imaging tandem mass spectrometry capabilities is a useful strategy, which may help to alleviate the problem of isobaric species and the lack of HRMS and IMS instrumentation. Taking stock from a recently published work from Clench's group (Cole et al. 2013), we have adapted this strategy to latent fingermarks and recently reported the results in a Spectroscopy Europe issue (Bradshaw and Francese 2014). Here, as an example, two species having the same nominal mass, namely protonated cocaine ions (monoisotopic *m/z* 304.1548) and dimethylbenzylammonium ions (DBA), which can be found in many toiletry products (monoisotopic *m/z* 304.3004). It is clear that erroneously claiming the presence of cocaine in the defendant's fingermarks (whilst in fact the species in question is just a common bactericide found in many hygiene products) may have profound and unacceptable consequences in a court case. Thus, the claim must be sound and based on solid evidence. In our study, the sample preparation included

Table 6 Parameters for the preparation and MSI analysis of DBA- and cocaine-containing marks

Preparation of cocaine- and DBA-contaminated marks for MALDI MSI analysis			
Matrix and standards application		Instrumental parameters	
Matrix	CHCA solution in 70:30 ACN: 0.5% TFA	Mass spectrometer	Q-star Pulsar *i* (Applied Biosystems)
Application mode	Spray-coating using SunCollect autosprayer	Mode/laser repetition rate	Continuous raster imaging/positive ion mode
Number of layers and speed	5 at 2 μL/min	Declustering potential 2/focusing potential	40/15 a.u.
Standards	Each standard separately mixed 1:1 with the matrix and 0.5 μL spotted in two distinct locations at the side of each contaminated mark	Collision energy	25
		Accumulation time	0.5 s
		Spatial resolution	150 μm × 150 μm

'controlled' contamination of fingertips and the use of *natural* fingermarks to mimic real conditions as far as possible (Bradshaw and Francese 2014). Fingertip contamination with the two substances was achieved by (a) rubbing a fingertip against a glass slide containing a known dry residue amount of cocaine (5 μg) and (b) wiping a fingertip using Dettol® antibacterial surface wipes containing DBA, before depositing separate marks onto aluminum sheets. Preparation and analysis of the marks were undertaken using the method illustrated in Table 6.

As seen in Fig. 17, simple MS data acquisition cannot help in the discrimination of the isobaric species.[19] However, if MS/MS images are interrogated specifically for cocaine product ions (*m/z* 182.1, 150.1, 82.1), only the cocaine standard spot appears (Fig. 17a). Similarly, DBA-specific product ions (*m/z* 212.2, 91.0, 58.0) generate images of both the fingermark and the standard DBA spot with the standard cocaine spot no longer visible (Fig. 17b). In the latter case, a highly informative and visually impactful image was also obtained by superimposing the images from the DBA product ion at *m/z* 212.2 and the cocaine product ion at *m/z* 182.1, demonstrating that cocaine is absent in the mark and only present in the standard. In conclusion, this strategy may be adopted for molecular confirmation and does not require HRMS or IMS using reference standards.

[19] Note that another limiting factor can be the bin size of the imaging software used (e.g., BioMap; http://www.maldi-msi.org/index.php?option=com_content&view=article&id=14&Itemid=32). Even if the data were acquired in high resolution, software often cannot cope with a narrow bin sizes, thus generating (depending on the decimal places) the same uncertainties in discriminating between two species with the same nominal mass.

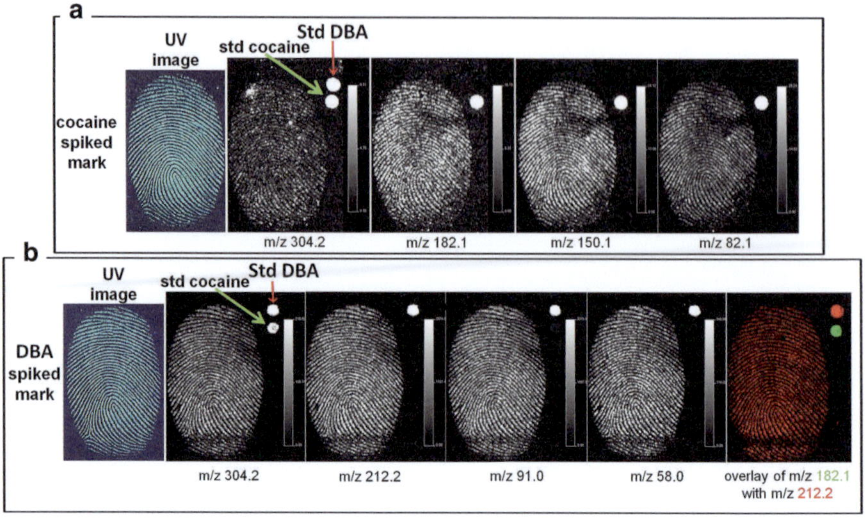

Fig. 17 MALDI MS and MS/MS imaging using reference standards for the discrimination of the isobaric species cocaine and DBA. (**a**) UV and MS image (*m/z* 304.2) for a cocaine-spiked fingermark and the corresponding MS/MS images of the cocaine product ions. As only the cocaine standard spot appears in the MS/MS images, it is concluded that cocaine and not DBA is the species present in the mark. (**b**) UV and MS image (*m/z* 304.2) for a DBA-spiked mark and the corresponding MS/MS images of the DBA product ions. As only the DBA standard spot appears in the MS/MS images, it is concluded that DBA and not cocaine is present in the mark (Bradshaw and Francese, Spectroscopy Europe 2014, **26**, 6-8, reproduced with permission from John Wiley and Sons)

Acknowledgements My previous students Robert Bradshaw and Leesa Ferguson (now PhDs) and my current Ph.D. student Ekta Patel are gratefully acknowledged for the very hard work put into these pioneering studies and for sharing a huge enthusiasm towards this research. The Home Office and Dr. Steve Bleay are thanked for both intellectual and financial support to this research. Mr. Neil Denison is gratefully acknowledged together with West Yorkshire Police for the huge support in developing the technology in the field. The Biomedical Research Centre at Sheffield Hallam University is heartily thanked for supporting and believing in this emerging research area.

References

Bandey HL, Bleay SM, Bowman VJ, Downham RP, Sears VG. Bandey HL (eds.) Fingermark visualisation manual (2014) Home Office, London. ISBN 978-1-78246-234-237

Bailey, M. J. et al. Enhanced imaging of developed fingerprints using mass spectrometry imaging. Analyst. 138, 6246–50 (2013).

Benton M, Rowell F, Sundar L, Jan M (2010a) Direct detection of nicotine and cotinine in dusted latent fingermarks of smokers by using hydrophobic silica particles. Surf Interface Anal 42:378–385

Benton M, Chua MJ, Gua F, Rowell F, Ma J (2010b) Environmental nicotine contamination in latent fingermarks from smoker contacts and passive smoking. Forensic Sci Int 200:28–34

Bradshaw R, Francese S (2014) Matrix–assisted laser desorption ionisation tandem mass spectrometry imaging of small molecules from latent fingermarks. Spectroscopy Europe 26:6–8

Bradshaw R, Rao W, Wolstenholme R, Clench MR, Bleay S, Francese S (2012) Separation of overlapping fingermarks by matrix assisted laser desorption ionisation mass spectrometry imaging. Forensic Sci Int 222:318–326

Bradshaw R, Bleay S, Wolstenholme R, Clench MR, Francese S (2013a) Towards the integration of MALDI MSI into the current fingermark examination workflow. Forensic Sci Int 232: 111–124

Bradshaw R, Wolstenholme R, Ferguson LS, Sammon C, Mader K, Claude E, Blackledge R, Clench MR, Francese S (2013b) Spectroscopic imaging based approach for condom identification in condom contaminated fingermarks. Analyst 138:2546–2557

Bradshaw R, Bleay S, Clench MR, Francese S (2014) Direct detection of blood in fingermarks by MALDI MS profiling and imaging. Sci Justice 54:110–117

Bradshaw R, Cressein A, Francese S (2013) Technical Note #32|062013, Rapid MS imaging of fingermarks. http://www.htximaging.com/Content.aspx?type=LIB. Accessed 21 Jan 2015

Cole LM, Mahmoud K, Haywood-Small S, Tozer GM, Smith DP, Clench MR (2013) Recombinant 'IMS TAG' proteins- a new method for validating bottom-up matrix-assisted laser desorption/ionisation ion mobility separation mass spectrometry imaging. Rapid Commun Mass Spectrom 27:2355–2362

Drapel V, Becue A, Champod C, Margot P (2009) Identification of promising antigenic component in latent fingermark residues. Forensic Sci Int 184:47–53

Ferguson LS (2013) Analysis of the composition of latent fingermarks by spectroscopic imaging techniques. Ph.D. Thesis, Sheffield Hallam University

Ferguson L, Bradshaw R, Wolstenholme R, Clench MR, Francese S (2011) A novel two step matrix application for the enhancement and imaging of latent fingermarks. Anal Chem 83:5585–5591

Ferguson LS, Wulfert F, Wolstenholme R, Fonville JM, Clench MR, Carolan VA, Francese S (2012) Direct detection of peptides and small proteins in fingermarks and determination of sex by MALDI mass spectrometry profiling. Analyst 137:4686–4692

Ferguson L, Wolstenholme R, Francese S (2013) Improvements to MALDI MSI (Dry-wet matrix deposition). Patent no. GB2489215

Ferguson LS, Creasey S, Wolstenholme R, Clench MR, Francese S (2013b) Efficiency of the dry wet method for the MALDI-MSI analysis of latent fingermarks. J Mass Spectrom 48: 677–684

Fieldhouse S (2011) Consistency and reproducibility in fingermark deposition. Forensic Sci Int 207:96–100

Francese S, Wolstenholme R, Ferguson L, Wulfert F, Fonville JM (2011) Categorisation of biological deposits using matrix assisted laser desorption ionisation mass spectrometry UK Patent 1120533.3 28 Nov 2011; International Patent Application no. PCT/GB2012/051775, 24 July 2012

Francese S, Bradshaw R, Flinders B, Mitchell C, Bleay S, Cicero L, Clench MR (2013a) Curcumin: a multipurpose matrix for MALDI mass spectrometry imaging applications. Anal Chem 85:5240–5248

Francese S, Bradshaw R, Ferguson LS, Wolstenholme R, Bleay S, Clench MR (2013b) Beyond the ridge pattern: multi-informative analysis of latent fingermarks by MALDI MS. Analyst 138:4215–4228

Girod A, Ramotowski R, Weyermann C (2012) Composition of fingermark residue: a qualitative and quantitative review. Forensic Sci Int 223:10–24

Gruendling T, Weidner S, Falkenhagen J, Barner-Kowollik C (2010) Mass spectrometry in polymer chemistry: a state-of-the-art up-date. Polym Chem 1:599–617

Hanton SD (2001) Mass spectrometry of polymers and polymer surfaces. Chem Rev 101: 527–569

Hensel RR, King RC, Owens KG (1997) Electrospray sample preparation for improved quantitation in matrix-assisted laser desorption/ionization time-of-flight mass spectrometry. Rapid Commun Mass Spectrom 11:1785–1793

Ifa DR, Manicke NE, Dill AL, Cooks RG (2008) Latent fingerprint chemical imaging by mass spectrometry. Science 321:805

International Fingerprint Research Group (IFRG) (2014) Guidelines for the assessment of fingermark detection techniques. Journal of Forensic Identification 64:174–200

Knowles AM (1978) Aspects of physicochemical methods for the detection of latent fingerprints. J Phys E: Sci Instrum 11:713–721

McComb ME, Oleschuk RD, Chow A, Ens W, Standing KG, Perreault H, Smith M (1998) Characterization of hemoglobin variants by MALDI-TOF MS using a polyurethane membrane as the sample support. Anal Chem 70:5142–5149

Miklaszewska M (1968) Free amino acids of eccrine sweat. Method Pol Med J 7:617–623

Puolitaival SM, Burnum KE, Cornett DS, Caprioli RM (2008) Solvent free matrix dry-coating for MALDI imaging of phospholipids. J Am Soc Mass Spectrom 19:882–886

Ramotowski R (2001) Composition of latent print residue. In: Lee HC, Gaensslen RE (eds) Advances in fingerprint technology, 2nd edn. CRC, Boca Raton, pp 63–104

Strauss JS, Downing DT, Ebling FJ, Stewart ME (1991) Sebaceous glands. In: Goldsmith LA (ed) Physiology, biochemistry and molecular biology of the skin, 2nd edn. Oxford University Press, New York

Sugiura S, Shimma S, Setou M (2006) Two-step matrix application technique to improve ionization efficiency for matrix-assisted laser desorption/ionization in imaging mass spectrometry. Anal Chem 78:8227–8235

Sundar, L. & Rowell, F. Detection of drugs in lifted cyanoacrylate-developed latent fingermarks using two laser desorption/ionisation mass spectrometric methods. Analyst. 139, 633–642 (2014).

Wang HY, Liu CB, Wu WU (2011) A simple desalting method for direct MALDI mass spectrometry profiling of tissue lipids. J Lipid Res 52:840–849

Wolstenholme R, Bradshaw R, Clench MR, Francese S (2009) Study of latent fingermarks by matrix-assisted laser desorption/ionisation mass spectrometry imaging of endogenous lipids. Rapid Commun Mass Spectrom 23:3031–3039

Yang HG, Park KH, Shin S, Lee JH, Park S, Kim HS, Kim J (2013) Characterization of heme ions using MALDI-TOF MS and MALDI FT-ICR MS. Int J Mass Spectrom 343–344:37–44

(MA)LDI MS Imaging at High Specificity and Sensitivity

Aurélien Thomas, Nathan Heath Patterson, Martin Dufresne, and Pierre Chaurand

Abstract (Matrix-assisted) laser desorption/ionization ((MA)LDI) mass spectrometry imaging (MSI) has been driven by remarkable technological developments in the last couple of years. Although molecular information of a wide range of molecules including peptides, lipids, metabolites, and xenobiotics can be mapped, (MA) LDI MSI only leads to the detection of the most abundant soluble molecules in the cells and, consequently, does not provide access to the least expressed species, which can be very informative in the scope of disease research. Within a short period of time, numerous protocols and concepts have been developed and introduced in order to increase MSI sensitivity, including in situ tissue chemistry and solvent-free matrix depositions. In this chapter, we will discuss some of the latest developments in the field of high-sensitivity MSI using solvent-free matrix depositions and will detail protocols of two methods with their capability of enriching molecular MSI signal as demonstrated within our laboratory.

1 Introduction

In the last decade, the emergence of MSI has offered exciting new possibilities for bioanalytical imaging (Chaurand 2012). Compared to conventional MS approaches, imaging techniques also allow the molecular mapping of hundreds of potential biomarkers, including proteins, peptides, and lipids, in a tissue section while maintaining a high correlation with the tissue architecture (Reyzer and Caprioli 2007; Schwamborn and Caprioli 2010).

A. Thomas
Unit of Toxicology, CURML, University of Lausanne, Lausanne, Switzerland
e-mail: Aurelien.Thomas@chuv.ch

N.H. Patterson • M. Dufresne • P. Chaurand (✉)
Department of Chemistry, Université de Montréal, C.P. 6128, succursale Centre-ville, Montreal, QC, Canada, H3C 3J7
e-mail: heath.patterson@umontreal.ca; martin.dufresne@umontreal.ca; pierre.chaurand@umontreal.ca

R. Cramer (ed.), *Advances in MALDI and Laser-Induced Soft Ionization Mass Spectrometry*, DOI 10.1007/978-3-319-04819-2_7

The first step of an MSI experiment is the preparation of thin sections from a tissue block or biopsy and their mounting on a flat (conductive) surface (Thomas and Chaurand 2014). Some MSI approaches require the homogeneous deposition of a desorption/ionization agent on the sections. In the next step, a desorption/ionization probe such as a laser beam in MALDI, an ion beam in Secondary Ion Mass Spectrometry (SIMS) or a focused solvent spray in Desorption ElectroSpray Ionization (DESI) is used to raster a grid array of regularly spaced coordinates over the tissue section. A mass spectrum will be acquired at each array coordinate (or pixel) with a preset lateral resolution usually between 5 and 300 μm (Römpp and Spengler 2013). By using dedicated software, ion images will be generated by integrating the intensity of each detected peak as a function of spectrum coordinate (Chaurand 2011).

Although (MA)LDI MSI is extremely powerful and informative, the overall approach is not specific, and certain limitations exist (MacAleese et al. 2009; Chaurand et al. 2006; Thomas and Chaurand 2014). The common MSI strategies for the analysis of tissue sections can lead to the detection and localization of hundreds of molecules (metabolites, lipids, peptides, and proteins) over a mass range of typically up to 30,000 Da. Although much information can be obtained, this approach only leads to the detection of the most abundant and/or soluble compounds in the tissue, and consequently, dramatically lacks sensitivity (Chaurand et al. 2004). Generic specificity, meaning the general class of biomolecules to be analyzed, can be achieved with the proper choice of matrix. For example, to perform MSI measurements of intact proteins, the appropriate choice of matrix is sinapinic acid (Stoeckli et al. 2001). To investigate (phospho)lipids (Berry et al. 2011) and metabolites, other matrices such as α-cyano-4-hydroxycinnamic acid (CHCA) (Shanta et al. 2011), 1,5-diaminonaphtalene (DAN) (Thomas et al. 2012; Anderson et al. 2014) or 2,5-dihydroxybenzoic acid (DHB) (Guo et al. 2014) will be chosen. The mode of matrix deposition can also be critical. For example, matrix deposition from solutions (with a wet interface) is critical for successful MSI analysis of peptides and proteins, whereas some lipids and metabolites can be detected using dry matrix deposition approaches (Hankin et al. 2007; Puolitaival et al. 2008).

Among current efforts in (MA)LDI MSI, the improvement of specificity and sensitivity is particularly challenging for the investigation of certain classes of low-abundance molecules. Many such molecules play essential roles in cell regulation or disease onset and development. Although numerous efforts have been undertaken to develop methodologies and instrumentation, (MA)LDI MSI is still undergoing further development to push the technology forward and to make it more routinely accessible to users (Jungmann and Heeren 2012). The aim of this chapter is to present some of the latest methodologies developed in our laboratory to increase specificity and sensitivity in IMS of different classes of lipids. Among these, two proven protocols are detailed here providing the necessary information for readers for successful implementation in their laboratories.

2 Applications

2.1 *General Applications*

MSI has been used in a wide range of bioanalytical applications (Chughtai and Heeren 2010; Norris and Caprioli 2013). In particular, molecular expression and organization in diseases such as cancer and neurologic disorders have been studied (Cazares et al. 2009, 2011; Thomas et al. 2013; Lemaire et al. 2007; Morita et al. 2010; Wisztorski et al. 2008). MSI also has been used extensively in developmental and fundamental biology (Burnum et al. 2009; Grey et al. 2009; Minerva et al. 2008). Another application area of growing interest is the mapping of xenobiotics and related metabolites in tissues (Stoeckli et al. 2007; Sun and Walch 2013; Stauber 2012). Not only can the exact location and absolute quantity of the drug be determined within a tissue specimen, but its effect on the proteome or lipidome can be monitored as a function of time or dose (Reyzer et al. 2004). There is also currently a growing interest for the monitoring of the metabolomic and proteomic content of plant tissues by MSI (Debois et al. 2013; Cavatorta et al. 2009).

2.2 *Imaging of Lipids*

2.2.1 Employing Sublimation of DAN on Tissue Sections

MSI experiments at high-spatial resolution require homogeneous application of the matrix. Different approaches, including manual and automatic systems, have been developed to form homogeneous matrix crystals across the section and limit lateral migration of the molecules (Chen et al. 2008). In this regard, solvent-based matrix deposition processes carry the risk of introducing an inherent limitation for high-resolution MSI due to the risk of analyte delocalization. In some cases, well-controlled automated matrix deposition approaches have resulted in encouraging results (Chaurand et al. 2006). Matrix sublimation is a highly efficient procedure for MSI of lipids at high-spatial resolution by forming an extremely homogeneous micrometer-thick layer of matrix on the sections (Hankin et al. 2007; Thomas et al. 2012; Chaurand et al. 2011; Angel et al. 2012). Although different matrices such as DHB and CHCA can be sublimated, most of these are only efficient in one mode of ionization (Murphy et al. 2011), leading to an important loss of information since some phospholipids, sulfatides, ceramides, cardiolipins, and gangliosides as well as other classes of lipids are better ionized in the positive or negative ionization mode (Thomas et al. 2012; Woods and Jackson 2006).

To improve the information obtainable from existing protocols, we recently investigated several matrix candidates for their sublimation ability, sensitivity in both positive and negative ionization modes, lack of matrix cluster interferences, and stability under vacuum. Among investigated candidates, the use of a DAN matrix deposited by sublimation has been demonstrated to be of high efficiency for

MSI of lipids, providing rich information in both ionization modes (Thomas et al. 2012). By using a fixed offset in the *x* and *y* dimensions of the grid array, ion images can be acquired serially from a single sample preparation in both positive and negative ion mode. As detailed below, we have developed a robust sample preparation protocol, which has to date been successfully applied to MSI analyses of large cohorts of clinical tissue biopsies.

2.2.2 Employing Silver Sputter Deposition on Thin Tissue Sections

Cholesterol, fatty acids, and other olefin-related compounds are widely investigated because of their important functional role in multiple diseases, including cancer and cardiovascular diseases (Maxfield and Tabas 2005; Lusis 2000). MSI of these compounds is however a challenging task due to their poor ionization efficiency with conventional (MA)LDI approaches. The addition of silver salts to homogeneous samples is a longstanding strategy to enhance ionization of olefin compounds for MS analyses (Schriemer and Li 1996). Recently, some promising approaches have been developed to make the process convenient with regard to MSI requirements. In one approach, a silver nitrate layer was deposited on a porous silicon waffle target prior to mounting the tissue on the surface (Patti et al. 2010a, b). For this specific LDI MS technique, called nanostructure-initiator mass spectrometry (NIMS), the deposition of a silver nitrate layer significantly enhanced MSI of cholesterol. In another approach, colloidal silver layer deposition also yielded improved results for various long-chain fatty acids and hydrocarbons using an LDI MSI approach (Jun et al. 2010). In these studies, the molecular ions were observed as silver cations.

These different techniques provided encouraging results for LDI MSI of cholesterol and other fatty acids of tissue sections. We further explored this potential and developed a novel solvent-free LDI MSI approach to image cholesterol and fatty acids with high specificity, sensitivity, and spatial resolution after silver cationization (Dufresne et al. 2013). Here, some aspects need to be considered when increasing the lateral resolution, including the homogeneity of the desorption/ionization agent, the minimum laser raster distance (minimum step size of the sample stage) and the minimum laser focus achievable on target. A smaller laser focus leads to a lower number of ions generated and thus a diminution in intensity of the corresponding ion signals. This last point can be partially overcome by increasing the laser power or by using a more efficient ionization approach such as the use of silver cationization as discussed above (Chaurand et al. 2011).

Due to the highly homogeneous matrix deposition with minimal lateral analyte migration, matrix sublimation is the technique of choice for achieving high-spatial-resolution MSI. Our work demonstrated that sputter deposition of metal yields a micrometer scale homogeneous layer (Nygren and Malmberg 2004), providing experimental conditions similar to those obtained from the matrix sublimation process. On the basis of previous findings involving silver cationization of cholesterol and fatty acids, we developed a robust silver sputtering deposition method for high-resolution LDI MSI through metal cationization of a variety of olefinic compounds from tissue sections. A protocol for applying on-tissue, silver sputtering to LDI MSI of cholesterol and other olefinic compounds is described below.

3 Materials and Protocols

3.1 MSI Employing Sublimation of DAN on Tissue Sections

3.1.1 Materials

The setup of a sublimation apparatus is given in Fig. 1. It is strongly recommended to install the complete sublimation system in a fume hood. All the glassware pieces (sublimation chamber and cold trap) should be fixed using static mounts. Great care is required in setting up the system to align the various joints between all pieces and ensure there are no leaks in the tubing. The area where tubing and glass joints meet can be wrapped tightly with Parafilm to further prevent air leaks. The cold trap of the apparatus is essential to maintain the vacuum and prevent sublimated matrix

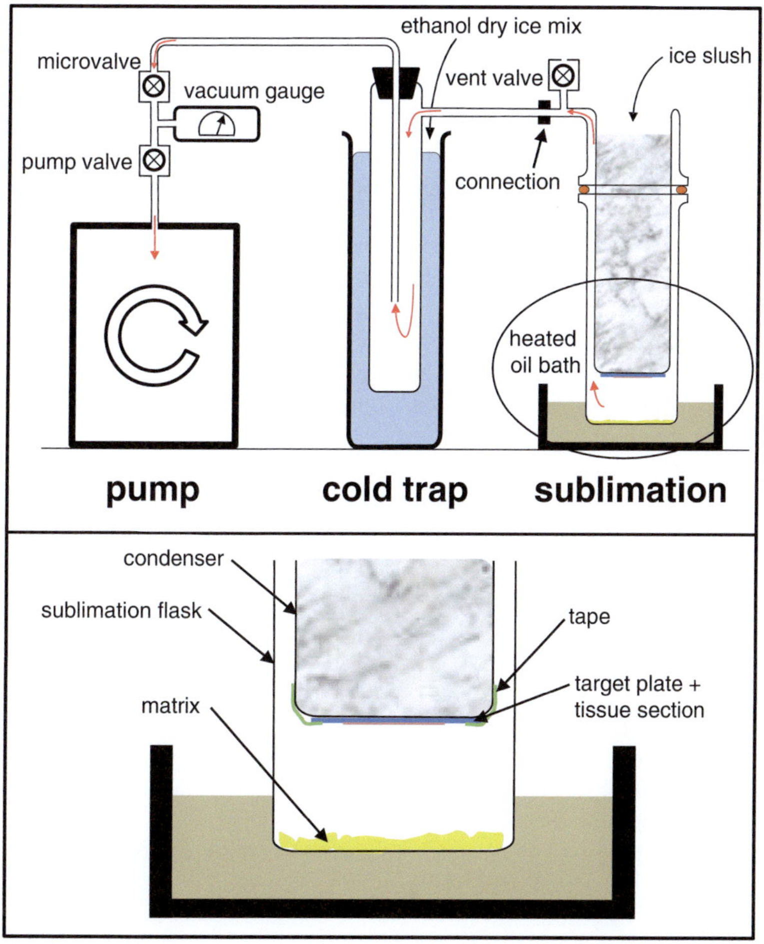

Fig. 1 Scheme of a matrix sublimation system. *Red arrows* indicate airflow. See text for details

from contaminating the vacuum gauge. It should be filled with laboratory grade ethanol and pieces of dry ice to cool the trap. The apparatus needs to be able to achieve a pressure of 0.05 Torr for DAN sublimation. An oil bath (sand bath or other heating element) with digital temperature control is recommended for heating the bottom of the sublimation chamber as it provides reproducible and even heating (Fig. 1). The oil bath should be installed on a sturdy height adjustable stand. WARNING: The oil bath is operated at high temperature and extra care must be taken when adjusting the oil bath height.

As the section to be imaged (sometimes with mounting tape, slide or even MALDI target) will be fitted onto the sublimation apparatus' condenser, it is necessary to obtain sublimation glassware that can accommodate the dimensions of the section, tape, slide, or MALDI target. Within our lab, we have two interchangeable sublimation chambers, one smaller diameter system (Product number CG-3038; Chemglass Life Sciences, Vineland, NJ, USA) that accommodates 45 mm × 45 mm square glass slides, and a larger diameter system (Product number 8023-55; Ace Glass Inc., Vineland, NJ, USA) that can accommodate two 75 mm × 25 mm microscope glass slides. For MSI experiments, it is recommended to use standard microscope glass slides coated with indium–tin oxide (ITO). All of our glass slides are purchased from Delta Technologies, Loveland, CO, USA, which come coated with a low-resistivity ($R \leq 200\ \Omega$) ITO film to insure conductivity across the slide for MALDI-TOF MSI analyses. The slides should be as clean as possible before use. Material on the surface can create 'hot spots' of sublimated matrix. We typically clean thoroughly with soap and water followed by rinsing thoroughly with methanol. The slides are then immediately dried using Kimwipes™.

The DAN matrix can be purchased from Sigma-Aldrich at 97% purity according to the manufacturer. It does not require recrystallization before its use for sublimation. WARNING: Caution should be taken when handling DAN due to possible carcinogenetic effects. Exposure to the chemical should be minimized and fume hoods should always be used.

MALDI MSI experiments were performed on an UltraFlextreme MALDI-TOF MS system (Bruker Daltonics, Billerica, MA, USA) using the reflectron geometry at +25 kV of accelerating potential under optimized laser energy and delayed extraction conditions. For MS and MSI data acquired on our MALDI-TOF MS system, we routinely obtain mass accuracies better than 5 ppm for single spectra and better than 10 ppm across entire images.

3.1.2 Parameters for DAN Sublimation

Parameters for sublimation will need to be adjusted for each apparatus including the amount of matrix to be added. The slides can be weighed before and after sublimation to determine the amount of matrix sublimated, and this can be correlated to the results obtained. We recommend 110 $\mu g/cm^2$ deposited on tissue sections.

Vacuum: The vacuum in the sublimation apparatus should be at least 0.05 Torr and be maintained at this value during the sublimation process.

Heating: DAN requires an oil bath (heating element) temperature of at least 145 °C for sublimation. In our experience, heating above this value is preferable as the temperature of the oil bath (heating element) will gradually decrease during sublimation. The depth of submersion of the sublimation chamber in the heat bath is also of importance.

3.1.3 Procedure

1. The heat bath should be heated to 160 °C with stirring.
2. The matrix should be weighed out, and an amount sufficient to completely cover the bottom of the sublimation chamber is used. For our larger diameter sublimation chamber, we use ~17 g of DAN. The matrix should be initially pulverized in a mortar and pestle to generate an ultrafine homogeneous powder. It should be spread evenly across the bottom to produce a homogeneous layer, minimizing valleys and peaks. This can be aided by applying pressure onto the matrix, for example, with the bottom of a flask or beaker that fits into the chamber.
3. The slide (including the thaw-mounted tissue section) will then be fixed to the bottom of the condenser (see Fig. 1). Optimal results will be obtained when the cooling is homogeneous across the surface. As even contact across the surface cannot be guaranteed without some pressure applied to push the slide onto the glassware, we strongly recommend taping the slide onto the condenser with single-sided tape. Typically, we tightly tape the shorter edge of the microscope slide, and then tightly tape the opposite side. After the first layer of tape, we add a second for reinforcement and to ensure a tight contact. The bottom of our condenser has a small height disparity between the middle and the outside areas and with our taping method, we force the slide to bend very slightly onto the surface and ensure good contact despite an uneven surface. Alternatively, double-sided tape along the entire length of the slide can be used. This approach can however be problematic if the binding is too strong as it may lead to breaking the slide when dismounting it from the condenser.
4. Once the slide is taped in place, the sublimation chamber can be closed and sealed. After this, a vacuum of at least 0.05 Torr will be established in the apparatus. We recommend creating the vacuum before putting ice and water in the condenser to cool the slides, preventing condensation on the tissue surface (Fig. 1). In general, a handful of ice is added, and just enough water to make the ice float is needed. This prevents cold spots from forming in areas where the ice is touching the glassware.
5. After 10–15 min of vacuum exposure without heating, the DAN matrix is heated by immersing the sublimation chamber into the heat bath. If using an oil bath, marking the oil's displacement on the outside of the oil bath's bowl can improve reproducibility between experiments. The sublimation time will depend on pressure, temperature, amount of matrix, and depth of chamber submersion. With our setup using the larger chamber, the sublimation time is approximately 15 min.

6. After sublimation, the heating element is removed and the vacuum pump is isolated (using the pump valve, cf. Fig. 1). Immediately after isolating the vacuum pump, it is recommended to maintain vacuum pressure inside the sublimation apparatus. With the sublimation chamber still under vacuum, the majority of the ice water is removed from the condenser and replaced by a large volume of room temperature (tap) water to prevent water condensation on the slide when releasing the vacuum. After the room temperature water is added to the condenser, the apparatus should rest at room temperature for ~5 min to warm the sample to room temperature. The vacuum is then slowly released (using the vent valve, cf. Fig. 1), avoiding any disturbance of the leftover matrix at the bottom of the flask. The condenser and flask can now be separated and the water in the condenser can be discarded. WARNING: Great care must be taken when manipulating the sublimation apparatus. Shattering the apparatus while under vacuum may lead to serious injuries.
7. Next, the slide is carefully removed from the bottom of the condenser and set aside for analysis. Ideally, 2–3 mg of matrix should be deposited on a 75 mm × 25 mm ITO-coated glass slide. This can be easily checked by weighing the slide before and after the sublimation process. If samples are going to be analyzed at a later date after matrix deposition, it is important to store them under freezing conditions to prevent lipid degradation (Patterson et al. 2014), preferably in a small plastic bag filled with nitrogen to prevent condensation when removing from the freezer.
8. Excess matrix deposited on the surface of the condenser can be removed with Kimwipes™ and methanol. If multiple sublimations are to be performed, it is recommended to repack the matrix and ensure again a level layer of matrix.

3.1.4 Troubleshooting

To insure proper functioning of the complete setup, it is recommended to perform a dummy sublimation at the optimized timing and heating parameters to assess and ensure that the expected amount of matrix is deposited per unit surface area.

3.1.5 An Example: MALDI MSI of a Mouse Brain Sagittal Tissue Section After DAN Sublimation

Figure 2a shows the averaged mass spectra acquired from a mouse brain imaged at 75 μm resolution in both positive and negative ion mode. Negative ion mode MSI was achieved on the same section using a 37-μm offset in the *x* and *y* coordinates. The inset of Fig. 2a shows a scan of the ablation pattern on the section after MSI analysis. The black and red spots indicate where positive and negative ion mode spectra were acquired. The small area of the inset is part of the hippocampus (labeled c in Fig. 2 biv). The average spectra demonstrate that dual ionization

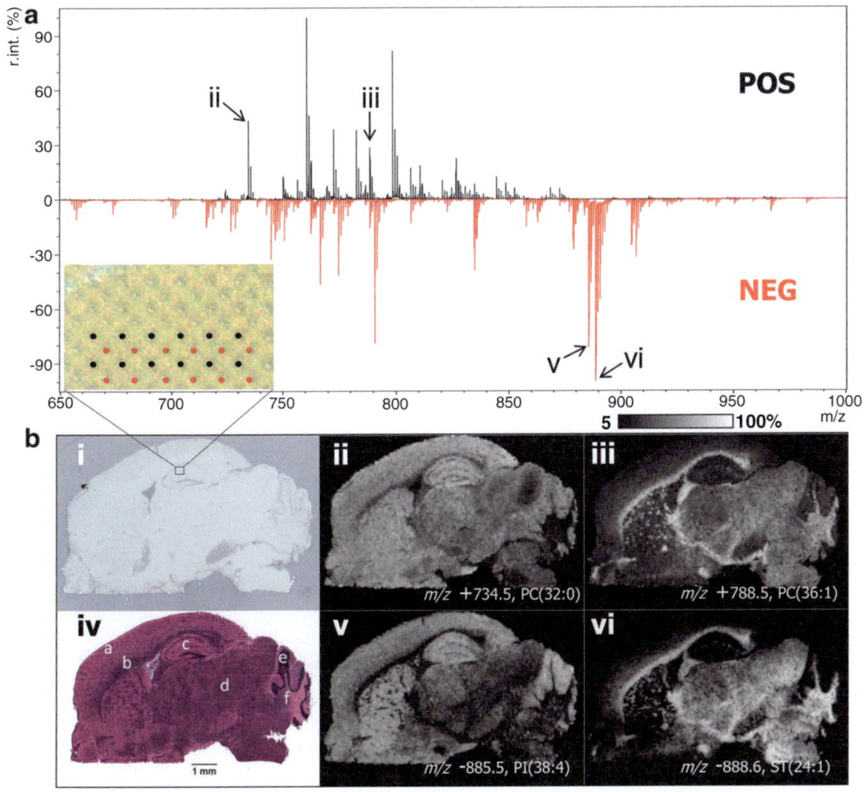

Fig. 2 MALDI MSI of a sagittal mouse brain tissue section acquired using DAN as MALDI matrix at 75 μm-spatial resolution with representative ion images. (**a**) *Top* and *bottom*—average mass spectra displaying the *m/z* range of 650–1000 from the MSI analysis of the section for the positive (*black*) and negative (*red* and *inverted*) ionization mode. *Inset*—photograph of the laser ablation pattern after MSI data acquisition with black and red spots highlighting the positive and negative acquisitions, respectively. (**b**) (*i*) digital optical scan of the tissue section after DAN matrix deposition by sublimation prior to MSI; (*ii–iii*) representative ion images from the positive ionization mode for *m/z* 734.55 (PC(32:0)) and *m*/z 788.61 (PC(36:1)), respectively; (*iv*) H&E staining image of a serial section for histological comparison, labels *a–f* indicate the cerebral cortex, forebrain *white matter*, hippocampus, midbrain, cerebellum, and cerebellum *white matter*, respectively; (*v–vi*) representative ion images from negative ionization mode for *m/z* 885.55 (PI(38:4)) and *m*/z 888.62 (ST(24:1)), respectively

imaging using an offset gives access to many more lipid species per sample analysis. Figure 2b shows the optical image of the tissue section covered with sublimated DAN (i), two positive ion mode images (ii–iii), an H&E staining image of a serial section (iv) and two negative ion mode images (v–vi). In this demonstration we show images of correlated phosphotidylcholines (PC) from positive ion mode acquisitions and phosphatidylinositols (PI) and sulfatides (ST) imaged by negative ionization mode. The images for PC(32:0) (ii) and PI(38:4) (v) highlight several regions of the brain, including the cerebral cortex, hippocampus, midbrain (less intense in PI(38:4)), and some areas of the cerebellum (annotated in Figure 2biv as a, c, d, and e, respectively). Conversely, PC(36:1) (iii) and ST(24:1) (vi) are distributed in different areas of the brain including the white matter in the forebrain and cerebellum (annotated in Figure 2biv as b and f, respectively). MSI clearly highlights major brain histologies, and dual ionization mode MSI allows correlation of phospholipid species from a wider variety of classes.

3.2 *MSI Employing Silver Sputter Deposition on Thin Tissue Sections*

3.2.1 Materials

The sample preparation system used in our laboratory for metallic silver deposition on thin tissue sections is the magnetron sputtering system (308R sputter coater, Cressington Scientific Instruments Ltd, Watford, England). Silver and chrome targets for the sputtering system (size depends on the system used) are required with at least 99.5% purity (ESPI Metals, Ashland OR, USA). Standard microscope glass slides (25 mm × 75 mm) can be purchased from most generic scientific vendors. LDI MSI experiments were performed on a MALDI-TOF MS system (UltraFlextreme, Bruker Daltonics, Billerica MA, USA) using the reflectron geometry at +25 kV of accelerating potential under optimized laser energy and delayed extraction conditions with similar mass accuracies as described in Sect. 3.1.1.

3.2.2 Procedure

1. *Preparation of the silver-coated slides*: Microscope glass slides are used to prepare silver-coated slides. These are first rinsed with deionized water followed by a methanol wash. The excess solvent is removed by wiping the slides with a low particle- and fiber-generating tissue wipe such as the BEMCOT PS-2 (Asahi Kasei, Tokyo, Japan) to insure no particles are left on the slides prior to the metal sputtering. A thin film of chrome of about 2–3 nm thickness is initially deposited to increase the adhesion of silver to the slides. Silver is then deposited over the thin chrome film until a thickness of 100 nm is reached. These two steps can be carried out using either a sputtering system or a thermal evaporator (requires venting and evacuating the chamber twice if chrome is used) to achieve the desired thickness. For the sputtering system more details will be given in step 3.
2. *Tissue deposition*: Freshly frozen or formalin-fixed tissue samples can be cut at thicknesses between 5 and 40 μm using either a cryostat (frozen) or a vibratome (formalin-fixed). The layer of silver subsequently added onto the sections establishes conductivity over the section which reduces any isolating effect of thicker tissue cuts.

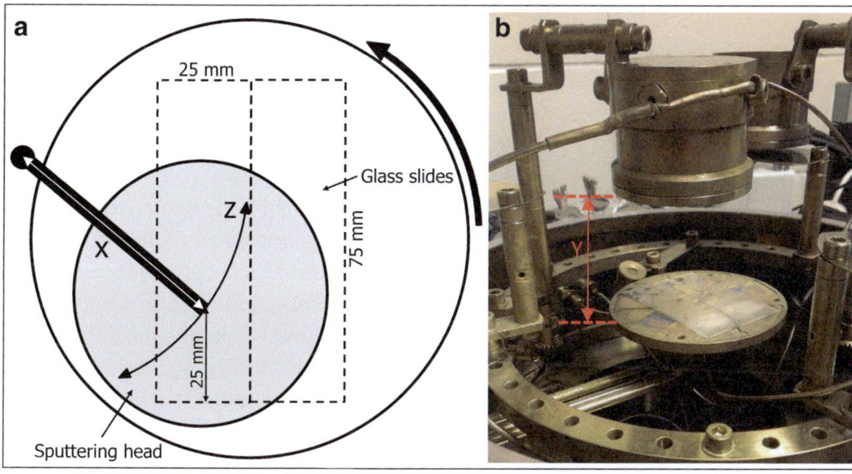

Fig. 3 (**a**) Scheme (*top view*) and photograph (*side view*) of a silver sputter system. See text for details

3. *Silver sputtering on tissue sections*:

(a) Thickness control and calibration curve: Sputtering systems have five parameters, which need to be considered, and a calibration method will be given below if your sputtering system does not allow for all those parameters to be adjusted. The head position and the distance from the sample can affect the amount and homogeneity of the silver layer being produced. The system we use offers x (horizontal head support)/y (vertical head support)/z (head rotation) positioning (see Fig. 3). For reproducibility purposes, all dimensions should initially be optimized and kept constant. Fixed values for x (~9 cm for our system) and z positioning offset the sputtering head from the center of the system (see below). The distance y between the sputtering head and the target glass slides is also maintained constant (~7.5 cm). In our system, to generate nanometer-thick homogeneous layers of silver over a 25 mm×75 mm slide, we first center the edge of one sample slide onto the center of the sample stage (see Fig. 3). This allows for two sample slides to be positioned within the system. We then center the sputtering head horizontally over the first third of a slide (see Fig. 3). This, coupled with the sample stage rotation, will give a similar exposure to the sputtering plume for the entire surface of both slides. The sample stage rotation is ~25 rounds per minute. The argon partial pressure necessary for the sputter process controls the roughness of the silver surface. In the system used, a pressure value of 0.02 mbar was the optimal value for best silver layer homogeneity. The value for the current used to form the argon plasma for the sputtering was set to the maximum value of our system, which is 80 mA. The last parameter to control the silver layer thickness is the deposition time. As mentioned above if you cannot change the above parameters on your system, a calibration curve can be constructed for the sputtering system by plotting the silver layer thickness as a function of deposition time. To make the curve you will need an instrument to measure the silver layer thickness such as an atomic force microscopy system. To generate a curve, a small glass cover slip (22×22 mm^2) can be placed on a microscope glass slide before silver deposition. After silver deposition, the cover slip is removed and the thickness is measured on the edges around the area the cover slip occupied. Repeat these steps for multiple deposition times and you should now have a calibration curve for your own sputtering parameters. It is important that the depositions are not performed in quick successions when you create the calibration curve as the silver target will start to heat up after longer periods of operations. Indeed, the hotter the silver target, the faster the silver deposition will occur. This means that the curve will not be linear with time if all data points are sequentially generated. For long deposition times (>1 min) this heating up process is unavoidable; enough time needs to be given to the system to cool down between each deposition. With the above procedure, one should be able to accurately generate a homogeneous silver layer of any thickness with a given set of sputtering parameters.

(b) Optimization of silver layer thickness: With the sputtering system we use and the parameters mentioned above, a deposition time of 30 s generates a 14 ± 2 nm-thick silver layer, which is used for non-fatty tissue sections mounted on silver slides. However, silver layer deposition on very fatty tissues can be more demanding since it is difficult to predict the amount of silver needed—more information will be given in the troubleshooting section. Minimizing the silver layer thickness is an important consideration for LDI MSI since silver still reflects a large proportion of the UV laser light at the typical wavelengths used. This means that thinner films (less silver) require less laser energy for the LDI process to take place while thicker films require higher laser energy, which can lead to analyte fragmentation. Some very fatty samples will require thicker silver films but in these cases the silver applied should be the minimum amount necessary for proper ionization as explained in Sect. 3.2.3.

(c) Serial silver coatings of multiple tissue sections: To prepare a batch of sections from the same or multiple tissue specimens, the above mentioned silver block heating issue needs to be taken into account. In the system we use, to avoid waiting for the silver target to cool down, the quartz balance inside the system is used to evaluate the thickness of the silver layer. The balance response needs however to be calibrated and is dependent on its distance from the deposition head. If this distance is fixed for all samples, the response of the balance should be consistent. We usually fix the position of the head for silver coating the first section and never touch it again for the coating of subsequent sections. This means that the balance will evaluate the amount of silver deposited between samples in a reproducible manner. We usually record the silver thickness the balance provides for the first three sections and then rely on the balance readings and not the deposition time for controlling the thickness for all following sections. If for any reason the position of the silver deposition head needs to be changed, one needs to go through the process of cooling down the silver target and the deposition head to room temperature, which can take several hours.

4. *Data acquisition*: Once the silver layer is deposited on the sample, sample analysis by LDI MSI can be started. If the sections cannot be analyzed on the same day as the preparation, we suggest storing the slides in a sealed bag under N_2 atmosphere in a freezer. The N_2 atmosphere will prevent further oxidation of the sample and water condensation when taking the sample out of the freezer. Step one is to define a suitable laser energy for obtaining good reproducible MS signals. Since no MALDI matrix is used, the laser fluence range on target can be very critical. As with most LDI techniques, the laser fluence range for obtaining intense MS signals from intact molecular ions is relatively small and the transition to more ion fragmentation is very sharp. The necessary laser energies are usually set around the same values as those necessary for MALDI MS using CHCA or DHB matrices. During MSI data acquisition, the detection of very distinct MS signals corresponding to the formation of multiple primarily odd-numbered silver clusters ranging from Ag_2^+ to Ag_9^+ may be noticed (see Fig. 4). The presence of these silver clusters in basically every mass spectrum can be very useful for both internal mass calibration and troubleshooting.

3.2.3 Troubleshooting

Very fatty tissue specimens are the most difficult to analyze using silver-assisted LDI MS and are a good example of how to troubleshoot the deposition of the silver layer. The best way to find the right silver layer thickness for fatty tissues is to start with the values mentioned above. When the silver layer of the sample is too thin, very few LDI MS analyte ion signals are observed in the spectrum and those observed will most likely be more intense than the signals for the $[Ag_2]^+$ and $[Ag_3]^+$ silver clusters. This is an indication that there is an insufficient amount of silver as ionization agent; $[Ag_2]^+$ and $[Ag_3]^+$ should be observed with dominant ion signals in the mass spectrum. At this point the deposition of an additional layer of silver on the section is recommended. We usually gradually add silver by increments of 5 nm (~10 s of sputtering time) and analyze the section again by LDI MS until the $[Ag_2]^+$ and $[Ag_3]^+$ peaks become the dominant signals in the mass spectrum. This simple procedure allows to quickly optimize the silver layer thickness for any type of tissue.

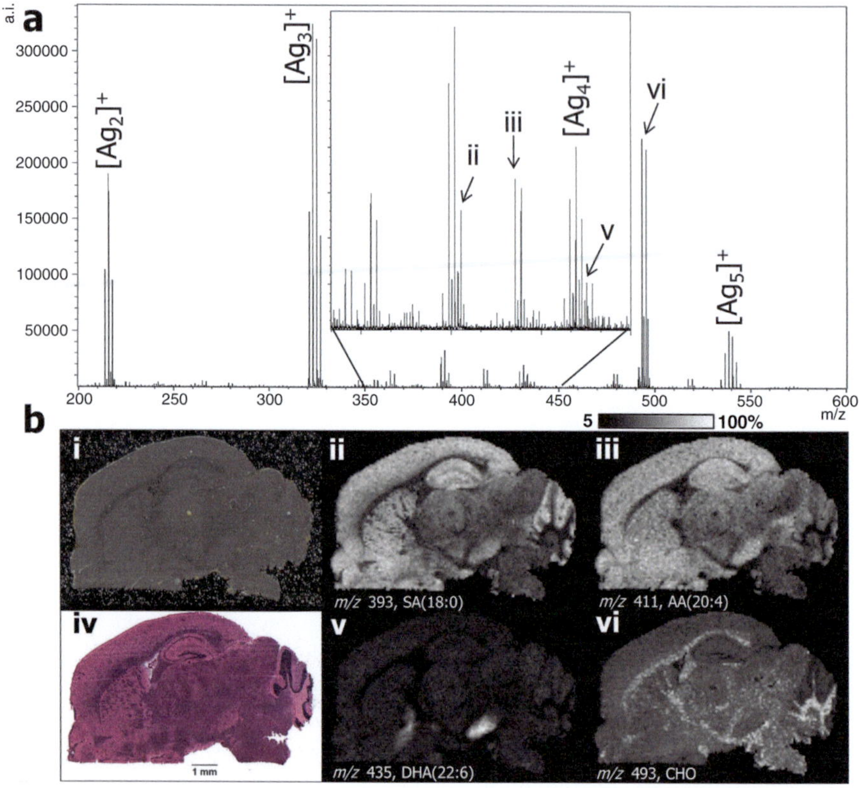

Fig. 4 Silver-assisted LDI MSI of a mouse brain section at 75 μm-lateral resolution with representative ion images. (**a**) LDI-TOF MS mass spectrum and (**b**) images from a section cut in the sagittal plane at around 1.2 mm from bregma; (*i*) scanned optical image of the brain section after coating with a 14 ± 2 nm silver layer; (*ii–iii*) MS ion images for $[M+Ag]^+$ ions of stearic acid (*ii*) and arachidonic acid (*iii*); (*iv*) H&E staining image of a serial section to highlight the major histological features; (*v–vi*) MS ion images for $[M+Ag]^+$ ions of docosahexaenoic acid (*v*) and cholesterol (*vi*)

Another problem that can occur is the loosening of the tissue sections from the silver slides. This occurs because the underlying silver layer detaches from the slides. This does not happen often and unfortunately cannot be predicted. However, some steps can be taken to reduce its frequency. For instance, as mentioned above, for better adhesion of the silver layer to the slides, the addition of a first layer of chrome is recommended. This step will significantly reduce the frequency of tissue section loosening. You can also reduce this phenomenon by using freshly prepared silver slides. Older silver-coated glass slides can display white spots on their surface due to the silver layer locally detaching from the surface. If spots are observed, the slide should not be used.

3.2.4 An Example: Silver-Assisted LDI MSI of a Mouse Brain Sagittal Tissue Section

Figure 4a shows a typical MS spectrum acquired from a mouse brain gray matter region where the presence of various silver-cationized fatty acid molecular ions in the range of *m/z* 330–450 as well as strong signals for the silver-cationized molecular ion of cholesterol at *m/z* 493 are clearly seen. Fatty acids and cholesterol always present two distinct peaks that originate from the natural abundance of the ^{107}Ag and ^{109}Ag silver isotopes with relative abundances of 52% and 48%, respectively. This makes all silver-cationized molecules appear as doublets with this very distinct isotopic pattern. However, this makes the spectra more complex and can in some instances become confusing. For instance, two fatty acids only differing by two hydrogen atoms due to a slightly different degree of saturation would lead to the formation of a triplet isotopic pattern system where the adducts of ^{109}Ag overlap with the adducts of ^{107}Ag of the more saturated molecule. Along with the photographs of the tissue section after silver deposition (i) and the matching histology image (H&E staining of a serial section, iv), Fig. 4b shows four typical MS ion images for three fatty acids and cholesterol in a sagittal section of a mouse brain (~1.2 mm from bregma). Steric acid (ii) shows a distribution throughout the gray matter of the brain and cerebellum, while cholesterol (vi) shows correlation with the white matter in both the cerebellum and forebrain regions. Arachidonic acid (iii) also shows good correlation with the gray matter of the brain but has a much lower signal in the cerebellum. The MS ion image for docosahexaenoic acid (v) shows a very strong correlation with two specific brain substructures, the *substantia innominata* and the *substantia nigra*. Most fatty acids that are typically detected in the mouse brain have carbon chain lengths between 16 and 22.

4 Concluding Remarks

Current (MALDI) MSI approaches only allow to broadly analyze classes of biomolecules. Based on the matrix employed and its mode of deposition on tissue sections, different classes of biomolecules (proteins, peptides, or lipids) can be analyzed. For example, current MSI methods for lipids do not allow specific targeting of a chosen class of lipids, which may be of high value for diagnosis, prognosis or as indicator of response to therapy. For MSI to become a competitive technology in the clinical setting, it has to be tailored to target specific sets of biomarkers with high specificity and sensitivity.

Some specificity can be introduced with the appropriate choice of matrix, the solvent system used, and the mode of matrix deposition onto tissue sections. DAN matrix deposition on tissue sections allows to specifically detect and image (phospho)lipids by MALDI MSI after DAN deposition by sublimation. One of the unique properties of this matrix is the possibility to analyze lipids with both ion modes, which allows the detection of all the major classes of phospholipids. The high homogeneity of the matrix coating allows MSI analyses to be performed at very high (≤10 μm) spatial resolution, in the range of cellular dimensions (Thomas et al. 2012). Efforts by our group and others to develop novel MALDI approaches and methods to implement various degree of specificity in MSI are ongoing.

Based on their structure, some biocompounds have high affinities for various metal cations. This unique property can be used to selectively ionize and detect such biocompounds by MSI even when present in complex mixtures. As presented and detailed here, we have recently developed a tissue section preparation protocol to analyze cholesterol, fatty acids, as well as other olefin-containing molecules present within tissue sections by silver-assisted LDI MSI with high specificity and sensitivity. The high homogeneity of the silver coating even allows MSI analyses to be performed at very high (≤5 μm) spatial resolution, in the range of cellular dimensions (Dufresne et al. 2013). Other metals, metal oxides, and cations have interesting chelation properties with various classes of lipids that are currently explored by our group to introduce a further degree of specificity and sensitivity to MSI.

References

Anderson DM, Ablonczy Z, Koutalos Y, Spraggins J, Crouch RK, Caprioli RM, Schey KL (2014) High resolution MALDI imaging mass spectrometry of retinal tissue lipids. J Am Soc Mass Spectrom 25(8):1394–1403. doi:10.1007/s13361-014-0883-2

Angel PM, Spraggins JM, Baldwin HS, Caprioli R (2012) Enhanced sensitivity for high spatial resolution lipid analysis by negative ion mode matrix assisted laser desorption ionization imaging mass spectrometry. Anal Chem 84(3):1557–1564. doi:10.1021/ac202383m

Berry KA, Hankin JA, Barkley RM, Spraggins JM, Caprioli RM, Murphy RC (2011) MALDI imaging of lipid biochemistry in tissues by mass spectrometry. Chem Rev 111(10):6491–6512. doi:10.1021/cr200280p

Burnum KE, Cornett DS, Puolitaival SM, Milne SB, Myers DS, Tranguch S, Brown HA, Dey SK, Caprioli RM (2009) Spatial and temporal alterations of phospholipids determined by mass spectrometry during mouse embryo implantation. J Lipid Res 50(11):2290–2298. doi:10.1194/jlr.M900100-JLR200

Cavatorta V, Sforza S, Mastrobuoni G, Pieraccini G, Francese S, Moneti G, Dossena A, Pastorello EA, Marchelli R (2009) Unambiguous characterization and tissue localization of Pru P 3 peach allergen by electrospray mass spectrometry and MALDI imaging. J Mass Spectrom 44(6):891–897. doi:10.1002/jms.1562

Cazares LH, Troyer D, Mendrinos S, Lance RA, Nyalwidhe JO, Beydoun HA, Clements MA, Drake RR, Semmes OJ (2009) Imaging mass spectrometry of a specific fragment of mitogen-activated protein kinase/extracellular signal-regulated kinase kinase kinase 2 discriminates cancer from uninvolved prostate tissue. Clin Cancer Res 15(17):5541–5551. doi:10.1158/1078-0432.CCR-08-2892

Cazares LH, Troyer DA, Wang B, Drake RR, Semmes OJ (2011) MALDI tissue imaging: from biomarker discovery to clinical applications. Anal Bioanal Chem 401(1):17–27. doi:10.1007/s00216-011-5003-6

Chaurand P (2011) Imaging mass spectrometry: current performance and upcoming challenges. Curr Trends Mass Spectrom 30–37. http://www.chromatographyonline.com/imaging-mass-spectrometry-current-performance-and-upcoming-challenges

Chaurand P (2012) Imaging mass spectrometry of thin tissue sections: a decade of collective efforts. J Proteomics 75:4883–4892. doi:10.1016/j.jprot.2012.04.005

Chaurand P, Schwartz SA, Capriolo RM (2004) Profiling and imaging proteins in tissue sections by MS. Anal Chem 76(5):87A–93A

Chaurand P, Norris JL, Cornett DS, Mobley JA, Caprioli RM (2006) New developments in profiling and imaging of proteins from tissue sections by MALDI mass spectrometry. J Proteome Res 5(11):2889–2900. doi:10.1021/pr060346u

Chaurand P, Cornett DS, Angel PM, Caprioli RM (2011) From whole-body sections down to cellular level, multiscale imaging of phospholipids by MALDI mass spectrometry. Mol Cell Proteomics 10(2):O110.004259. doi:10.1074/mcp.O110.004259

Chen Y, Allegood J, Liu Y, Wang E, Cachon-Gonzalez B, Cox TM, Merrill AH Jr, Sullards MC (2008) Imaging MALDI mass spectrometry using an oscillating capillary nebulizer matrix coating system and its application to analysis of lipids in brain from a mouse model of Tay-Sachs/Sandhoff disease. Anal Chem 80(8):2780–2788. doi:10.1021/ac702350g

Chughtai K, Heeren RM (2010) Mass spectrometric imaging for biomedical tissue analysis. Chem Rev 110(5):3237–3277. doi:10.1021/cr100012c

Debois D, Ongena M, Cawoy H, De Pauw E (2013) MALDI-FTICR MS imaging as a powerful tool to identify Paenibacillus antibiotics involved in the inhibition of plant pathogens. J Am Soc Mass Spectrom 24(8):1202–1213. doi:10.1007/s13361-013-0620-2

Dufresne M, Thomas A, Breault-Turcot J, Masson JF, Chaurand P (2013) Silver assisted laser desorption ionization for high spatial resolution imaging mass spectrometry of olefins from thin tissue sections. Anal Chem 85(6):3318–3324. doi:10.1021/ac3037415

Grey AC, Chaurand P, Caprioli RM, Schey KL (2009) MALDI imaging mass spectrometry of integral membrane proteins from ocular lens and retinal tissue. J Proteome Res 8(7):3278–3283. doi:10.1021/pr800956y

Guo S, Wang Y, Zhou D, Li Z (2014) Significantly increased monounsaturated lipids relative to polyunsaturated lipids in six types of cancer microenvironment are observed by mass spectrometry imaging. Sci Rep 4:5959. doi:10.1038/srep05959

Hankin JA, Barkley RM, Murphy RC (2007) Sublimation as a method of matrix application for mass spectrometric imaging. J Am Soc Mass Spectrom 18(9):1646–1652. doi:10.1016/j.jasms.2007.06.010

Jun JH, Song Z, Liu Z, Nikolau BJ, Yeung ES, Lee YJ (2010) High-spatial and high-mass resolution imaging of surface metabolites of Arabidopsis thaliana by laser desorption-ionization mass spectrometry using colloidal silver. Anal Chem 82(8):3255–3265. doi:10.1021/ac902990p

Jungmann JH, Heeren RM (2012) Emerging technologies in mass spectrometry imaging. J Proteomics. doi:10.1016/j.jprot.2012.03.022

Lemaire R, Menguellet SA, Stauber J, Marchaudon V, Lucot JP, Collinet P, Farine MO, Vinatier D, Day R, Ducoroy P, Salzet M, Fournier I (2007) Specific MALDI imaging and profiling for biomarker hunting and validation: fragment of the 11S proteasome activator complex, Reg alpha fragment, is a new potential ovary cancer biomarker. J Proteome Res 6(11):4127–4134. doi:10.1021/pr0702722

Lusis AJ (2000) Atherosclerosis. Nature 407(6801):233–241. doi:10.1038/35025203

MacAleese L, Stauber J, Heeren RM (2009) Perspectives for imaging mass spectrometry in the proteomics landscape. Proteomics 9(4):819–834. doi:10.1002/pmic.200800363

Maxfield FR, Tabas I (2005) Role of cholesterol and lipid organization in disease. Nature 438(7068):612–621. doi:10.1038/nature04399

Minerva L, Clerens S, Baggerman G, Arckens L (2008) Direct profiling and identification of peptide expression differences in the pancreas of control and ob/ob mice by imaging mass spectrometry. Proteomics 8(18):3763–3774. doi:10.1002/pmic.200800237

Morita Y, Ikegami K, Goto-Inoue N, Hayasaka T, Zaima N, Tanaka H, Uehara T, Setoguchi T, Sakaguchi T, Igarashi H, Sugimura H, Setou M, Konno H (2010) Imaging mass spectrometry of gastric carcinoma in formalin-fixed paraffin-embedded tissue microarray. Cancer Sci 101(1):267–273. doi:10.1111/j.1349-7006.2009.01384.x

Murphy RC, Hankin JA, Barkley RM, Zemski Berry KA (2011) MALDI imaging of lipids after matrix sublimation/deposition. Biochim Biophys Acta 1811(11):970–975. doi:10.1016/j.bbalip.2011.04.012

Norris JL, Caprioli RM (2013) Analysis of tissue specimens by matrix-assisted laser desorption/ionization imaging mass spectrometry in biological and clinical research. Chem Rev 113(4):2309–2342. doi:10.1021/cr3004295

Nygren H, Malmberg P (2004) Silver deposition on freeze-dried cells allows subcellular localization of cholesterol with imaging TOF-SIMS. J Microsc 215(Pt 2):156–161. doi:10.1111/j.0022-2720.2004.01374.x

Patterson NH, Thomas A, Chaurand P (2014) Monitoring time-dependent degradation of phospholipids in sectioned tissues by MALDI imaging mass spectrometry. J Mass Spectrom 49(7):622–627. doi:10.1002/jms.3382

Patti GJ, Shriver LP, Wassif CA, Woo HK, Uritboonthai W, Apon J, Manchester M, Porter FD, Siuzdak G (2010a) Nanostructure-initiator mass spectrometry (NIMS) imaging of brain cholesterol metabolites in Smith-Lemli-Opitz syndrome. Neuroscience 170(3):858–864. doi:10.1016/j.neuroscience.2010.07.038

Patti GJ, Woo H-K, Yanes O, Shriver L, Thomas D, Uritboonthai W, Apon JV, Steenwyk R, Manchester M, Siuzdak G (2010b) Detection of carbohydrates and steroids by cation-enhanced nanostructure-initiator mass spectrometry (NIMS) for biofluid analysis and tissue imaging. Anal Chem 82(1):121–128. doi:10.1021/ac9014353

Puolitaival SM, Burnum KE, Cornett DS, Caprioli RM (2008) Solvent-free matrix dry-coating for MALDI imaging of phospholipids. J Am Soc Mass Spectrom 19(6):882–886. doi:10.1016/j.jasms.2008.02.013

Reyzer ML, Caprioli RM (2007) MALDI-MS-based imaging of small molecules and proteins in tissues. Curr Opin Chem Biol 11(1):29–35. doi:10.1016/j.cbpa.2006.11.035

Reyzer ML, Caldwell RL, Dugger TC, Forbes JT, Ritter CA, Guix M, Arteaga CL, Caprioli RM (2004) Early changes in protein expression detected by mass spectrometry predict tumor response to molecular therapeutics. Cancer Res 64(24):9093–9100. doi:10.1158/0008-5472.CAN-04-2231

Römpp A, Spengler B (2013) Mass spectrometry imaging with high resolution in mass and space. Histochem Cell Biol. doi:10.1007/s00418-013-1097-6

Schriemer DC, Li L (1996) Detection of high molecular weight narrow polydisperse polymers up to 1.5 million Daltons by MALDI mass spectrometry. Anal Chem 68(17):2721–2725. doi:10.1021/ac960442m

Schwamborn K, Caprioli RM (2010) Molecular imaging by mass spectrometry—looking beyond classical histology. Nat Rev Cancer 10(9):639–646. doi:10.1038/nrc2917

Shanta SR, Zhou LH, Park YS, Kim YH, Kim Y, Kim KP (2011) Binary matrix for MALDI imaging mass spectrometry of phospholipids in both ion modes. Anal Chem 83(4):1252–1259. doi:10.1021/ac1029659

Stauber J (2012) Quantitation by MS imaging: needs and challenges in pharmaceuticals. Bioanalysis 4(17):2095–2098. doi:10.4155/bio.12.187

Stoeckli M, Chaurand P, Hallahan DE, Caprioli RM (2001) Imaging mass spectrometry: a new technology for the analysis of protein expression in mammalian tissues. Nat Med 7(4):493–496. doi:10.1038/86573

Stoeckli M, Staab D, Schweitzer A (2007) Compound and metabolite distribution measured by MALDI mass spectrometric imaging in whole-body tissue sections. Int J Mass Spectr 260(2–3):195–202. doi:10.1016/j.ijms.2006.10.007

Sun N, Walch A (2013) Qualitative and quantitative mass spectrometry imaging of drugs and metabolites in tissue at therapeutic levels. Histochem Cell Biol 140(2):93–104. doi:10.1007/s00418-013-1127-4

Thomas A, Chaurand P (2014) Advances in tissue section preparation for MALDI imaging MS. Bioanalysis 6(7):967–982. doi:10.4155/bio.14.63

Thomas A, Charbonneau JL, Fournaise E, Chaurand P (2012) Sublimation of new matrix candidates for high spatial resolution imaging mass spectrometry of lipids: Enhanced information in both positive and negative polarities after 1,5-diaminonapthalene deposition. Anal Chem 84(4):2048–2054. doi:10.1021/ac2033547

Thomas A, Patterson NH, Marcinkiewicz MM, Lazaris A, Metrakos P, Chaurand P (2013) Histology-driven data mining of lipid signatures from multiple imaging mass spectrometry analyses: application to human colorectal cancer liver metastasis biopsies. Anal Chem 85(5):2860–2866. doi:10.1021/ac3034294

Wisztorski M, Croix D, Macagno E, Fournier I, Salzet M (2008) Molecular MALDI imaging: an emerging technology for neuroscience studies. Dev Neurobiol 68(6):845–858. doi:10.1002/dneu.20623

Woods AS, Jackson SN (2006) Brain tissue lipidomics: direct probing using matrix-assisted laser desorption/ionization mass spectrometry. AAPS J 8(2):E391–E395. doi:10.1208/aapsj080244

Microprobe MS Imaging of Live Tissues, Cells, and Bacterial Colonies Using LAESI

Bindesh Shrestha, Callee M. Walsh, Gregory R. Boyce, and Peter Nemes

Abstract Laser ablation electrospray ionization (LAESI) is an ambient ionization technique for mass spectrometry that is capable of performing direct spatial imaging on biological specimens with minimal-to-no sample preparation, ensuring experimental conditions that can maintain viability. Mass spectrometry imaging (MSI) by LAESI has accomplished utility in mapping the spatial distribution of small molecules including metabolites and lipids in a wide variety of biological samples under ambient conditions, ranging from sectioned animal and live plant tissues to living microbial colonies and small cohorts of cells. In this chapter, we provide a brief introduction to LAESI and offer practical guidance on performing MSI using this technique. The focus here is to discuss the main steps of custom-building a LAESI setup, to perform multidimensional imaging of tissues and cells, and to demonstrate the utility of LAESI MSI under native or native-like conditions. As recent commercialization has extended this new analytical resource to a broader user base, we anticipate two- and three-dimensional MSI by LAESI to benefit basic and applied research.

1 Introduction

Laser ablation electrospray ionization (LAESI) is an ambient ionization technique for mass spectrometry (MS) that is well suited for the investigation of biological tissues and cells (Nemes and Vertes 2007). As the underlying principles and operation of the technology have been the topic of various recent reviews (Nemes and Vertes 2012; Wu et al. 2013; Monge et al. 2013), visualized experiments (Nemes and Vertes 2010a; Shrestha and Vertes 2010), and book chapters (Vertes et al. 2013;

B. Shrestha • P. Nemes (✉)
Department of Chemistry, W. M. Keck Institute for Proteomics Technology and Applications, The George Washington University,
800 22nd Street, NW, Suite 4000, Washington, DC, USA
e-mail: bindesh@gwu.edu; petern@gwu.edu

C.M. Walsh • G.R. Boyce
Protea Biosciences, Morgantown, WV, USA
e-mail: callee.walsh@proteabio.com; gregory.boyce@proteabio.com

R. Cramer (ed.), *Advances in MALDI and Laser-Induced Soft Ionization Mass Spectrometry*, DOI 10.1007/978-3-319-04819-2_8

Nemes and Vertes 2010b, 2015; Li et al. 2015b), the discussion here is limited to a brief introduction of the method followed by protocols to allow for performing LAESI MS imaging (MSI). using a custom-built system as well as the LAESI DP-1000, the commercialized platform.

LAESI is designed to utilize the substantial amount of water that is natively present in or exogenously added to the sample of interest such as a biological specimen. Sampling is accomplished by fast mid-infrared (mid-IR) ablation using 2940 nm as the excitation wavelength. At this wavelength, light energy is efficiently coupled into water-rich samples via resonant excitation of the O–H vibrations, predominantly in water, typically the most abundant component of biological samples. Based on fast imaging experiments (Apitz and Vogel 2005), the sudden deposition of energy gives rise to ablation at atmospheric pressure, which proceeds via three overlapping stages. In the first few hundred nanoseconds, fast surface evaporation and phase explosion produce an ablation plume that rapidly expands away from the surface of the sample. As expansion of this plume is slowed by collisions with ambient gas molecules, the plume eventually halts and collapses back onto the sample, causing the buildup of significant pressure locally in the sample surface. The last stage of the process is driven by the relaxation of this pressure through the ejection of neutral material, projecting particulate matter over hundreds of microseconds and several millimeters to centimeters above the sample surface.

Immediately after ablation, these ejected neutral projectiles are captured in charged droplets generated by an electrospray source to convert molecules of the sample to gas-phase ions that can be analyzed by a mass spectrometer. Based on the formation of multiply charged droplets and little-to-no fragmentation, the process by which ions are generated from the charged droplets in LAESI is thought to be similar to those in electrospray ionization (ESI) (Nemes et al. 2012; Nemes and Vertes 2007). These mechanisms lend utility to convert a broad range of biomolecules to intact molecular ions, ranging from small metabolites to large proteins (Shrestha et al. 2013). As LAESI-generated ions have been found to have indistinguishable internal energy to those produced by traditional ESI (Nemes et al. 2012), laser ablation practically extends classical, soft ESI to microprobe operation with in situ and in vivo operation.

Furthermore, sampling by mid-IR laser ablation and ionization by ESI allows for flexibility in meeting select analytical and biological research needs. There is flexibility in the physical dimensions of interrogation, the resolution of a spatial imaging experiment, and the types of biomolecules of interest that can be pursued. By adjusting the laser light fluence and number of laser pulses, the extent of ablation is readily tunable usually with cylindrical voxel dimensions between ~500 μm diameter × ~500 μm depth (multiple laser shots), or ~100 nL, and ~20 μm diameter × ~20 μm depth (single laser pulse), or ~5 pL, using classical light-focusing optical elements ('conventional' LAESI) and sharpened optical fibers (fiber-optic LAESI), respectively. These metrics have fostered applications in the domains of spatial profiling (Nemes and Vertes 2015), lateral imaging in two dimensions (Nemes et al. 2010; Li et al. 2015a; Nemes et al. 2008) and with single-cell resolution (Shrestha et al. 2010b; Shrestha et al. 2011), as well as subcellular interrogations (Stolee and Vertes 2013; Stolee et al. 2012). Furthermore, repeated ablation at a particular location (pixel) on the sample enables depth profiling of the chemical composition in tissues (Nemes et al. 2008). The combination of depth profiling and two-dimensional (2D) MSI laid the foundation for three-dimensional (3D) MSI (Nemes et al. 2009) in situ without the need for tissue sectioning and subsequent 2D imaging of the sections, a feature that is unique to LAESI (Nemes and Vertes 2012).

MSI using mid-IR LAESI offers complementary performance to contemporary MS technologies. The use of water as the natural light-absorbing medium minimizes signals in the low-mass region of the LAESI mass spectrum ($<m/z$ 500), extending classical ultraviolet matrix-assisted laser ablation desorption/ionization (MALDI) to small molecules, particularly primary and secondary metabolites, lipids, and drugs. Besides LAESI, various other ambient ionization sources have demonstrated a continually expending array of success for MSI. Representative examples include but are not limited to desorption electrospray ionization (DESI), electrospray-assisted laser desorption ionization (ELDI), and liquid extraction surface analysis (LESA) mass spectrometry. For these and other ambient ionization MSI techniques, readers are referred to recent reviews in the field (Wu et al. 2013; Nemes and Vertes 2012). In what follows, we discuss the use of LAESI for spatial interrogations of biomolecules in samples with no or limited sample preparation.

2 Applications

The integration of three different technologies—these are laser ablation, electrospray ionization, and mass spectrometry—harbors broad operational space in LAESI, fostering niche applications in MSI. These include analysis with in situ and in vivo operation, microprobe-type chemical analysis for lateral and depth profiling, single-cell measurements, and 3D MSI (see references earlier). Representative applications are summarized in Table 1. Here, we review methods and provide useful comments, or tips, toward a successful LAESI MSI experiment.

2.1 *LAESI MSI Using Classical Optics*

In the conventional configuration, LAESI employs classical light-diffraction elements to steer and focus the mid-IR laser beam and an electrospray source to ionize ablated biomolecules. The schematics of the LAESI MSI platform are shown in Fig. 1. The relative positioning of the ablation plume and electrospray source determines the efficiency of their intersection, which in turn controls the efficiency of ion production. The positioning of this intersection volume relative to the inlet of the mass spectrometer defines the ion collection efficiency for mass analysis, evaluated as the signal-to-noise ratio in the LAESI mass spectrum. Hence, LAESI experiments typically begin with the careful initialization of the configuration, encompassing various interconnected variables for ionization. Critical variables are the angle of laser ablation (θ_{LA}; cf. Fig. 1) and absolute distances between the sample

Table 1 Representative applications for LAESI MSI in two (2D) and three dimensions (3D) and single-cell imaging

LAESI system	Imaging modality	Spatial resolution	Specimen (biological)	Reference
Custom-built[a]	2D	350 μm	Zebra plant (*A. squarrosa*) leaf	Nemes et al. (2008)
			Peace lily (*S. lynise*) leaf	Vertes et al. (2008)
Custom-built[a] (Peltier-cooled)		200 μm	Rat brain (*R. norvegicus*)	Nemes et al. (2010)
DP-1000; Custom-built[a]		250 μm	Mouse brain (*M. musculus*); Geranium (*P. peltatum*) leaf	Li et al. (2014)
DP-1000		300 μm	Mouse lung sections (9 CFW-1)	Kiss et al. (2015a)
Custom-built[a]	3D	300 μm lateral × 30–40 μm depth	Zebra plant (*Aphelandra squarrosa*) and peace lily leaf	Nemes et al. (2009)
DP-1000	2D; 3D	200–400 μm lateral × 40–50 μm depth	Rose leaf, orange slice, lemon slice, ergot body from rye (*S. cereal*), tip cap of a maize kernel	Nielen and van Beek (2015a)
Custom-built: Fiber-optic	Cell-by-cell	30 μm	Onion (*A. cepa*) epidermal cells	Shrestha et al. (2011)
			Easter lily (*L. longiflorum*), garlic (*A. ampeloprasum*), and onion epidermal cells	Li et al. (2015)

For a review on LAESI configurations, please refer to reference (Nemes and Vertes 2015)
[a]Using conventional light-focusing elements in reflection geometry

and electrospray ($d_{S\text{–}ES}$), laser ablation plume and the electrospray plume ($d_{LA\text{–}ES}$), and the laser ablation plume and the MS inlet ($d_{LA\text{–}MS}$). Spray potential, electrospray flow rate, and angle of ion collection settings are secondary variables that are adjusted afterward. To accomplish a required set of performance metrics, iterative refinement is recommended for these parameters.

Molecular imaging by LAESI relies on the coordination of spatially resolved microsampling and mass spectrometric analysis of the generated ions. Typically, the sample is rastered in the focal point of the mid-IR laser beam across a pre-defined interrogation area in the *X*–*Y* directions ($D_{X/Y}$), desired spatial resolution ($R_{X/Y/Z}$), and accumulation times (dwell time, $\tau_{X/Y}$) in the *X* and *Y* lateral directions during 2D imaging. An additional set of variables in the *Z* direction extends imaging to three dimensions. Hence, the total amount of time required for imaging is given by $t_{2D} = (D_X D_Y \tau_X \tau_Y)/(R_X R_Y)$ in two dimensions and by $t_{3D} = (D_X D_Y D_Z \tau_X \tau_Y \tau_Z)/(R_X R_Y R_Z)$ in three dimensions. Synchronization between these variables spatially and temporally has been automated in the LAESI

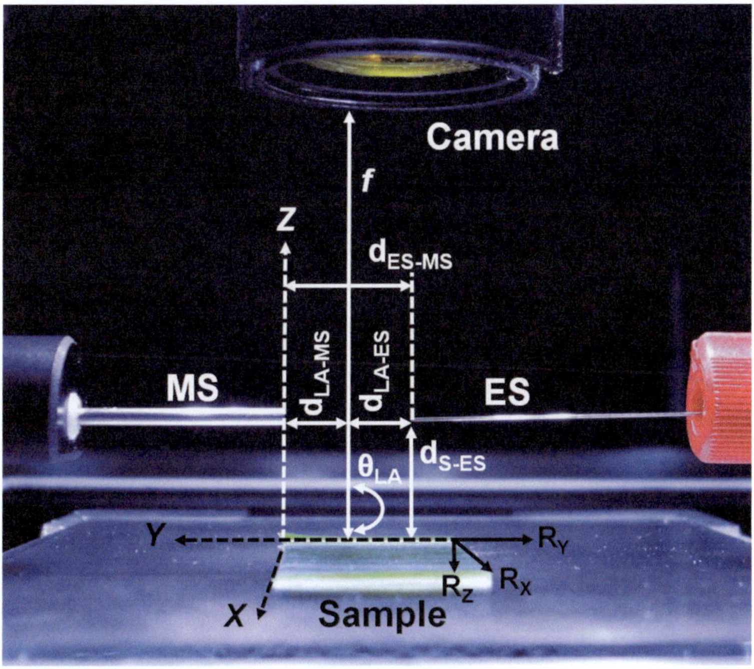

Fig. 1 LAESI ion source for a successful MSI experiment. The schematic highlights the primary experimental variables, overlaid on the picture of the LAESI DP-1000 ion source (Protea Biosciences). *Key*: MS, mass spectrometer inlet (extension tube); d_{LA-MS}, distance between laser ablation plume and mass spectrometer inlet; d_{LA-ES}, distance between laser ablation plume and electrospray emitter tip; d_{ES-MS}, distance between electrospray tip and mass spectrometer inlet; d_{S-ES}, distance between sample and electrospray axis; *X–Y–Z*, independent axes of translation; *f*, working distance of mid-IR light-focusing optical element; θ_{LA}, angle of incidence for mid-IR light from the sample surface; and $R_{X/Y/Z}$, imaging resolution in the *X*/*Y*/*Z* direction

DP-1000® (Protea Biosciences, Morgantown, WV, USA), the commercialized LAESI platform that houses all components necessary for mid-IR laser ablation, electrospray ionization, sample translation, equipment for cooling/heating, and specialized software analysis tools. These instrumental components are fine-controlled by a dedicated software (LAESI Desktop Software; Protea Biosciences) to allow for thorough optimization of the experimental parameter space. LAESI DP-1000 is capable of interfacing to various types of mass spectrometers using an adapter assembly (see MS extension tube in Fig. 1), and the resulting data sets are analyzed using ProteaPlot (Protea Biosciences), a dedicated imaging software.

LAESI MS demonstrates increasing utility in mapping the distribution of metabolites and lipids from different tissues (plant to animal) and bacterial and fungal species (Lin et al. 2014; Walsh et al. 2012; Razunguzwa et al. 2014; Boyce et al. 2014) such as *Bacillus cereus*, *Staphylococcus aureus*, *Pseudomonas aeruginosa*, *Aspergillus fumigatus*, and *Arbuscular mycorrhizal* fungi spores, many of which are pathogenic for humans.

2.2 High-Resolution MSI with Optical Fibers

To accomplish different imaging resolutions for particular applications, as shown in Fig. 2, the size of the ablation crater is tuned depending on the physicochemical properties of the tissue (e.g., water content and tensile strength) and the laser (e.g., fluence and pulse width). Conventional optics for mid-IR light-focusing such as CaF_2 or ZnSe lenses have successfully led to imaging resolutions between ~150–500 μm, capable of deciphering major anatomical regions in tissues. In cases, where finer features such as cell-to-cell differences need to be distinguished, higher resolution is needed; a resolution of ~25–50 μm is required to resolve neighboring cells in many epidermal tissues of plants. Furthermore, because cells are the fundamental units of organization in these samples, it is logical to use cells as the actual pixels for MSI. To realize imaging with 10–50 μm resolution, germanium oxide (GeO_2) optical fibers have been used as the mid-IR light-focusing element in a near-field configuration. The optical fiber was etched down to a few micrometers and positioned proximally to the cell of interest to ablate areas smaller than single cells

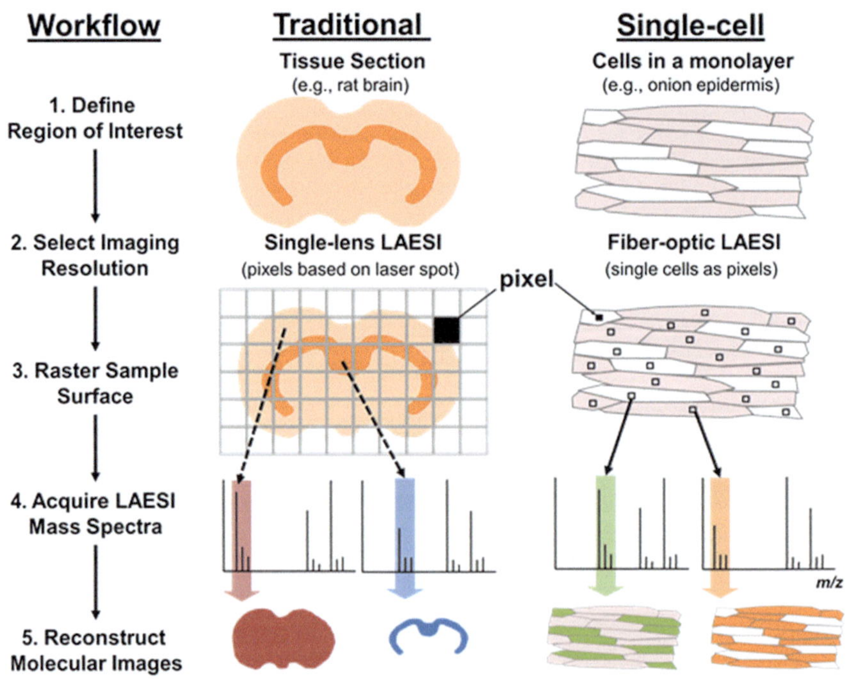

Fig. 2 Experimental workflow for MSI using LAESI (*Left*). Key workflow steps (*Middle*). In the traditional setting, conventional mid-IR light-focusing optical elements are used to obtain resolutions between ~150–500 μm (pixel size). Typical samples in this setting range from plant and animal tissues to bacterial colonies (*Right*). To extend the resolution to single cells, the mid-IR light-focusing optic is replaced with a fiber optic, allowing for local ablation with ~10–50 μm resolution. By utilizing the coordinates of each cell to define pixels of the raster pattern, the entire tissue is imaged cell by cell

(Shrestha and Vertes 2009). Consecutive ablation of single cells in a preselected area using this fiber allowed for cell-by-cell imaging using LAESI (Shrestha et al. 2011; Li et al. 2015). Figure 2 provides the workflow and decision process for two applications demanding different resolutions.

2.3 Ion Mobility Separation Coupled to LAESI MSI

Recent studies demonstrated that the addition of a separation step prior to mass analysis of LAESI-generated ions enhances the signal-to-noise ratio and facilitates molecular identifications (Li et al. 2015a; Shrestha and Vertes 2014). Ion mobility mass spectrometry (IMS) has sufficiently high spectral resolution and speed to be able to analyze the complex mixture of ions that is generated within seconds of a LAESI experiment. The combination of LAESI and IMS extends measurements to a new domain (LAESI IMS MSI), where it becomes possible to distinguish isobaric ions and structural isomers that are not separated by mass-to-charge ratio (*m*/*z*) in the traditional approach. In combination with IMS, LAESI MSI raises an opportunity to identify and map the spatial distribution of isobaric molecules that can have very different biological functions (Nielen and van Beek 2014).

3 Materials and Protocols

In this section, we describe the essential steps to custom-build a LAESI interface and to utilize it for 2D and 3D MSI of tissues of animals and plants. Representative examples are shown for coronal brain sections of the rat (*Rattus norvegicus*) and the zebra plant (*Aphelandra squarrosa*) in Fig. 3. We present progressive advances in refining the resolving power to the level of individual cells. Using the LAESI DP-1000, we demonstrate that MSI is significantly simplified using various automated features and software that are provided by this platform.

3.1 2D and 3D LAESI MSI

3.1.1 Materials

1. Set up an electrospray source as follows. Supply the electrospray solution at 200–400 nL/min through a tapered-tip stainless steel emitter (e.g., MT320-50-5-5; New Objective, Woburn, MA, USA) using a syringe pump (e.g., Physio 22; Harvard Apparatus, Holliston, MA, USA). Connect the emitter to a stable high-voltage power supply capable of providing regulated ~3000 V (e.g., PS350; Stanford Research Systems, Sunnyvale, CA, USA).
2. Position the tip of the electrospray emitter ~10 mm on-axis from the inlet of a mass spectrometer ($d_{\text{MS-ES}}$) using a three-axis translation stage.

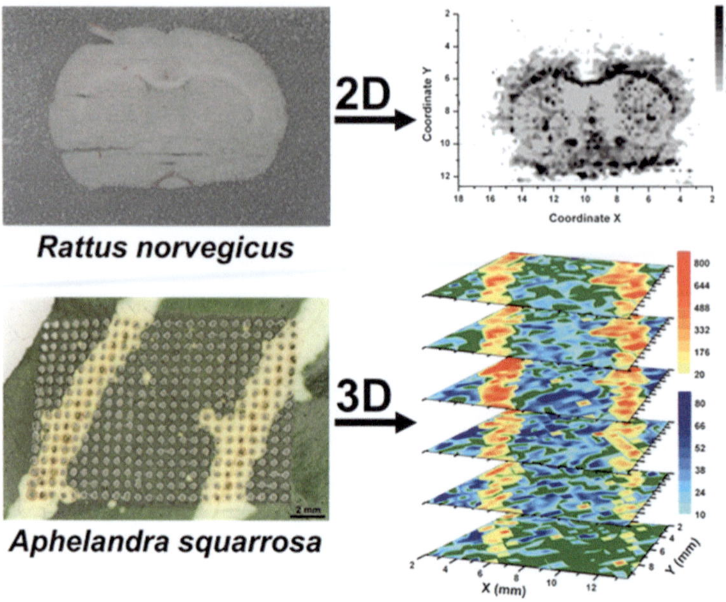

Fig. 3 Examples of MSI in 2D and 3D using a custom-built LAESI system (*Top*). 2D imaging of a coronal rat brain section (*R. norvegicus*) revealed heterogeneous distribution of glycerophosphoethanolamine 36:2 (*m/z* 728.559) across the tissue with accumulation in the *corpus callosum* (*Bottom*). 3D imaging of the zebra plant leaf (*A. squarrosa*) uncovered different molecular distributions laterally and cross-sectionally in the tissue. For example, chlorophyll *a* (*green-blue*, *m/z* 893.546) was more abundant in the mesophyll layer, whereas kaempferol/luteolin (*yellow-red*, *m/z* 287.049) accumulated in the veins of the plant with higher concentration in the second-to-third layer of the tissue (Images were adapted with permission from (Nemes et al. 2009). Copyright 2009 American Chemical Society.)

3. Mount the sample on a motorized three-axis translation stage (e.g., LTA-HS; Newport, Irvine, CA, USA) ~13 mm below the electrospray emitter ($d_{S\text{-}ES}$). Ensure that the sample surface is appreciably uniform in topography (flat), water content, and tissue strength across the surface that is desired to be imaged.

- **TIP**: Note the locations of fibrous veins in plants, bones in animal tissues, and dried areas as these areas will likely generate little-to-no signal. This information can be helpful when interpreting results.

4. Focus a laser beam of 2940 nm wavelength, <100 ns pulse duration, and 2–20 Hz frequency on the sample surface by steering it through a plano-convex lens using appropriate mirrors. In our experiments, we used a Nd:YAG laser with an optical

parametric oscillator to generate the light beam (e.g., Opolette 100; Opotek Inc., Carlsbad, CA, USA) and gold-coated mirrors (Thorlabs, Newton, NJ, USA) to focus the laser light through a CaF_2 lens with $f=50$ mm (Thorlabs) onto the sample.[1]

WARNING: Direct exposure of Class IV laser beams can cause permanent eye damage. Always wear appropriate protective eyewear that block the beam at the operating wavelength and perform these experiments in appropriate and designated laser areas. Consult your local safety officer and laser health and safety rules and regulations for details.

- **TIP**: Set the distance between the lens and sample to the focal length of the lens before focusing the laser beam to establish a good starting point for fine-tuning ablation in later steps.

5. Generate the electrospray by applying ~3000 V to the emitter and supplying 50% methanol (or acetonitrile) with 0.1% acetic/formic acid for the positive ion mode and 50% methanol (or acetonitrile) with 0.1% ammonium acetate (or ammonium hydroxide) for the negative ion mode at the above stated flow rate using a syringe pump.

   WARNING: High voltage poses electric shock hazard. Properly shield all conductive surfaces, and ensure appropriate d_{ES-MS} (at least ~10 mm) to avoid electrical breakdown to the mass spectrometer.

   WARNING: Follow standard safety protocols to work with solvents and reagents. As a general guide, avoid direct contact or inhalation of solvents to prevent irritation to the skin, eyes, and the respiratory tract.

6. Interface the LAESI ion source to a mass spectrometer that has an atmospheric pressure (AP) interface. We have successfully adapted multiple custom-built LAESI systems and different LAESI DP-1000s to various types of mass spectrometers ranging from time-of-flight and ion trap instruments to instruments incorporating ion mobility separation (refer to publications earlier).
7. (Optional) In preparation for MSI of animal tissues, utilize a cryostat microtome (e.g., CM1800; Leica Microsystems Inc., Buffalo Grove, IL, USA) to prepare ~20–150 μm-thick sections of tissue and immediately freeze these section onto glass slides to avoid the loss of native water.

   WARNING: Always handle cold surfaces using a pair of appropriate protective gloves.

[1] Er:YAG lasers are widely used to produce this type of laser beam, e.g., in IR-MALDI MS applications.

3.1.2 Setup Initialization

1. Coarse-adjust the optical beam path to yield ablation marks of desired size on the sample. The dimension of the ablation area will determine the maximal imaging resolution.

- **TIP**: Start with a thermosensitive (burn) paper as the sample (e.g., S-18642; Uline, Pleasant Prairie, WI, USA) to monitor the optical path and establish conditions yielding ablation marks of ~250 μm circular diameter.[1] Inspect the ablation marks for symmetry (circular vs. ellipsoidal burn marks). Afterward, the thermosensitive paper should be replaced with the sample to refine the setup (e.g., for an animal tissue section).

2. Obtain a LAESI mass spectrum for the sample, and select a biomolecular ion of interest which will be used to optimize the setup.

- **TIP**: It is useful to select two different ions (*m/z* peaks), one with high initial abundance to coarse-adjust the setup, and the other with lower abundance for fine tuning. Inspect the ablation marks to avoid obtaining increasing *S/N* at the expense of lower resolution, unless acceptable otherwise.

3. Fine-tune all parameters of the LAESI setup (including the source geometry, voltages, and optics) to maximize ion production for the biomolecular ions of interest. During this step, it is important to utilize a pristine tissue area at each *X–Y–Z* coordinate between consecutive conditions.

- **TIP**: Continuously raster the sample in the *X–Y* dimensions manually or by using the motorized translation stage (line scan function).

4. For 3D MSI, determine the resolution of depth profiling for the tissue by inspecting the depth of the ablation crater under a microscope (upright or stereo). The depth of the ablation crater determines the depth resolution (R_Z) during 3D MSI.

- **TIP**: Fine-adjust the laser fluence while inspecting the ablation crater in the lateral (*X–Y*) and the cross-sectional (*Z*) directions using the microscope.

[1] 1 Today's supermarket receipts are often made of thermosensitive paper and therefore inexpensive alternatives

3.1.3 Perform 2D/3D LAESI MSI

1. Mount a fresh sample on the translation stage.

- **TIP**: For plant tissues, use tape to immobilize the sample on a microscope glass slide, and mount the slide into a slide holder held by the translation stage. For animal tissues, thaw-mount the tissue by allowing it to thaw for a few seconds followed by prompt refreezing on a substrate cooled by a Peltier stage.[2]

2. Determine the area of interest on the sample, and calculate the amount of time required for imaging, t_{2D} for 2D and t_{3D} for 3D imaging. Based on this time, calculate the amount of electrospray solvent that will be necessary to perform the imaging experiment, and also determine the size of MS data that will be generated.
3. To ensure that the experiment will be performed without obstruction, inspect that the translation stage has a free travel range in the *X*, *Y*, and *Z* directions, the syringe supplying the electrospray solvent contains a sufficient volume of the solvent and the corresponding piston has adequate translation range, and the PC controlling the mass spectrometer has sufficient disk space. Ensure that the MS instrument is able to collect data for the entire time.
4. (Important) Depending on the sample, ablated particles may deposit on the electrospray emitter, at which point direct ESI may elevate the background signal for an extended amount of time. This can be readily avoided by refining the laser pulse energy, increasing $d_{S\text{-}ES}$, and/or occasionally cleaning the surface of the electrospray emitter.
5. To perform MSI in two dimensions, execute the following steps in sync: turn on the mid-IR laser, start rastering the sample surface in the *X*–*Y* plane, and simultaneously acquire MS data while delivering multiple laser shots at each location, i.e., *X*–*Y* coordinate (pixel).
6. To perform MSI in three dimensions, utilize single laser pulses and register the corresponding mass spectra individually for each laser pulse.

- **TIP**: Utilize at least 2–3-times faster data acquisition rates than the repetition rate of the mid-IR laser source. In our experiments, we found 0.2 Hz repetition rate and 1–2 Hz mass spectrometric data acquisition rate particularly useful during the development of 3D MSI. Note that the LAESI DP-1000 provides a commercial source for 3D LAESI MSI.

[2]Instructions on building a Peltier stage are available elsewhere (Shrestha et al. 2010a

7. To finish the experiment, turn off the mid-IR laser, the data collection, the electrospray source, and the syringe pump.

- **TIP**: Rinse the electrospray emitter with water–methanol mixture.

8. Visualize the molecular distribution of detected ions by correlating their selected-ion peak areas to the coordinates of analysis, *X–Y* in 2D MSI and *X–Y–Z* in 3D MSI, where *Z* corresponds to individual laser pulses delivered in the *Z* direction.

- **EXAMPLE**: Representative imaging results are provided in Fig. 3 using the coronal section of the rat brain for a 2D and the leaf of the zebra plant (*A. squarrosa*) for a 3D experiment. Comparison of the optical image with the 2D ion image reveals that glycerophosphoethanolamine was accumulated in the *corpus callosum* of the brain. The example for 3D imaging demonstrates that chlorophyll *a* and kaempferol/luteolin have a vastly different lateral and cross-sectional distribution in the leaf.

3.2 *Imaging Bacterial Colonies Using the LAESI DP-1000*

The following protocol outlines molecular imaging of living microbial colonies on agar using the LAESI DP-1000 system.

3.2.1 Materials

1. Install the LAESI DP-1000 system on a compatible mass spectrometer (Thermo Fisher Scientific and Waters mass spectrometers), and activate the data acquisition and analysis software packages.
2. Pour agar close to the brim of the Petri dish. After solidification of the agar, plate the bacteria and incubate them until desired amount of growth has occurred.

3.2.2 Optimization of the Ion Source

1. Place the bacterial culture plate on the sample tray of the DP-1000. Using interactive mode, adjust the stage in the *Z* dimension to $d_{\text{S–ES}} = \sim 15$ mm.
2. Set $d_{\text{MS–ES}}$ to 7 mm. At this distance, the $d_{\text{LA–ES}}$ measurement will be 2 mm.

- **TIP**: Good starting parameters for the electrospray are: 4000 V spray potential; 1 μL/min flow rate; 50% methanol with 0.1% acetic acid in positive ion mode.

3. Use the interactive mode to fine-tune the focusing distance (*f*, Fig. 1).

- **TIP**: It is recommended to empirically determine the focal distance by translating the focusing lens until the sample is in visual focus (in line with the camera) and by monitoring the mass spectrometric signal intensity upon delivering the laser pulses.

4. Select the appropriate laser fluence.

- **TIP**: Recommended starting parameters for the laser include 80% energy (~800 μJ) and 10 pulses at 10 Hz repetition rate for 2D imaging of bacterial colonies.

3.2.3 Data Acquisition

1. To start an imaging experiment, set up a project in the control software (LAESI Desktop Software) of the LAESI DP-1000. Take a photograph of the sample using the onboard wide-angle camera and define a desired imaging area by using a grid.
2. Set the spatial resolution of the grid using the LAESI Desktop Software. A setting of ~200 μm provides a robust initial estimate.
3. Save experimental parameters including analysis locations, electrospray conditions, stage temperature, and laser settings, in the project for archive use. After saving the project, start the analysis. MS data can be acquired manually or automatically with the LAESI DP-1000. When the analysis is complete, the LAESI DP-1000 will automatically turn off the LAESI subsystems by default. The data acquisition can be stopped automatically (e.g., after a pre-defined time period in the software) or terminated manually.

3.2.4 Data Analysis

1. Use ProteaPlot to visualize molecular images for the detected ions of interest.

- **TIP**: Experiment with background subtraction, photographic and molecular image overlay, spectral comparison, and data export capabilities in ProteaPlot to highlight features of interest and enhance data analysis.
- **EXAMPLE**: A representative example is provided for the human pathogen *Serratia marcescens* in Fig. 4. For this experiment, a bacterial colony of *S. marcescens* was grown on LBK agar and imaged in two dimensions using the LAESI DP-1000. The positive ion mode mass spectrum (*m*/*z* 150–2000) shows a complex molecular makeup with hundreds of metabolites. The molecular images were constructed for six selected molecular species in the colony using the ProteaPlot software.

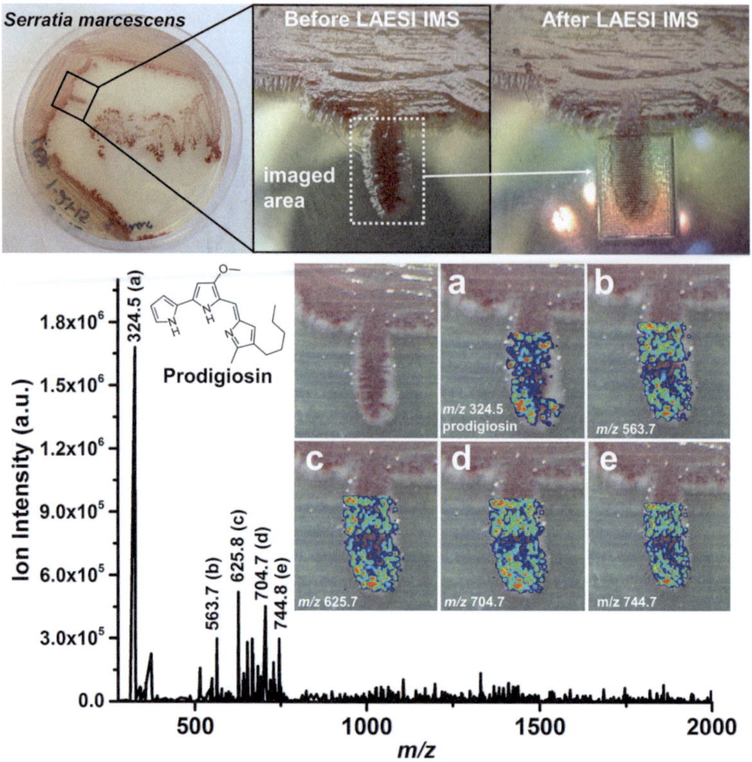

Fig. 4 MSI of bacterial colonies using the LAESI DP-1000 (*Top*). Photographs of *Serratia marcescens* bacterial colony growing on LBK agar area before and after LAESI. Ablation craters are visible after the analysis (*Bottom*). A representative mass spectrum acquired during LAESI MSI of the colony. *Insets* (**a**) to (**e**) correspond to the molecular image of various mass-selected ions that were detected in this spectrum (see *m/z* values in spectrum). Ion intensity increases from *blue* to *red* color. Using accurate mass and tandem MS, the signal labeled (**a**) at *m/z* 324.5 was identified as prodigiosin, a secondary metabolite pigment that provides the characteristic red color of this bacterial species

3.3 Imaging Cell by Cell

Last, this section describes a cell-by-cell imaging protocol for a monolayer of single cells using a custom-built LAESI system that is equipped with an optical fiber (cf. Fig. 2). A representative image of the optical fiber setup is shown in Fig. 5.

3.3.1 Materials

1. Set up an electrospray source, mid-IR laser, three-axis translation stage, and mass spectrometer in function as described in Sect. 3.1.
2. Use a germanium oxide (GeO_2) optical fiber transmitting in the mid-IR spectral range (e.g., 450 μm core diameter HP fiber; Infrared Fiber Systems Inc., Silver

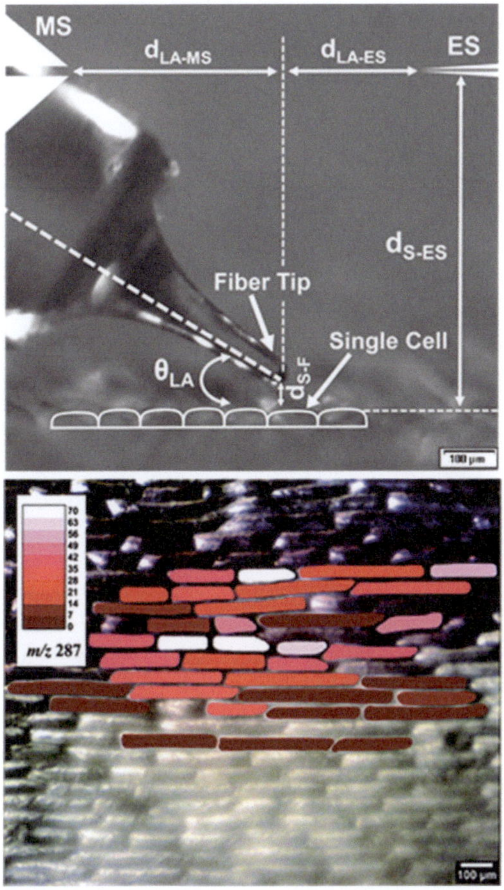

Fig. 5 LAESI MSI with single-cell resolution (*Top*). The tip of an etched optical fiber transmitting mid-IR light is brought proximally to individual cells of interest to perform local ablation. A two-camera system measures distances and angles of incidence (not shown). Critical parameters requiring careful optimization are labeled. Key: d_{S-F}, distance between sample (cell) and fiber tip (*Bottom*). To obtain molecular images on a cohort of cells, the LAESI analysis is performed on every cell of the population, one cell at a time, using the sharpened optical fiber. As an example, the ion image of cyanidin (*m*/*z* 287) is overlaid on the optical image of an onion epidermal tissue, demonstrating that this ion correlates with purple pigmentation (Images were adapted with permission from (Shrestha and Vertes 2009; Shrestha et al. 2011). Copyrights 2009 and 2011 American Chemical Society)

Spring, MD, USA) and sapphire blade (e.g., KITCO Fiber Optics, Virginia Beach, VA, USA) for cleaving this optical fiber.

3. Use a fiber micromanipulator (e.g., MN-151; Narishige, Tokyo, Japan) and other miscellaneous optical components (Thorlabs) as needed.
4. For visualization of the tissue in perpendicular directions, employ a long-working-distance microscope system consisting of two independent cameras (e.g., K1 CentriMax; Edmund Optics Inc., Barrington, NJ, USA).

5. Prepare a mock sample such as the abaxial or adaxial epidermis layers peeled off from a leaf (e.g., Peace lily), mounted on a clean microscope slide, and attached to the three-axis translation stage.

3.3.2 Preparing the Optical Fiber

1. Remove the plastic coat of the optical fiber by dipping it into heated 1-methyl-2-pyrrolidinone (reagent grade) until the coat turns soft, allowing to peel off the coat with a lint-free tissue. Minor debris needs to be washed off by rinsing with methanol. Score the ends of the fiber using a sapphire blade, and gently snap the ends off to attain clean, blunt tip ends.

- **TIP**: For the next step, it is important to ensure that the optical fiber has smooth ends. Repeat this step if needed. Inspect the quality of the fiber ends under a microscope.

2. Chemically sharpen (etch) one end of the fiber using 1–2% nitric acid (reagent grade) as follows: To produce a fiber with ~20 μm tip diameter, immerse the end of the fiber vertically in the nitric acid solution ~0.3–0.5 mm below the meniscus until the fiber tip automatically detaches from the acid surface. Remove the sharpened tip and rinse it with deionized water prior to use. The other end of the fiber should be left unmodified.

3.3.3 Setup for LAESI Single-Cell MSI

1. Mount the fiber on a fiber chuck (e.g., BFT-5; Siskiyou Corporation, Grants Pass, OR, USA) using a bare fiber chuck (e.g., BFC300; Siskiyou). A schematic of the setup is shown in Fig. 5 (top panel).
2. Couple the mid-IR laser beam into the fiber by focusing it onto the blunt end of the fiber using a plano-convex CaF_2 lens.
3. Position the sharpened end of the fiber a few millimeters from the sample surface at an incidence angle θ_{LA} of ~30°.

- **TIP**: Use a dual-camera system to aid positioning of the fiber and to prevent accidental damages to the fiber tip. Should the tip break off, dismount the optical fiber, and re-etch its end to the desired tip dimension.

4. Deliver the mid-IR laser pulses at high repetition rate (e.g., 10 Hz) to individual cells using the sharpened optical fiber to cause localized ablation, usually with dimensions <50 μm in diameter. Gently lower the fiber tip to a distance from the

surface of the tissue (d_{S-F}) of ~20 μm to ablate the cell of interest. It is common that rupture of the cell wall occurs after a few tens of laser pulses.

- **TIP**: Optimize the position of the fiber tip with respect to the mass spectrometer to maximize the signal-to-noise ratio for biomolecules of interest.

5. Maneuver the sharpened end of the fiber tip in front of the mass spectrometer orifice by using a micromanipulator at an angle θ_{LA} of ~45° and adjust the distance/position between the optical fiber tip and the cell surface (d_{S-F}) to obtain the best signal. Be careful not to break the tip.
6. Turn on the electrospray source, the mass spectrometer, and the mid-IR laser to begin data collection.

- **TIP**: Step-by-step instructions on single-cell analysis using LAESI are discussed in greater detail elsewhere (Shrestha et al. 2010a).

3.3.4 Cell-by-Cell LAESI MSI

1. Perform successive ablations on individual selected cells across a pre-defined area of the tissue by positioning the sharpened optical fiber tip proximally to the cells of interest and coupling the mid-IR light into the optical fiber.

- **TIP**: This process can be performed manually. The sample can be readily positioned using a manual translation stage while observing single cells with the long-working-distance microscope. This process can also be automated based on the coordinates of the single cells.

2. Optimize the repetition rate of the laser and its energy output to obtain the highest signal-to-noise (*S*/*N*) ratio from each sample pixel at a desired dwell time (e.g., 4 s).

- **TIP**: (1) For single-cell analysis, make sure to tune the laser parameters to obtain the best mass spectrum with minimal damage to the neighboring cells. (2) For automated cell-by-cell imaging, identify coordinates around the centroids of single cells using an image processing software (e.g., MetaMorph; Olympus Corporation, Center Valley, PA, USA) and supply the coordinate data to a stage-control program to sequentially position the cells under the optical fiber tip for analysis. A protocol for automated cell-by-cell imaging is described in detail elsewhere (Li et al. 2015b).

3. To finish the experiment, turn off all the subsystems and stop the data acquisition.
4. Visualize the molecular distribution of the detected ions across the cell population, cell by cell, by correlating peak intensities/areas to the coordinates of each analyzed cell. Normalizing ion intensities to a common ion or the total ion signal aids the interpretation of the results. Molecular images are constructed by overlaying false colors representing peak intensities on the microscopic image of cells.

- **EXAMPLE**: Figure 5 demonstrates the cell-by-cell imaging of *A. cepa* epidermal cells using a sharpened optical fiber for LAESI. Upon overlaying the distribution of the cyanidin ion (*m/z* 287) with the optical image of the epidermis, a correlation is apparent between this compound and cells that have purple pigmentation.

References

Apitz I, Vogel A (2005) Material ejection in nanosecond Er: YAG laser ablation of water, liver, and skin. Appl Phys A 81(2):329–338

Boyce G, Walsh C, Seeley E, Morton J, Kilby G (2014) Fatty acid and lipid profiling of Arbuscular mycorrhizal fungi with LAESI-MS. In: The annual meeting of the American Society for Mass Spectrometry, Baltimore

Kiss A, Smith DF, Reschke BR, Powell MJ, Heeren RMA (2014) Top-down mass spectrometry imaging of intact proteins by laser ablation ESIFT-ICR MS. Proteomics 14(10):1283–1289. doi:10.1002/pmic.201300306

Li H, Smith BK, Márk L, Nemes P, Nazarian J, Vertes A (2015a) Ambient molecular imaging by laser ablation electrospray ionization mass spectrometry with ion mobility separation. Int J Mass Spectrom 375:681

Li H, Smith BK, Shrestha B, Márk L, Vertes A (2015b) Automated cell-by-cell tissue imaging and single-cell analysis for targeted morphologies by laser ablation electrospray ionization mass spectrometry. In: He L (ed) Methods in molecular biology, Mass spectrometry imaging of small molecules. Humana Press, New York, pp 117–127

Lin D, H G, King J, Cichewicz R (2014) Investigation of an unprecedented natural non-enzymatic reaction with laser ablation electrospray ionization (LAESI) mass spectrometry technology. In: The Annual Meeting of the American Society of Mass Spectrometry, Baltimore

Monge ME, Harris GA, Dwivedi P, Fernandez FM (2013) Mass spectrometry: recent advances in direct open air surface sampling/ionization. Chem Rev 113(4):2269–2308. doi:10.1021/cr300309q

Nemes P, Vertes A (2007) Laser ablation electrospray ionization for atmospheric pressure, in vivo, and imaging mass spectrometry. Anal Chem 79(21):8098–8106. doi:10.1021/ac071181r

Nemes P, Vertes A (2010a) Atmospheric-pressure molecular imaging of biological tissues and biofilms by LAESI mass spectrometry. J Vis Exp 43, e2097

Nemes P, Vertes A (2010b) Laser ablation electrospray ionization for atmospheric pressure molecular imaging mass spectrometry. In: Rubakhin SS, Sweedler JV (eds) Mass spectrometry imaging, Principles and protocols. Humana, New York, pp 159–171. doi:10.1007/978-1-60761-746-4_9

Nemes P, Vertes A (2012) Ambient mass spectrometry for in vivo local analysis and in situ molecular tissue imaging. TrAC Trends Anal Chem 34:22–34

Nemes P, Vertes A (2015) Laser ablation electrospray ionization mass spectrometry: mechanisms, configurations, and imaging applications. In: Domin M, Cody R (eds) Ambient ionization mass spectrometry. Royal Society of Chemistry, Cambridge, p 348

Nemes P, Barton AA, Li Y, Vertes A (2008) Ambient molecular imaging and depth profiling of live tissue by infrared laser ablation electrospray ionization mass spectrometry. Anal Chem 80(12):4575–4582

Nemes P, Barton AA, Vertes A (2009) Three-dimensional imaging of metabolites in tissues under ambient conditions by laser ablation electrospray ionization mass spectrometry. Anal Chem 81(16):6668–6675

Nemes P, Woods AS, Vertes A (2010) Simultaneous imaging of small metabolites and lipids in rat brain tissues at atmospheric pressure by laser ablation electrospray ionization mass spectrometry. Anal Chem 82(3):982–988

Nemes P, Huang H, Vertes A (2012) Internal energy deposition and ion fragmentation in atmospheric-pressure mid-infrared laser ablation electrospray ionization. Phys Chem Chem Phys 14(7):2501–2507

Nielen MWF, van Beek TA (2014) Macroscopic and microscopic spatially-resolved analysis of food contaminants and constituents using laser-ablation electrospray ionization mass spectrometry imaging. Anal Bioanal Chem 406(27):6805–6815. doi:10.1007/s00216-014-7948-8

Razunguzwa TT, Henderson HD, Reschke BR, Walsh CM, Powell MJ (2014) Laser-ablation electrospray ionization mass spectrometry (LAESIs-MS): ambient ionization technology for 2D and 3D molecular imaging. In: Domin M, Cody R (eds) Ambient ionization mass spectrometry. Royal Society of Chemistry, Cambridge, pp 462–481

Shrestha B, Vertes A (2009) In situ metabolic profiling of single cells by laser ablation electrospray ionization mass spectrometry. Anal Chem 81(20):8265–8271. doi:10.1021/ac901525g

Shrestha B, Vertes A (2010) Direct analysis of single cells by mass spectrometry at atmospheric pressure. J Vis Exp 43, e2144. doi:10.3791/2144

Shrestha B, Vertes A (2014) High-throughput cell and tissue analysis with enhanced molecular coverage by laser ablation electrospray ionization mass spectrometry using ion mobility separation. Anal Chem 86(9):4308–4315

Shrestha B, Nemes P, Nazarian J, Hathout Y, Hoffman EP, Vertes A (2010a) Direct analysis of lipids and small metabolites in mouse brain tissue by AP IR-MALDI and reactive LAESI mass spectrometry. Analyst 135(4):751–758. doi:10.1039/b922854c

Shrestha B, Nemes P, Vertes A (2010b) Ablation and analysis of small cell populations and single cells by consecutive laser pulses. Appl Phys A 101(1):121–126

Shrestha B, Patt JM, Vertes A (2011) In situ cell-by-cell imaging and analysis of small cell populations by mass spectrometry. Anal Chem 83(8):2947–2955

Shrestha B, Javonillo R, Burns JR, Pirger Z, Vertes A (2013) Comparative local analysis of metabolites, lipids and proteins in intact fish tissues by LAESI mass spectrometry. Analyst 138(12):3444–3449. doi:10.1039/c3an00631j

Stolee JA, Vertes A (2013) Toward single-cell analysis by plume collimation in laser ablation electrospray ionization mass spectrometry. Anal Chem 85(7):3592–3598. doi:10.1021/ac303347n

Stolee JA, Shrestha B, Mengistu G, Vertes A (2012) Observation of subcellular metabolite gradients in single cells by laser ablation electrospray ionization mass spectrometry. Angew Chem Int Ed 51(41):10386–10389. doi:10.1002/anie.201205436

Vertes A, Nemes P, Shrestha B, Barton AA, Chen Z, Li Y (2008) Molecular imaging by Mid-IR laser ablation mass spectrometry. Appl Phys A 93(4):885–891

Vertes A, Shrestha B, Nemes P (2013) Direct metabolomics from tissues and cells: new approaches for small molecule and lipid characterization. In: Wevers R, Lutz N, Sweedler JV (eds) Methodologies for metabolomics, Experimental strategies and techniques. Cambridge University Press, Cambridge, pp 140–158

Walsh C, Henderson H, Boyce G, Sexstone A, Gibson L, Powell M (2012) Mass spectrometric profiling of bacterial and mammalian cells with LAESI-MS. In: The Annual meeting of the American Society for Mass Spectrometry, Vancouver

Wu CP, Dill AL, Eberlin LS, Cooks RG, Ifa DR (2013) Mass spectrometry imaging under ambient conditions. Mass Spectrom Rev 32(3):218–243. doi:10.1002/mas.21360

MALDESI: Fundamentals, Direct Analysis, and MS Imaging

Milad Nazari and David C. Muddiman

Abstract Mass spectrometry (MS) has become one of the most important tools in analytical and bioanalytical fields. One of the areas of research in MS is the development of ambient ionization methods that allow for the analysis of samples with little to no sample preparation. Matrix-assisted laser desorption electrospray ionization (MALDESI) is a hybrid ionization method combining attributes from matrix-assisted laser desorption/ionization (MALDI) and electrospray ionization (ESI). The vast experimental space governing the MALDESI process has been investigated thoroughly, while its applications in direct analysis of biomolecules were explored.

MALDESI has evolved to perform mass spectrometry imaging (MSI) of biological and forensic samples. Using ice as the energy-absorbing matrix, a mid-IR laser is used to desorb material from the sample. The desorbed material then partitions into the charged electrospray droplets and are ionized in an ESI-like process. Ion images of endogenous and exogenous analytes in biological tissues have been generated using an IR-MALDESI MSI source. Another application of IR-MALDESI MSI is in the field of forensics, where dyes and fibers can be directly analyzed. This chapter discusses the fundamentals, mechanisms, and some applications of IR-MALDESI. Finally, a detailed protocol for performing a quantitative MSI (QMSI) experiment using the IR-MALDESI source is provided.

1 Introduction

Mass spectrometry (MS) has become an irreplaceable analytical tool due to its sensitivity, specificity, and versatility. The advent of 'soft' ionization techniques such as electrospray ionization (ESI) (Fenn et al. 1989) and matrix-assisted laser desorption/ionization (MALDI) (Karas and Hillenkamp 1988; Tanaka et al. 1988) arguably revolutionized the field of MS by allowing the analysis of large biomolecules, and further advancing the utility of MS in analytical and biological applications (Aebersold and Mann 2003). One of the rapidly evolving areas of MS has been mass

M. Nazari • D.C. Muddiman (✉)
Department of Chemistry, W.M. Keck FTMS Laboratory for Human Health Research, North Carolina State University, Raleigh, NC 27695, USA
e-mail: mnazari@ncsu.edu; david_muddiman@ncsu.edu

R. Cramer (ed.), *Advances in MALDI and Laser-Induced Soft Ionization Mass Spectrometry*, DOI 10.1007/978-3-319-04819-2_9

spectrometry imaging (MSI), which involves generating mass spectra from discrete locations in an array over the surface of a sample, such as a biological tissue section. Using the exact position where each mass spectrum was collected and the abundance of any ion in the spectrum, a heat map can be generated that displays the distribution of that ion within the sample.

Even though the concept of MSI was initially conceived by the introduction of secondary ion mass spectrometry (SIMS) (Castaing and Slodzian 1962), pioneering work in this field using MALDI was demonstrated by Caprioli et al. in 1997 (Caprioli et al. 1997). The softness of the ionization process in MALDI and the capability to generate ions from a specific position make MALDI a prime candidate for MSI analysis. Indeed, MALDI is the most common ionization method used for MSI analysis with a variety of applications such as in proteomics (Burnum et al. 2008), metabolomics (Reyzer and Caprioli 2007), lipidomics (Berry et al. 2011), as well as pharmacokinetics/pharmacodynamics (Castellino et al. 2011). Vast improvements have been made to lasers and sample preparation methods utilized, as well as data acquisition methods and data/image processing algorithms since the introduction of MALDI MSI. However, the general demand for vacuum and relatively extensive sample preparation steps place some restrictions on the type of samples that can be analyzed by MALDI MSI (Laiko et al. 2000; Goodwin 2012).

In order to circumvent the limitations mentioned above, development of ambient ionization methods has been the focus of many groups. Atmospheric pressure MALDI (AP-MALDI) overcame the vacuum requirements (Laiko et al. 2000); however, the sample preparation steps in AP-MALDI were similar to those in conventional vacuum MALDI. The introduction of desorption electrospray ionization (DESI) (Takáts et al. 2004) signified the trend toward native sample analysis. DESI allows for direct analysis of samples from a surface with little to no sample preparation, and has been used for MSI of biological tissues (Wiseman et al. 2008; Eberlin et al. 2011) and plants (Esquenazi et al. 2009; Lane et al. 2009), as well as forensic applications (Morelato et al. 2013). Since the introduction of DESI, many new ambient ionization techniques including direct analysis in real time (DART) (Cody et al. 2005), atmospheric pressure solids analysis probe (ASAP) (McEwen et al. 2005), electrospray-assisted laser desorption/ionization (ELDI) (Shiea et al. 2005), and matrix-assisted laser desorption electrospray ionization (MALDESI) (Sampson et al. 2006) have been introduced.

Introduced in 2006, the MALDESI source was the first hybrid ionization source combining resonant laser desorption and electrospray post-ionization. Figure 1 shows the schematic of a typical MALDESI source. In summary, an ultraviolet (UV) or infrared (IR) laser can be utilized to resonantly excite the matrix. The term 'matrix' refers to any molecule, endogenous or exogenous, that is present in large excess and facilitates the desorption of neutral molecules from the sample by absorbing the energy of the laser. The plume of desorbed material partitions into the charged droplets of electrospray, where ions are generated in an ESI-like process and are sampled by the mass spectrometer. Characterization of the fundamentals of the ionization process (Sampson et al. 2006, 2007, 2008a, b, 2009; Dixon and Muddiman 2010) and optimizing the source parameters (Barry and Muddiman 2011), along with appli-

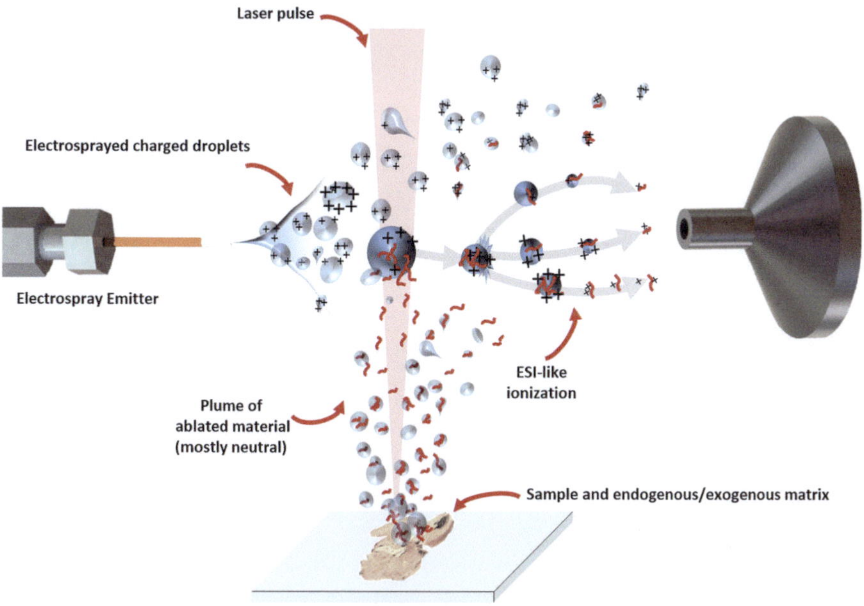

Fig. 1 Schematic of a typical MALDESI source

cations of the source in direct analysis of biological tissues (Robichaud et al. 2013a, 2014; Barry et al. 2014, 2015; Bokhart et al. 2014; Nazari and Muddiman 2014) and forensic analysis (Cochran et al. 2013, 2014) have been the focus of our group. Figure 2 shows an overview of the development and application areas of MALDESI source from investigation of fundamentals to its applications in direct analysis of samples.

2 Applications

2.1 *Intact Characterization and Direct Analysis of Biomolecules*

As the acronym suggests, MALDESI combines attributes from both MALDI and ESI. Naturally, many of the initial studies were focused on investigating and understanding the ionization mechanism. One of the applications of MALDESI is direct analysis of peptides and proteins from sample surfaces. The original MALDESI source utilized a 337-nm pulsed nitrogen laser and UV-absorbing organic matrices to desorb neutral peptides and proteins from the surface, where they subsequently partitioned into the charged electrospray droplets and were ionized. The sample preparation steps for these analyses are similar to those in MALDI. However, the

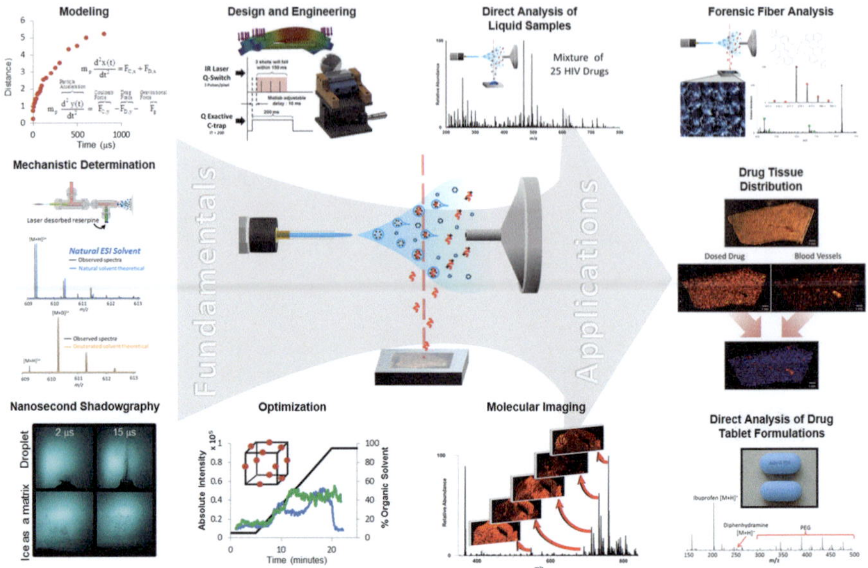

Fig. 2 Development and application areas of MALDESI from its introduction in 2006 up to the present date

multiple charging (lower *m/z*) afforded by ESI provides significant advantages for the analysis of macromolecules using high resolving power the analyzers such as Fourier transform ion cyclotron resonance (FT-ICR) or Orbitrap analyzers since the resolving power in these instruments has an inverse relationship with *m/z*. Figure 3 shows MALDESI-FT-ICR mass spectra obtained for multiply charged peptides and proteins, providing further evidence for the proposed ESI-like charging mechanism in MALDESI (Sampson et al. 2006, 2008b).

One of the major advantages of MALDESI is its inherent versatility. In 2009, a mid-IR laser ($\lambda = 2940$ nm, 10 Hz repetition rate) was implemented in place of the UV laser used in earlier experiments. Using an IR laser allows endogenous or exogenous water, in liquid or solid form, to be used as the energy-absorbing matrix by resonantly exciting the O–H stretching mode of water. Several advantages of using water as a matrix include its ubiquitous presence in nearly all biological samples and the fact that it will not result in matrix-related interferences typically observed in lower *m/z* range. The IR-MALDESI source was used for direct analysis of biomolecules with no sample preparation, using endogenous water as the energy-absorbing matrix (Fig. 4)[1]. The signal abundance was increased by a factor of ~15 when an IR laser was used, which can be attributed to the increased number of neutral molecules desorbed by IR laser ablation (Kampmeier et al. 1997).

Statistical design of experiments (DOE) can be used to explore the vast experimental space, and optimize the source parameters such as stage height, sample-inlet

[1]LAESI (using an IR laser – cf. Chapter 'Microprobe MS Imaging of Live Tissues, Cells, and Bacterial Colonies Using LAESI') and IR-MALDESI using endogenous water as the energy-absorbing matrix are conceptually the same ionization technique.

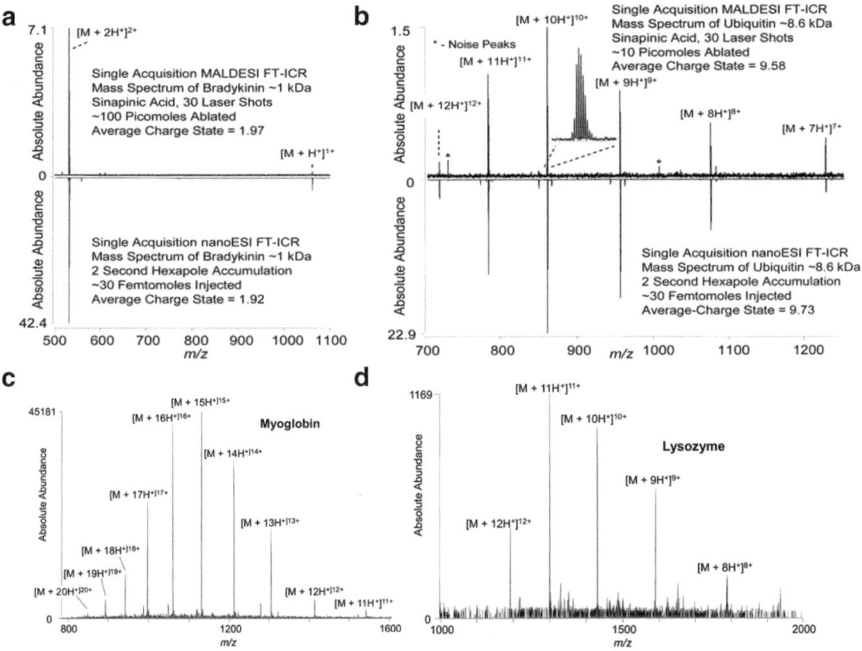

Fig. 3 (**a**, **b**) UV-MALDESI (*top*) and nano-ESI (*bottom*) FT-ICR mass spectra of bradykinin and ubiquitin, respectively. Adapted with permission from (Sampson et al. 2006) © 2006, Springer. (**c, d**) UV-MALDESI-FT-ICR spectra of a dried spot of myoglobin and lysozyme, respectively, mixed 1:1 (vol:vol) with 150 mg/mL 2,5-dihydroxybenzoic acid (DHB). Adapted with permission from (Sampson et al. 2008b) © 2008, Springer

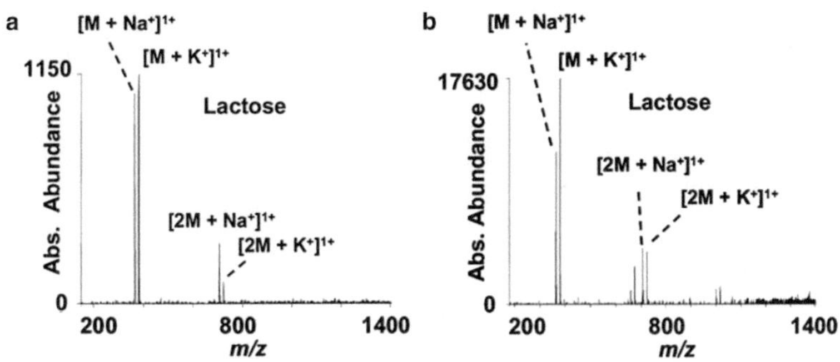

Fig. 4 MALDESI mass spectra of bovine milk with no sample preparation using (**a**) UV laser desorption, (**b**) IR laser desorption. Adapted with permission from (Sampson et al. 2009). © 2009, Springer

distance, and laser fluence in order to significantly improve the sensitivity. DOE inherently recognizes that the optimization of one variable at a time cannot be achieved due to factor–factor interactions. Successful completion of the DOE approach using liquid MALDESI allowed the detection of attomole amounts of intact protein (Barry and Muddiman 2011).

2.2 *Biological and Chemical MSI*

Introduced in 2013, the versatile IR-MALDESI MSI source can be utilized in direct analysis of endogenous and exogenous compounds from biological tissues (Robichaud et al. 2013a, 2014; Barry et al. 2014, 2015; Bokhart et al. 2014; Nazari and Muddiman 2014), as well as analysis of fibers and dyes in forensic applications (Cochran et al. 2013, 2014). Typically, in IR-MALDESI MSI experiments, a thin layer of ice is formed over the top of the sample as the energy-absorbing matrix. A mid-IR laser ($\lambda = 2940$ nm, 20 Hz repetition rate) is then used to facilitate the desorption of neutral material by resonantly exciting the O–H stretching mode of water present within the sample as well as the exogenous ice matrix. The neutral molecules are then post-ionized in an ESI-like process and analyzed by MS (Robichaud et al. 2013a, 2014). Parameters such as solvent composition, ESI laser spot distance, ESI tip diameter, and sample height were optimized in a DOE. Serial mouse brain sections were analyzed (Fig. 5) using the newly optimized parameters,

a

Parameter	Optimized settings for Tissue imaging	Optimized settings for Liquid MALDESI	Optimized settings proposed by others
Spot-Inlet Dist.	3 mm	8 mm	11 mm
ESI tip Diameter	30	30	100
Solvent Comp.	50/50 H_2O/MeOH	50/50 H_2O/ACN	50/50 H_2O/MeOH
Sample Height	5 mm	5 mm	15 mm
Solvent Flow	3 μL/min	0.5 μL/min	0.3 μL/min
ESI-Spot Dist.	1 mm	1 mm	3 mm

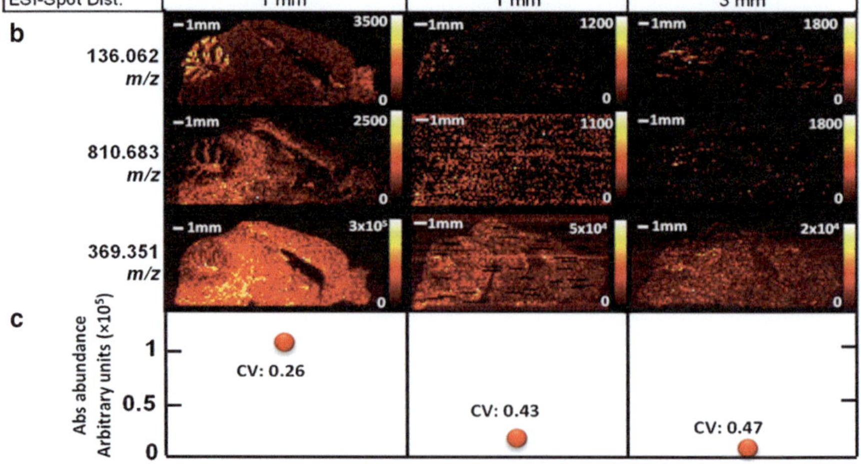

Fig. 5 Comparison of newly optimized settings for tissue imaging, the previously optimized parameters for liquid IR-MALDESI, and those parameters found optimal by others. (**a**) Summary of studied parameters, (**b**) comparison of sets of parameters in imaging experiments performed on serial 50 μm-thick mouse brain sagittal sections, (**c**) average abundance and coefficient of variation (CV) of cholesterol (*m/z* 369.351 ± 5 ppm, $[M\text{-}H_2O+H]^+$) over a region of interest where the molecule was found uniformly. Adapted with permission from (Robichaud et al. 2015). © 2015, Springer

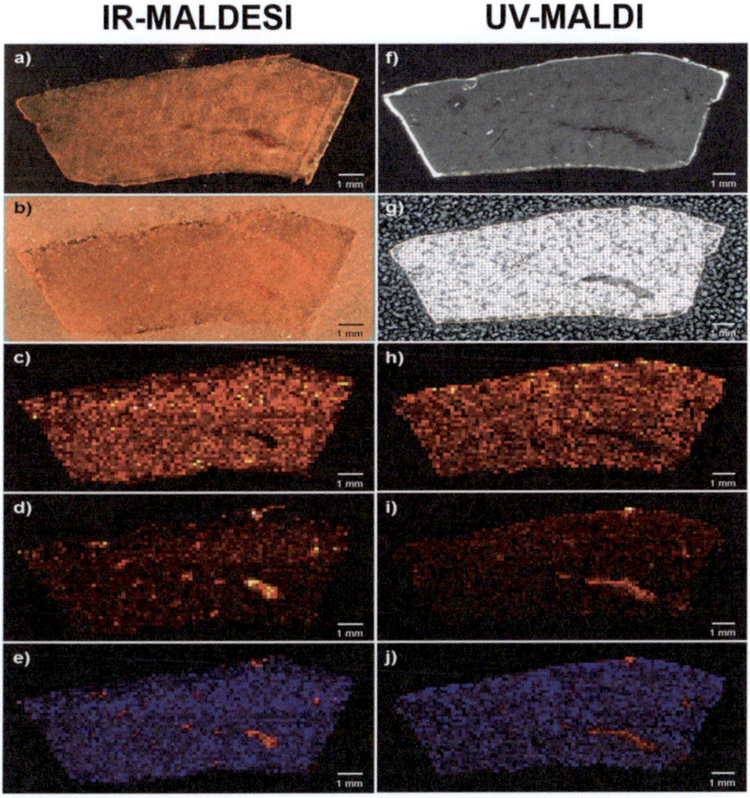

Fig. 6 Optical images of tissue sections analyzed by IR-MALDESI and UV-MALDI before (**a**, **f**) and after (**b**, **g**) matrix deposition, along with the ion images of lapatinib (*m/z* 581.142 ± 5 ppm $[M+H^+]^+$ (**c**, **h**), and SM(34:0) or PE-Cer(37:1) (*m/z* 725.558 ± 5 ppm $[M+H^+]^+$) (**d**, **i**) showing the distribution of the blood vessels. In (**e**, **j**) *blue* corresponds to lapatinib and *red* shows the distribution of blood vessels. Adapted with permission from (Barry et al. 2015). © 2015, Elsevier

the parameters previously found for liquid IR-MALDESI (Barry and Muddiman 2011), and those parameters found optimal by others in similar setups (Nemes et al. 2010). IR-MALDESI was empirically compared with UV-MALDI for imaging of biological tissue sections (Fig. 6), and it was demonstrated that both techniques offer similar information (Barry et al. 2015). However, there are inherent benefits and drawbacks for each method. For instance, IR-MALDESI reduces sample preparation methods involved and circumvents the restrictions imposed by UV-MALDI analysis under vacuum. However, the number of ions that are sampled in the mass spectrometer in IR-MALDESI (and other ambient ionization methods) is often a fraction of the total number of ions generated. Also, since IR-MALDESI is an ambient ionization method, many ambient ions can interfere with the analyte ion signal of interest. These ambient ions, combined with the sheer complexity of biological molecules present in tissue, require coupling of the ionization technique with high mass accuracy and high resolving power instruments in order to generate accurate and detailed images. For this, the IR-MALDESI source can be coupled to a Q Exactive Plus (Thermo Fisher Scientific, Bremen, Germany) mass spectrometer.

One of the areas of active research in the MSI community is obtaining quantitative information from the analytes of interest. A quantitative approach can be an invaluable tool in analyzing tissues since it provides information about the distribution of analyte and potential metabolites, as well as their concentrations in tissue. Although MSI has proven its utility in providing important spatial information of analytes of interest in tissues, obtaining quantitative information from MSI data with greater accuracy has proven to be a difficult task. Many factors such as the need for organic matrices, analysis under high vacuum, and pixel-to-pixel variability play a part in the difficulty of achieving quantitative information from MSI data. A quantitative IR-MALDESI method along with a normalization procedure to reduce voxel-to-voxel variability was recently used in order to quantify the content of an antiretroviral drug in incubated human cervical tissue. A visualization of the protocol used is presented in Fig. 7, while the details of the protocol are discussed in the next section.

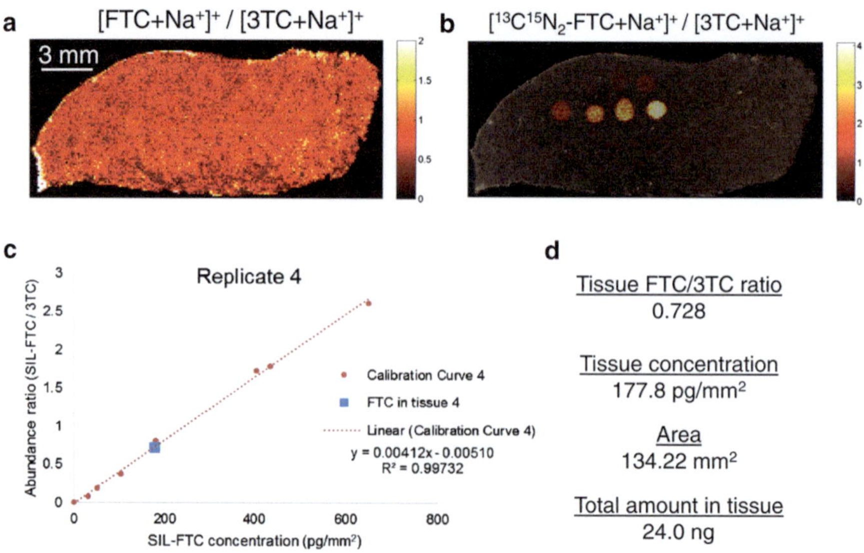

Fig. 7 Summary of the quantitative mass spectrometry imaging (QMSI) protocol for IR-MALDESI imaging. (**a**) The abundance of the analyte is normalized to that of a structural analogue previously applied to the slide in order to reduce variability, (**b**) Series of calibration spots with a stable-isotope labeled (SIL) compound are used to generate a calibration curve, (**c**) Resulting calibration curve generated using SIL molecules, along with the calculated tissue concentration, (**d**) Summary of values used to calculate the total abundance of drug present in tissue. FTC (emtricitabine) and 3TC (lamivudine) are nucleoside reverse transcriptase inhibitors (NRTIs) that are used in treatment of HIV infection. 3TC was used as the structural analogue of FTC in this study. Adapted with permission from (Bokhart et al. 2014). © 2014, Springer

IR-MALDESI has also been utilized in forensic applications in order to directly analyze fibers and dyes from surfaces. Traditional analysis of dyes by MS requires extraction of the dye from the fabric and separation of dye components by chromatography prior to MS analysis. Direct analysis of dyes using IR-MALDESI shortens the sample preparation steps involved in the extraction, thereby significantly reducing the sample analysis time (Cochran et al. 2013). A 'real-world' sample extraction was simulated by using a tape-lift method. The dye and fiber polymers were observed from individual pixels (Fig. 8), indicating that IR-MALDESI coupled to a high-resolution instrument is sensitive enough to detect these molecules from a 10-μm diameter fiber that is as small as 100 μm in length (Cochran et al. 2014).

Current studies are focused on fundamental and new applications of IR-MALDESI. One such application is using a polarity switching method to map out the distribution of lipids in ovary tissues in order to investigate the lipid metabolism and its alterations in ovarian cancer. Fundamentals of the MALDESI source such as re-designing the geometry of the laser path and optics in order to reduce the focus diameter, as well as approaches for reducing the voxel-to-voxel variability in images are currently under investigation.

3 Materials and Protocols

A detailed protocol used in IR-MALDESI quantitative MSI (QMSI) of biological tissues is presented below. The steps denoted with an asterisk (*) are those that only pertain to quantitative MSI. These steps can be skipped when only the spatial distribution of the analyte(s) is of interest.

3.1 Tissue Preparation

1. The tissue is flash-frozen in dry isopentane immediately after animal sacrifice and then stored in a freezer at −80 °C until analysis.
*2. Before analysis, a carefully chosen internal standard (often a structural analogue) is sprayed uniformly on the glass microscope slide using an automated pneumatic sprayer (e.g., TM Sprayer, LEAP Technologies, Carrboro, NC, USA).
3. The frozen tissue is sectioned into the desired thickness at −20 °C using a cryostat (CM1950, Leica, Buffalo Grove, IL, USA). In QMSI experiments, the tissue is thaw-mounted onto the glass slide sprayed with the internal standard. Since the laser in typical IR-MALDESI experiments completely ablates all tissue material from each location, the internal standard placed under the tissue is still effectively sampled.
*4. A dilution series of a stable-isotope labeled (SIL) analyte is prepared. 100 nL of each analytical standard is spotted directly on top of the tissue. Critical attention is paid to ensure that the spots do not overlap.

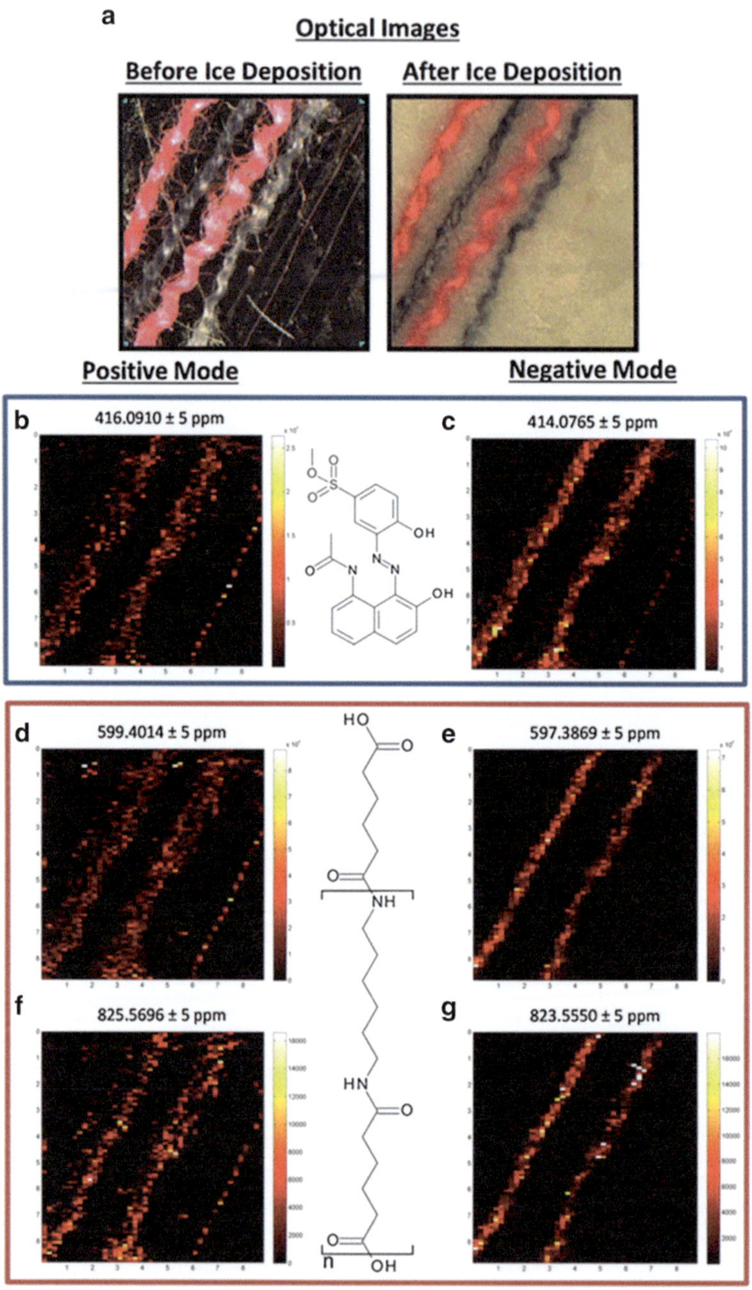

Fig. 8 Results from the polarity switching IR-MALDESI MSI of fibers. Optical images of the thread, fiber bundle, and single fibers for basic violet 16 and acid black 58 in nylon 6,6 are shown before and after deposition of ice matrix (**a**). Ion images of the protonated and deprotonated monomer of acid black 58 dye (**b**, **c**, respectively), along with the protonated (**d**, **f**) and deprotonated (**e**, **g**) form of the nylon 6,6 polymer (**d**, **e** are $n=2$ and **f**, **g** are $n=3$) are shown. Adapted with permission from (Cochran et al. 2014). © 2014, Springer

3.2 MSI Data Acquisition

1. The glass slide is placed on a water-cooled Peltier plate that is within a custom-built enclosure and mounted on an *X–Y* motion-controlled stage. The details of the different components of an IR-MALDESI MS source are discussed elsewhere (Robichaud et al. 2013a).
2. The enclosure is then purged with dry nitrogen in order to reduce the relative humidity inside, which is monitored throughout the experiment using a temperature/relative humidity sensor. The Peltier plate is turned on once the relative humidity is below 1%, in order to prevent condensation of water in the air on the surface of the tissue.
3. The stage is cooled to −9 °C, and the tissue slide is allowed to come to thermal equilibrium (approximately 10 min).
4. The nitrogen flow to the enclosure is turned off, and the tissue is exposed to the ambient environment. A thin layer of ice is deposited on the surface of the tissue due to condensation of water from the air. If the relative humidity is below 10–15%, a beaker of water can be placed inside the enclosure to facilitate the formation of an ice layer.
5. The enclosure is purged with nitrogen again in order to maintain a relative humidity of $10 \pm 2\%$, providing a consistent ice matrix layer throughout the experiment (Robichaud et al. 2014).[1]
6. Next, a region of interest (ROI) is chosen.[2] Other parameters such as the number of laser shots, repetition rate of the laser, as well as the delay between the laser trigger and MS acquisition can be controlled by the operator.
7. The oversampling method (Jurchen et al. 2005) is used in all imaging experiments.[3] A step size of 100 μm is used in most experiments, resulting in complete ablation of the tissue region analyzed. This also ensures that the amount of material desorbed from each voxel stays constant. The content of single cells can be assayed by using a step size of 10 μm (Nazari and Muddiman 2014).
8. Parameters such as solvent flow rate, capillary temperature, and the ionization mode can be chosen according to the application. Positive-ion mode is used in pharmaceutical analyses, whereas a polarity switching method is used in lipidomics and forensic analyses in order to gain a more comprehensive overview of the analytes involved.

---

[1] This sample preparation takes less time than a typical MALDI matrix spraying process, and simplifies the process since matrix solution preparation and application processes such as spraying or sublimation are not needed. More importantly, in contrast with MALDI where the choice of matrix depends on the analyte(s) of interest, this process is universal for all tissues and does not depend on the analyte(s) of interest (i.e., the ice matrix is used for small molecules as well as large biomolecules such as peptides and proteins.)

[2] In our lab, we use an in-house developed graphical user interface built in MATLAB (MathWorks, Natick, MA, USA) and control the *X–Y* stage with this.

[3] In the oversampling method, the sample spot will be fully desorbed before the laser focus is moved to its new desorption position by a distance, which is less than the focal diameter. Thus, a spatial resolution that is below the diameter of the laser focus can be achieved.

3.3 Data Analysis

1. The .raw files generated by the instrument are converted to imaging file formats such as mzXML (Kessner et al. 2008) or imzML (Schramm et al. 2012) using open-source converters. These files are then subsequently loaded into MSiReader (Robichaud et al. 2013b), which is an open-source software developed for analyzing high-resolution MSI data.
2. Ion maps of analytes of interest can be generated using MSiReader. Built-in features such as peak picking, optical image overlay, co-localization of ion maps, and normalization/interpolation can be utilized for further data analysis.

*3. Using the normalization feature in MSiReader, the abundance ratio of analyte to internal standard is calculated in each voxel. The average value of this ratio is then calculated over the entire area of the analyzed tissue region.

*4. An MSI calibration curve is generated by using the signal abundance and concentration per area of the SIL analyte (Fig. 7c).

*5. The average ratio calculated in step *3 is then used along with the calibration curve to calculate the average concentration of analyte per unit area in the tissue. The concentration per area can then be used to obtain the total amount of analyte in the tissue. This quantitative approach has been validated using LC-MS/MS of a serial tissue section (Bokhart et al. 2014).

Acknowledgements The authors would like to thank Dr. Jeremy Barry, Dr. Guillaume Robichaud, Dr. Elias Rosen, Mark Bokhart, and Kenneth Garrard for their contributions in developing the IR-MALDESI source. The authors also gratefully acknowledge the financial support received from the National Institutes of Health (R01GM087964), the W.M. Keck Foundation, and North Carolina State University.

References

Aebersold R, Mann M (2003) Mass spectrometry-based proteomics. Nature 422:198–207. doi:10.1038/nature01511

Barry JA, Muddiman DC (2011) Global optimization of the infrared matrix-assisted laser desorption electrospray ionization (IR MALDESI) source for mass spectrometry using statistical design of experiments. Rapid Commun Mass Spectrom 25:3527–3536. doi:10.1002/rcm.5262

Barry JA, Robichaud G, Bokhart MT et al (2014) Mapping antiretroviral drugs in tissue by IR-MALDESI MSI coupled to the Q Exactive and comparison with LC-MS/MS SRM assay. J Am Soc Mass Spectrom 25:2038–2047. doi:10.1007/s13361-014-0884-1

Barry JA, Groseclose MR, Robichaud G et al (2015) Assessing drug and metabolite detection in liver tissue by UV-MALDI and IR-MALDESI mass spectrometry imaging coupled to FT-ICR MS. Int J Mass Spectrom 377:448–455. doi:10.1016/j.ijms.2014.05.012

Berry KA, Hankin JA, Barkley RM et al (2011) MALDI imaging of lipid biochemistry in tissues by mass spectrometry. Chem Rev 111:6491–6512. doi:10.1021/cr200280p

Bokhart MT, Rosen E, Thompson C et al (2014) Quantitative mass spectrometry imaging of emtricitabine in cervical tissue model using infrared matrix-assisted laser desorption electrospray ionization. Anal Bioanal Chem. doi:10.1007/s00216-014-8220-y

Burnum KE, Frappier SL, Caprioli RM (2008) Matrix-assisted laser desorption/ionization imaging mass spectrometry for the investigation of proteins and peptides. Annu Rev Anal Chem 1:689–705. doi:10.1146/annurev.anchem.1.031207.112841

Caprioli RM, Farmer TB, Gile J (1997) Molecular imaging of biological samples: localization of peptides and proteins using MALDI-TOF MS. Anal Chem 69:4751–4760

Castaing R, Slodzian G (1962) Microanalysis by secondary ionic emission. J Microsc 1:395–410

Castellino S, Groseclose MR, Wagner D (2011) MALDI imaging mass spectrometry: bridging biology and chemistry in drug development. Bioanalysis 3:2427–2441. doi:10.4155/bio.11.232

Cochran KH, Barry JA, Muddiman DC, Hinks D (2013) Direct analysis of textile fabrics and dyes using infrared matrix-assisted laser desorption electrospray ionization mass spectrometry. Anal Chem 85:831–836. doi:10.1021/ac302519n

Cochran KH, Barry JA, Robichaud G, Muddiman DC (2014) Analysis of trace fibers by IR-MALDESI imaging coupled with high resolving power MS. Anal Bioanal Chem. doi:10.1007/s00216-014-8042-y

Cody RB, Laramée JA, Durst HD (2005) Versatile new ion source for the analysis of materials in open air under ambient conditions. Anal Chem 77:2297–2302. doi:10.1021/ac050162j

Dixon RB, Muddiman DC (2010) Study of the ionization mechanism in hybrid laser based desorption techniques. Analyst 135:880–882. doi:10.1039/b926422a

Eberlin LS, Ferreira CR, Dill AL et al (2011) Desorption electrospray ionization mass spectrometry for lipid characterization and biological tissue imaging. Biochim Biophys Acta 1811: 946–960. doi:10.1016/j.bbalip.2011.05.006

Esquenazi E, Dorrestein PC, Gerwick WH (2009) Probing marine natural product defenses with DESI-imaging mass spectrometry. Proc Natl Acad Sci U S A 106:7269–7270. doi:10.1073/pnas.0902840106

Fenn JB, Mann M, Meng CK et al (1989) Electrospray ionization for mass spectrometry of large biomolecules. Science 246:64–71

Goodwin RJA (2012) Sample preparation for mass spectrometry imaging: small mistakes can lead to big consequences. J Proteomics 75:4893–4911. doi:10.1016/j.jprot.2012.04.012

Jurchen JC, Rubakhin SS, Sweedler JV (2005) MALDI-MS imaging of features smaller than the size of the laser beam. J Am Soc Mass Spectrom 16:1654–1659. doi:10.1016/j.jasms.2005.06.006

Kampmeier J, Dreisewerd K, Schurenberg M, Strupat K (1997) Investigations of 2, 5-DHB and succinic acid as matrices for IR and UV MALDI. Part 1: I UV and IR laser ablation in the MALDI process. Int J Mass Spectrom Ion Process 176:31–41

Karas M, Hillenkamp F (1988) Laser desorption ionization of proteins with molecular masses exceeding 10,000 Daltons. Anal Chem 60:2299–2301

Kessner D, Chambers M, Burke R et al (2008) ProteoWizard: open source software for rapid proteomics tools development. Bioinformatics 24:2534–2536. doi:10.1093/bioinformatics/btn323

Laiko VV, Baldwin MA, Burlingame AL (2000) Atmospheric pressure matrix-assisted laser desorption/ionization mass spectrometry. Anal Chem 72:652–657

Lane AL, Nyadong L, Galhena AS et al (2009) Desorption electrospray ionization mass spectrometry reveals surface-mediated antifungal chemical defense of a tropical seaweed. Proc Natl Acad Sci U S A 106:7314–7319. doi:10.1073/pnas.0812020106

McEwen CN, McKay RG, Larsen BS (2005) Analysis of solids, liquids, and biological tissues using solids probe introduction at atmospheric pressure on commercial LC/MS instruments. Anal Chem 77:7826–7831. doi:10.1021/ac051470k

Morelato M, Beavis A, Kirkbride P, Roux C (2013) Forensic applications of desorption electrospray ionisation mass spectrometry (DESI-MS). Forensic Sci Int 226:10–21. doi:10.1016/j.forsciint.2013.01.011

Nazari M, Muddiman DC (2014) Cellular-level mass spectrometry imaging using infrared matrix-assisted laser desorption electrospray ionization (IR-MALDESI) by oversampling. Anal Bioanal Chem. doi:10.1007/s00216-014-8376-5

Nemes P, Woods AS, Vertes A (2010) Simultaneous imaging of small metabolites and lipids in rat brain tissues at atmospheric pressure by laser ablation electrospray ionization mass spectrometry. Anal Chem 82:982–988. doi:10.1021/ac902245p

Reyzer ML, Caprioli RM (2007) MALDI-MS-based imaging of small molecules and proteins in tissues. Curr Opin Chem Biol 11:29–35. doi:10.1016/j.cbpa.2006.11.035

Robichaud G, Barry JA, Garrard KP, Muddiman DC (2013a) Infrared matrix-assisted laser desorption electrospray ionization (IR-MALDESI) imaging source coupled to a FT-ICR mass spectrometer. J Am Soc Mass Spectrom 24:92–100. doi:10.1007/s13361-012-0505-9

Robichaud G, Garrard KP, Barry JA, Muddiman DC (2013b) MSiReader: an open-source interface to view and analyze high resolving power MS imaging files on Matlab platform. J Am Soc Mass Spectrom 24:718–721. doi:10.1007/s13361-013-0607-z

Robichaud G, Barry JA, Muddiman DC (2014) IR-MALDESI mass spectrometry imaging of biological tissue sections using ice as a matrix. J Am Soc Mass Spectrom 25:319–328. doi:10.1007/s13361-013-0787-6

Sampson JS, Hawkridge AM, Muddiman DC (2006) Generation and detection of multiply-charged peptides and proteins by matrix-assisted laser desorption electrospray ionization (MALDESI) Fourier transform ion cyclotron resonance mass spectrometry. J Am Soc Mass Spectrom 17:1712–1716. doi:10.1016/j.jasms.2006.08.003

Sampson JS, Hawkridge AM, Muddiman DC (2007) Direct characterization of intact polypeptides by matrix-assisted laser desorption electrospray ionization quadrupole Fourier transform ion cyclotron resonance mass spectrometry. Rapid Commun Mass Spectrom 21:1150–1154. doi:10.1002/rcm

Sampson JS, Hawkridge AM, Muddiman DC (2008a) Development and characterization of an ionization technique for analysis of biological macromolecules: liquid matrix-assisted laser desorption electrospray ionization. Anal Chem 80:6773–6778. doi:10.1021/ac8001935

Sampson JS, Hawkridge AM, Muddiman DC (2008b) Construction of a versatile high precision ambient ionization source for direct analysis and imaging. J Am Soc Mass Spectrom 19: 1527–1534. doi:10.1016/j.jasms.2008.06.013

Sampson JS, Murray KK, Muddiman DC (2009) Intact and top-down characterization of biomolecules and direct analysis using infrared matrix-assisted laser desorption electrospray ionization coupled to FT-ICR mass spectrometry. J Am Soc Mass Spectrom 20:667–673. doi:10.1016/j.jasms.2008.12.003

Schramm T, Hester A, Klinkert I et al (2012) ImzML - A common data format for the flexible exchange and processing of mass spectrometry imaging data. J Proteomics 75:5106–5110. doi:10.1016/j.jprot.2012.07.026

Shiea J, Huang M-Z, Hsu H-J et al (2005) Electrospray-assisted laser desorption/ionization mass spectrometry for direct ambient analysis of solids. Rapid Commun Mass Spectrom 19: 3701–3704. doi:10.1002/rcm.2243

Takáts Z, Wiseman JM, Gologan B, Cooks RG (2004) Mass spectrometry sampling under ambient conditions with desorption electrospray ionization. Science 306:471–473. doi:10.1126/science.1104404

Tanaka K, Waki H, Ido Y et al (1988) Protein and polymer analyses up to m/z 100 000 by laser ionization time-of-flight mass spectrometry. Rapid Commun Mass Spectrom 2:151–153

Wiseman JM, Ifa DR, Zhu Y et al (2008) Desorption electrospray ionization mass spectrometry: imaging of drugs and metabolites in tissues. Proc Natl Acad Sci U S A 105:18120–18125

Part IV
MS Profiling of Clinical Samples

Disease Profiling by MALDI MS Analysis of Biofluids

Stephane Camuzeaux and John F. Timms

Abstract There is an urgent need for accurate biomarkers of disease. The low-molecular weight proteome of blood serum or other biological fluids may be an ideal source of such biomarkers, although its analysis requires high-throughput strategies to enrich and quantify peptides and small proteins with biomarker potential. Herein, serum samples from cancer cases and controls are compared using a workflow of robotic reversed-phase extraction and clean-up, followed by automated MALDI MS spectral acquisition and analysis of the low-molecular weight peptidome. The aim of the presented methodology is to facilitate the discovery of candidate serum biomarkers of cancer using MALDI MS profiling, although the method is applicable to any comparative proteomic analysis of any biofluid.

1 Introduction

There is an urgent need for accurate disease biomarkers, which may provide novel approaches to diagnosis and screening. The blood serum/plasma proteome may be an ideal source of such biomarkers, although its complexity necessitates novel strategies to enrich and quantify low-abundance protein species with biomarker potential. The low-molecular weight proteome or 'peptidome' of serum/plasma is also a source of potential disease biomarkers (Geho et al. 2006; Hortin 2006; Liotta and Petricoin 2006; Petricoin et al. 2006; Villanueva et al. 2006a). Low-molecular weight proteins and peptides occur endogenously within the bloodstream, whilst protein fragments may be derived in vivo or ex vivo through the action of proteases on both blood-borne and cell/tissue-derived proteins. Both MALDI and its derivative SELDI MS are well suited for profiling the peptidome of biofluids and are particularly suited for high-throughput analyses, i.e. profiling of 100s or 1000s of samples in a single experiment. For comparative analyses in biomarker discovery, peptide peak intensities (peak areas or heights) from spectra acquired from multiple samples are compared between different clinical and control groups using dedicated

S. Camuzeaux • J.F. Timms (✉)
Cancer Proteomics Laboratory, Institute for Women's Health,
University College London, London, UK
e-mail: john.timms@ucl.ac.uk

R. Cramer (ed.), *Advances in MALDI and Laser-Induced Soft Ionization Mass Spectrometry*, DOI 10.1007/978-3-319-04819-2_10

analysis software tools. Differences in peak intensities are typically reported as ratios between the groups with an associated statistic to evaluate the significance of any differences. Since the comparisons are multivariate, correction for multiple testing must be applied. Any peaks of interest must then be identified, usually using orthogonal methods such as tandem MS or antibody-based methods. The onus is on proving that a peak identified by such methods is exactly the same species as the differential peak identified from the (MALDI) MS profiling. This is not always trivial, particularly given the lower mass accuracy of typical MALDI time-of-flight (TOF) instruments. Ideally, tandem MS identification is performed within the same experiment as the profiling.

Much criticism has been levelled at low-molecular weight serum/plasma profiling for biomarker discovery using MALDI and SELDI MS (Diamandis 2004a, 2006; Davis et al. 2010). This largely concerns experimental bias introduced during pre-analytical sample handling. It is without doubt the case that proteolysis has a huge influence on the peptidome patterns of biofluids, particularly serum. Any difference in handling (e.g. clotting time, temperature, storage) between samples may affect proteolysis and has been shown to influence the resulting peptidome patterns (Karsan et al. 2005; Banks et al. 2005; Timms et al. 2007; Baggerly et al. 2004). Essentially, this gives rise to technical variation that can mask true biological variation and increase the false positive rate. Thus, in any profiling study it must be ensured that all samples are collected, handled and stored as identically as possible to avoid such bias. Concerns have also been raised over assay reproducibility and the robustness of class-discriminating algorithms used for MALDI MS profiling biomarker discovery (Diamandis 2004b; Baggerly et al. 2005). Thus, monitoring and reporting of platform reproducibility is obligatory, whilst the robustness of algorithms must be assessed through proper training and test set validation. Finally, it has been argued that these high-throughput methods lack the sensitivity to detect low-abundance species, with coverage limited to abundant small proteins and fragments of coagulation proteins. This appears to be the case as evidenced by identifications assigned to MALDI MS spectral peaks recorded from serum samples, where fibrinogen fragments were the predominant species (Tiss et al. 2010). However, this is not to say that these small proteins, endogenous peptides or proteolytic fragments do not have potential as disease biomarkers as many seemingly well-controlled studies have reported. Indeed, it has been proposed that fragments of abundant coagulation proteins are surrogate peptide markers of cancer, generated *ex vivo* during coagulation through the action of tumour-specific exopeptidases (Villanueva et al. 2006c). This may also hold true in other diseases, where low-abundance, disease-associated proteases may generate specific patterns of protein fragments with diagnostic potential.

2 Applications

Herein, we describe a reproducible, high-throughput, semi-automated, MALDI MS profiling method similar to that used in (Timms et al. 2010) (see Fig. 1). The method describes the profiling of serum from case–control sample sets, although it is equally

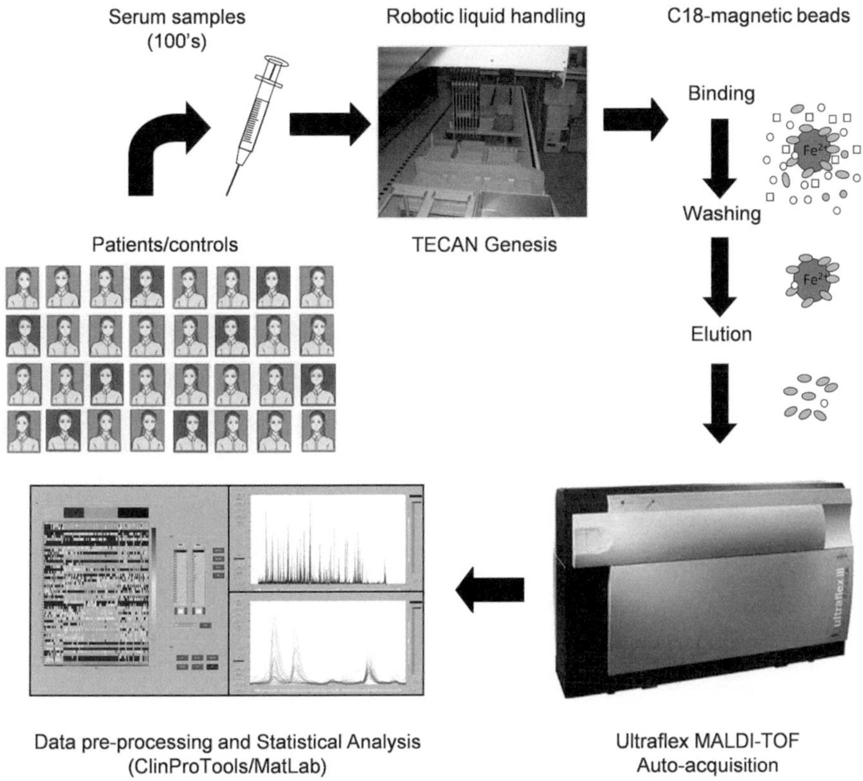

Fig. 1 Schematic workflow for MALDI MS profiling of biofluids

applicable to other biological fluids/clinical specimens such as plasma, cerebrospinal fluid, urine, ascites and saliva. We also describe the data analysis steps and a protocol for identification of discriminatory peaks with biomarker potential. The aim of the presented methodology is to facilitate the discovery of candidate serum biomarkers for the differential diagnosis of ovarian cancer.

This protocol and a slight variation of it with regard to the serum peptide extraction step (Tiss et al. 2007) have been used for biomarker discovery studies for the early detection of ovarian cancer (Timms et al. 2010, 2011) and differential diagnosis of biliary tract cancer (Sandanayake et al. 2014). Similar protocols have been used by other groups for diagnostic biomarker discovery in a variety of cancers (Villanueva et al. 2006b, c; Pietrowska et al. 2009) and other diseases/infections (Conraux et al. 2013; Li et al. 2012; Terracciano et al. 2011; Teunissen et al. 2011; Xiao et al. 2011), for predictive and prognostic biomarker discovery (Vafadar-Isfahani et al. 2010; Taguchi et al. 2007), and to identify blood-borne markers of ageing (Lu et al. 2012).

3 Materials and Protocols

3.1 Materials

3.1.1 Serum Collection and Storage

1. 8.5-mL BD Vacutainer® SST™ Advance Tubes (Becton Dickinson Diagnostics, New Jersey, USA).
2. Bench top centrifuge.
3. Cryovials.
4. −80 °C freezer.
5. Quality control human serum (cat. no.: H4522; Sigma-Aldrich Company Ltd., Gillingham, Dorset, UK).

3.1.2 Serum Polypeptide Extraction and Sample Preparation

1. 96-well Star PCR raised rim skirted plates (Starlab UK Ltd., Milton Keynes, UK).
2. RPC18 Dynabeads (Invitrogen Ltd., Paisley, UK).[1]
3. 96-well magnetic bead separator (Bruker UK, Coventry, UK).
4. Trifluoroacetic acid (TFA; 100% HPLC grade; Rathburn Chemicals Ltd., Walkerburn, Scotland).
5. Acetonitrile (ACN; 100% HPLC grade; Rathburn Chemicals Ltd.).
6. Methanol.
7. Genesis Freedom 200 liquid-handling work station (Tecan UK Ltd., Reading, UK).
8. α-Cyano-4-hydroxycinnamic acid (CHCA) matrix solution (6.2 mg/mL in 36%/56%/8% methanol/ACN/water; Agilent Technologies UK Ltd., Stockport, UK).
9. Ground steel MALDI target plates (Bruker UK).

3.1.3 MALDI-TOF MS Data Acquisition and Analysis[2]

1. Bruker Ultraflex II MALDI-TOF/TOF mass spectrometer (Bruker UK).
2. Flexcontrol v2.0 software (Bruker UK).[3]
3. Peptide calibrant standard II and protein calibrant standard I (Bruker UK).
4. ClinProTools v3.0 software (Bruker UK).

---

[1] RPC18 Dynabeads are paramagnetic, non-porous particles modified with hydrophobic C18 reversed phase chromatographic material. Other manufacturers and chromatographic materials can be used (e.g. C8 reversed phase, weak cation exchange, Cu^{2+}-IMAC), but we have found that the RPC18 Dynabeads gave good yields in terms of the numbers and intensities of the peaks detected.

[2] Other high-performance MALDI-TOF MS systems should be perfectly adequate, though might require additional data analysis software (cf. Sects. 3.2.3 and 3.2.4).

[3] Software versions higher (newer) than stated should also be adequate.

3.1.4 Peptide Identification by LC-MS/MS

1. Ultimate 3000 nano LC system coupled to an LTQ Orbitrap XL mass spectrometer (Thermo Fisher Scientific, Hemel Hempstead, UK) via a PicoView nanospray source (New Objective Inc., Woburn, MA, USA).[4]
2. C18 PepMap guard column (300 μm i.d. × 5 mm, 5 μm bead size, 100 Å pore size; LC Packings, Amsterdam, The Netherlands).
3. C18 PepMap nano LC column (75 μm i.d. × 150 mm, 3 μm bead size, 100 Å pore size; LC Packings).
4. Solvent A: 0.1% formic acid (FA) in HPLC grade water.
5. Solvent B: 0.1% FA, 99.9% HPLC grade ACN (Rathburn Chemicals Ltd.).
6. Xcalibur v2.0 (Thermo Scientific) and Mascot Server v2.4 (Matrix Science Ltd., London, UK) software[3].
7. Novex® Bolt™ Gel Electrophoresis System with pre-cast 10–20% gradient SDS-PAGE mini gels and running buffer (Life Technologies, Paisley, UK).
8. InstantBlue colloidal Coomassie blue protein stain (Expedeon, Cambridge, UK).
9. Siliconised Eppendorf tubes.
10. SpeedVac.

3.2 *Methods*

3.2.1 Serum Collection

1. Collect blood samples from consenting subjects by venepuncture into 8.5-mL BD Vacutainer® SST™ Advance tubes.[5]
2. Gently invert tubes five times and allow the blood to clot at room temperature for 60 min.
3. Centrifuge tubes at 2200 rpm at 4 °C for 10 min and aliquot 200 μL of serum supernatant into clearly labeled cryovials and freeze at −80 °C until further use. Record date of sampling and freezing.

3.2.2 Serum Polypeptide Extraction and Sample Preparation

1. Thaw one 200-μL aliquot of serum per subject/patient and distribute 50 μL into three replicate 96-well Star PCR raised rim skirted plates. Change the location of samples across each replica plate. Also add 50 μL of quality control (QC)

---

[4] Other instrumentation and modes of operation can be used for peptide identification.

[5] Blood collection requires informed consent from donors and studies require ethical approval from the relevant Committees on the Ethics of Human Research. Blood should be taken by a trained phlebotomist. Time of venepuncture should be recorded along with other subject/patient information relevant to the study, such as clinico-pathological features, treatment details, demographic data and any routine clinical assay results. Ideally, controls should be matched as closely as possible to cases. Ensure all samples are handled and processed identically (see Timms et al. 2010 for further information).

serum at 6–12 random positions on each plate which are used to monitor assay reproducibility. Record all sample positions. Re-store plates at −80 °C and run each plate on three separate days.
2. Wash magnetic beads in 0.1% TFA solution and resuspend to a concentration of 50 mg/mL in an Eppendorf tube.
3. Carry out the following steps on a Genesis Freedom 200 liquid-handling work station (or similar)[6] except where indicated.
4. Resuspend magnetic beads by pipetting up and down ten times and transfer 5 μL to the wells of a 96-well plate containing the 50 μL aliquots of serum.
5. Mix by pipetting up and down ten times and allow to stand for 1 min.
6. Pull the beads to one side of the wells using the magnetic bead separator and allow beads to settle on the side for 30 s.
7. Remove the supernatant and discard.
8. Add 200 μL of wash solution (0.1% TFA) and pull the beads left to right ten times using the magnetic bead separator and then allow to settle on one side for 30 s.
9. Remove the wash solution and repeat the wash step.
10. Spin the beads to the bottom of the wells by centrifugation at 2000 g for 2 min and remove remaining wash buffer.
11. Add 7 μL of elution solvent consisting of 50% ACN in 0.1% TFA and mix by pipetting up and down ten times. Leave the mixture to stand for 30 s.
12. Transfer the 96-well plate to the magnetic bead separator and pull the beads to one side for 30 s.
13. Add 35 μL of pre-prepared CHCA matrix solution to each well and mix by pipetting up and down five times.
14. Spot volumes of 1 μL of the eluate/matrix mix in quadruplicate onto a ground steel MALDI target plate and allow samples to dry at room temperature.[7] This generates 12 spotting replicates per sample.

3.2.3 MALDI-TOF MS Data Acquisition

1. Externally calibrate the MALDI-TOF mass spectrometer[8] in the linear positive ion mode using commercial peptide and protein calibration standards (in CHCA matrix) spotted onto the same target plate as the samples of interest. Use approximately 30 fmol of peptide and 500 fmol of protein standards per spot and use average masses for calibration.

---

[6] Other robotic liquid-handling platforms can be employed, although the protocol would need to be adjusted for other platforms. For example, a CyBi™-Disk liquid handling robot (CyBio AG, Jena, Germany) with pre-packed C4 and C18 ZipTips (Millipore, Watford, UK) was used successfully in previous work (Tiss et al. 2007).

[7] Note that drying at relative humidity below ~30–40% has been reported to be detrimental for MALDI MS (Tiss et al. 2007). Thus, ensure that the relative humidity is ≥35–40%.

[8] The Ultraflex II MALDI-TOF/TOF mass spectrometer is equipped with a 337 nm nitrogen laser, a gridless ion source, delayed-extraction (DE), a high-resolution timed ion selector and a 2 GHz digitizer. Other MALDI-TOF platforms can be used (see Footnote 2).

2. Set the following automated irradiation programme and data quality filtering using FlexControl's 'AutoXecute' function: each spectrum is the sum of 1000 laser shots per spotted sample delivered to ten different locations on the spot in ten sets of 100 shots (at a laser frequency of 10 Hz), after pre-irradiation with ten shots at 5% higher laser power to improve spectral quality; set evaluation parameters so that only spectra (of 100 shots) containing at least one peak with a resolving power of greater than 300 and a signal-to-noise ratio (S/N) > 10 in the m/z range of 700–4000 are accumulated.
3. Automatically acquire spectral profiles of the samples over a mass-to-charge (m/z) range of approximately 700–10,000 in the linear positive ion mode under 25 kV of ion acceleration, a delayed extraction (DE) potential difference of 1.4 kV, a lens potential of 5.9 kV and high gating strength to deflect ions below m/z 400. Apply DE (in our case at 80 ns) to give appropriate time-lag focusing after each laser shot. For signal detection, the detector gain is typically set to 7.5 and the sample rate to 2 GS/s.
4. For further analysis, only include data for an individual serum sample when at least three of the four spotting replicates had 1000 summed shots in at least two of the three run replicates. Typical averaged spectra for case control serum samples are shown in Fig. 2 with ~300 aligned peaks detected across the dataset (S/N>3.0).

3.2.4 Spectral Data Analysis[9]

1. Use ClinProTools v2.2 software for processing spectral data as follows: apply smoothing by averaging the intensities within a 5-point width moving window; apply baseline subtraction using an algorithm based on finding the lowest points between dominant local intensity maxima within a particular mass window; apply normalisation by dividing the intensity of each data point in a spectrum by the sum of all intensities in that spectrum; multiply intensities by a constant (e.g. 2×10^7) to give manageable output intensities; define peaks by identifying all local maxima in the spectra above a normalized intensity threshold of 0.2 and signal-to-noise ratio of 3; perform peak alignment to define common peaks using a mass window of 1500 ppm.
2. Determine average peak areas and standard deviations for QC serum samples and for each sample group (case, control, etc.). All aligned peaks from the QC samples can be used to determine intra-plate and inter-plate assay reproducibility which should be within 10% and 20%, respectively (see Fig. 3).
3. Determine distribution of peak areas and apply appropriate test of significance to define peaks, which discriminate between sample groups. Apply correction for multiple testing.
4. Split full dataset into training and test datasets. Construct multi-marker models (e.g. neural networks, support vector machines, genetic algorithms) from training data and determine classification performance on the test set, ideally in a blinded fashion.[10]

---

[9] See Footnote 2 in Sect. 3.1.3.

[10] Multi-marker model building and testing can be performed with the ClinProTools software. As before, other (classification) software can be used.

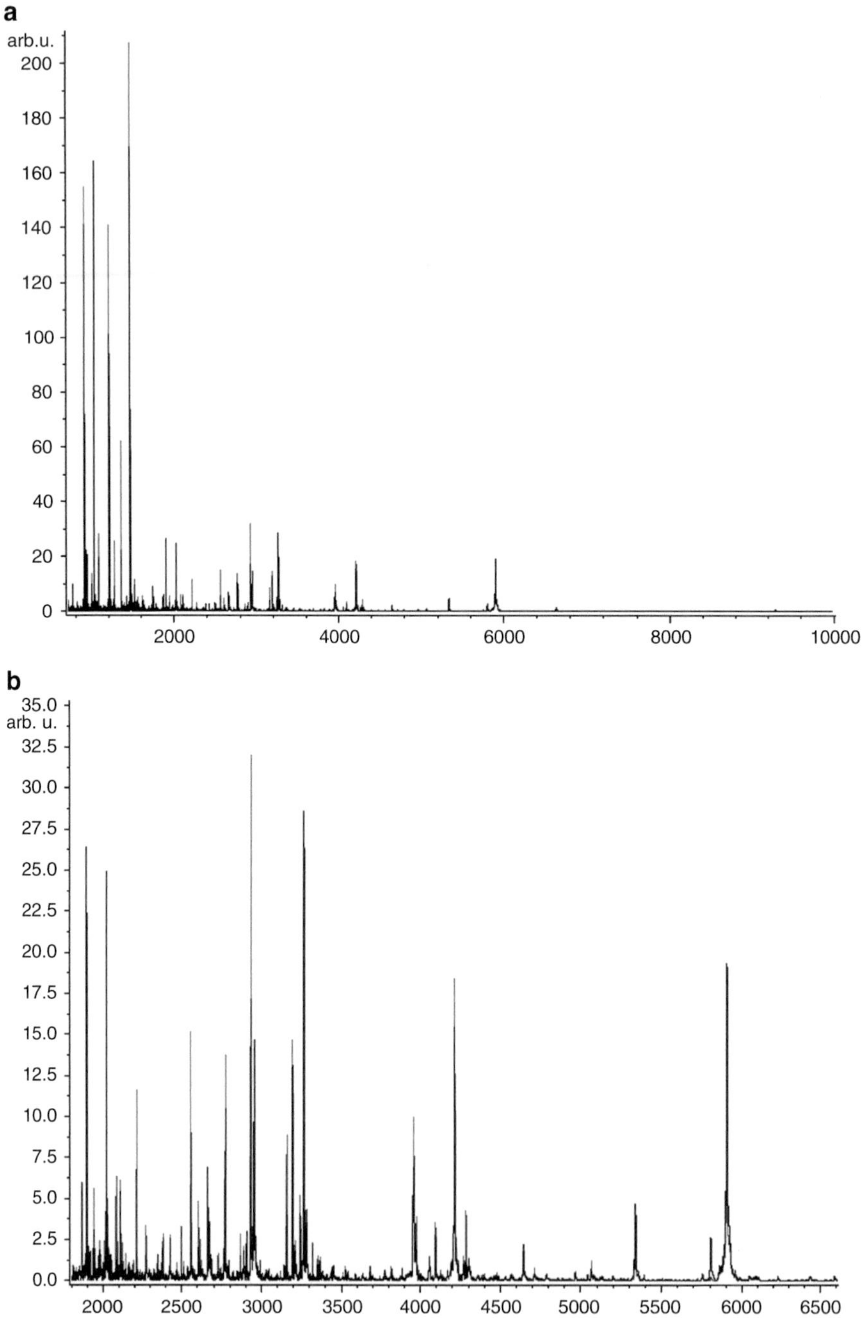

Fig. 2 (**a**) Average MALDI-TOF MS spectra for case (*dark grey*) and control (*light grey*) serum samples over the full scan range of *m/z* 700–10,000. (**b**) Zoomed MS spectra over the *m/z* range of 1700–6900

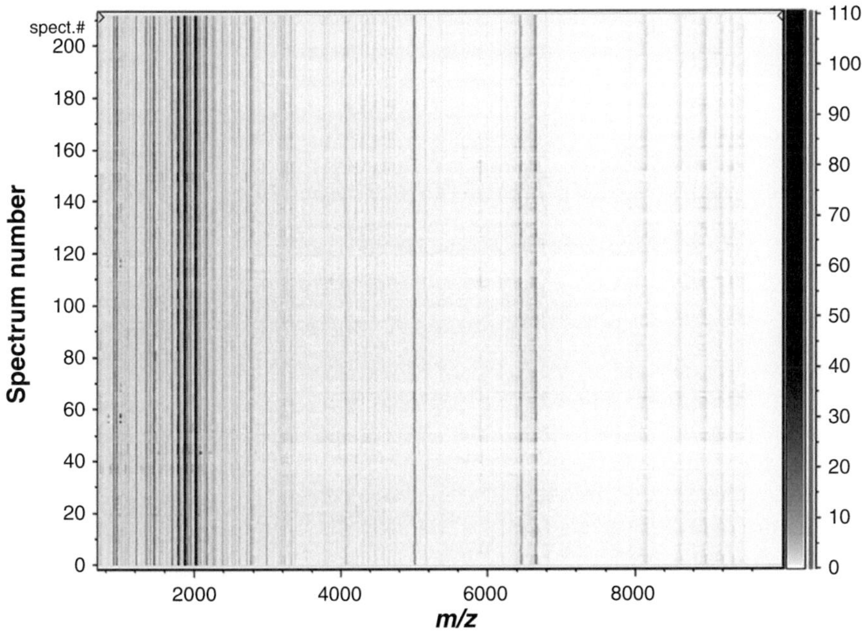

Fig. 3 Aligned MS spectra of QC serum replicates in pseudo-gel view (three replicate runs performed on different days with 12 samples per run each with six spotting replicates, i.e. total of 216 spectra). Using all peaks, intra-assay variation was 6.2% ± 4.8 (SD%) and inter-assay variation was 13.9% ± 7.6 (SD%)

3.2.5 Identification of MALDI-TOF Peaks by LC-MS/MS[11]

1. Prepare a fresh C18-extracted sample from 50 μL of serum without adding matrix solution (see Sect. 3.2.2; up to step 12). Using an Ultimate 3000 LC system, inject 5 μL of the extract from the autosampler onto a C18 PepMap guard column and wash for 3 min with 100% solvent A at a flow rate of 25 μL/min.
2. Switch to an analytical C18 PepMap nano LC column with 10% solvent B and apply a linear gradient of 10–50% B over 90 min, then to 100% B over 3 min. Continue with 100% B for 20 min and then reduce to 10% B over 0.5 min and continue for a further 20 min to re-equilibrate the column for the next injection.
3. Operate the mass spectrometer (LTQ Orbitrap XL) in the data-dependent mode for automated switching between MS and MS/MS acquisition. Acquire survey full scan MS spectra (from *m/z* 400–2000) in the orbitrap with a resolution of 60,000 at *m/z* 400. Select the 'top 6' most intense ions for CID. Select a target ion value of 1×10^6 and maximum scan time of 500 ms for the survey full scan in the orbitrap. Select target ion values of 1×10^4 and a scan time setting of

[11] See Footnote 4 in Sect. 3.1.4.

150 ms for CID. Dynamically exclude ions selected for MS/MS for 60 s. Enable the lock mass option for accurate mass measurement, using the polydimethylcyclosiloxane ion (*m/z* 455.1200) as an internal calibrant.

4. Use initial precursor ion monitoring to identify masses matching the peaks of interest and then target these by mass and retention time in subsequent runs using high-resolution MS/MS in the orbitrap.
5. Process raw spectra using Mascot Distiller and search against the human UniProtKB database. For searching, select 'no enzyme', set MS tolerance to ±10 ppm and MS/MS tolerance to 0.1 Da. Set oxidation (M), dehydration (N-term C) and deamidation (NQ) as variable modifications. Also search data in the same way using the 'de novo sequencing' function of Mascot Distiller.[12]
6. For larger peaks (*m/z*>5000), dry down the extracted sample in a SpeedVac, resuspend in sample buffer and resolve on a 10–20% gradient mini gel following the manufacturer's instructions. Stain the gel with InstantBlue and excise a gel piece in the molecular weight region of interest.
7. Destain the gel piece by shaking at room temperature for 30 min in 200 μL of 50% methanol/10% acetic acid and then wash the gel piece in 100 μL of 100% ACN with shaking for 15 min. Extract polypeptides in 200 μL of 50% formic acid, 25% ACN, 15% isopropanol by vigorous shaking for 2 h at room temperature. Centrifuge and recover the extract, dry down and resuspend in 0.1% formic acid. Analyse one fifth of this sample by MALDI-TOF MS (see Sect. 3.2.3) to verify the presence of peaks of interest. ZipTip the remaining sample and analyse by LC-MS/MS as described above in this section, starting with the first step.

Acknowledgements This work was funded by MRC grant G0301107 and was supported by the National Institute for Health Research University College London Hospitals Biomedical Research Centre.

References

Baggerly KA, Morris JS, Coombes KR (2004) Reproducibility of SELDI-TOF protein patterns in serum: comparing datasets from different experiments. Bioinformatics 20(5):777–785

Baggerly KA, Morris JS, Edmonson SR, Coombes KR (2005) Signal in noise: evaluating reported reproducibility of serum proteomic tests for ovarian cancer. J Natl Cancer Inst 97(4):307–309

Banks RE, Stanley AJ, Cairns DA, Barrett JH, Clarke P, Thompson D, Selby PJ (2005) Influences of blood sample processing on low-molecular-weight proteome identified by surface-enhanced laser desorption/ionization mass spectrometry. Clin Chem 51(9):1637–1649

Conraux L, Pech C, Guerraoui H, Loyaux D, Ferrara P, Guillemot JC, Meininger V, Pradat PF, Salachas F, Bruneteau G, Le Forestier N, Lacomblez L (2013) Plasma peptide biomarker discovery for amyotrophic lateral sclerosis by MALDI-TOF mass spectrometry profiling. PLoS One 8(11), e79733. doi:10.1371/journal.pone.0079733

[12] If the identity of the peak of interest is still ambiguous, it may be necessary to carry out fractionation of the scaled-up extract. For a detailed protocol, refer to (Tiss et al. 2010).

Davis MT, Auger PL, Patterson SD (2010) Cancer biomarker discovery via low molecular weight serum profiling–are we following circular paths? Clin Chem 56(2):244–247. doi:10.1373/clinchem.2009.127951
Diamandis EP (2004a) Analysis of serum proteomic patterns for early cancer diagnosis: drawing attention to potential problems. J Natl Cancer Inst 96(5):353–356
Diamandis EP (2004b) Mass spectrometry as a diagnostic and a cancer biomarker discovery tool: opportunities and potential limitations. Mol Cell Proteomics 3(4):367–378
Diamandis EP (2006) Serum proteomic profiling by matrix-assisted laser desorption-ionization time-of-flight mass spectrometry for cancer diagnosis: next steps. Cancer Res 66(11):5540–5541
Geho DH, Liotta LA, Petricoin EF, Zhao W, Araujo RP (2006) The amplified peptidome: the new treasure chest of candidate biomarkers. Curr Opin Chem Biol 10(1):50–55. doi:10.1016/j.cbpa.2006.01.008
Hortin GL (2006) The MALDI-TOF mass spectrometric view of the plasma proteome and peptidome. Clin Chem 52(7):1223–1237
Karsan A, Eigl BJ, Flibotte S, Gelmon K, Switzer P, Hassell P, Harrison D, Law J, Hayes M, Stillwell M, Xiao Z, Conrads TP, Veenstra T (2005) Analytical and preanalytical biases in serum proteomic pattern analysis for breast cancer diagnosis. Clin Chem 51(8):1525–1528
Li L, Li J, Jin H, Shang L, Li B, Wei F, Liu Q (2012) Detection of Leishmania donovani infection using magnetic beads-based serum peptide profiling by MALDI-TOF MS in mice model. Parasitol Res 110(3):1287–1290. doi:10.1007/s00436-011-2604-0
Liotta LA, Petricoin EF (2006) Serum peptidome for cancer detection: spinning biologic trash into diagnostic gold. J Clin Invest 116(1):26–30. doi:10.1172/JCI27467
Lu J, Huang Y, Wang Y, Li Y, Zhang Y, Wu J, Zhao F, Meng S, Yu X, Ma Q, Song M, Chang N, Bittles AH, Wang W (2012) Profiling plasma peptides for the identification of potential ageing biomarkers in Chinese Han adults. PLoS One 7(7), e39726. doi:10.1371/journal.pone.0039726
Petricoin EF, Belluco C, Araujo RP, Liotta LA (2006) The blood peptidome: a higher dimension of information content for cancer biomarker discovery. Nat Rev Cancer 6(12):961–967. nrc2011[pii]. doi:10.1038/nrc2011
Pietrowska M, Marczak L, Polanska J, Behrendt K, Nowicka E, Walaszczyk A, Chmura A, Deja R, Stobiecki M, Polanski A, Tarnawski R, Widlak P (2009) Mass spectrometry-based serum proteome pattern analysis in molecular diagnostics of early stage breast cancer. J Transl Med 7:60. doi:10.1186/1479-5876-7-60
Sandanayake NS, Camuzeaux S, Sinclair J, Blyuss O, Andreola F, Chapman MH, Webster GJ, Smith RC, Timms JF, Pereira SP (2014) Identification of potential serum peptide biomarkers of biliary tract cancer using MALDI MS profiling. BMC Clin Pathol 14(1):7. doi:10.1186/1472-6890-14-7
Taguchi F, Solomon B, Gregorc V, Roder H, Gray R, Kasahara K, Nishio M, Brahmer J, Spreafico A, Ludovini V, Massion PP, Dziadziuszko R, Schiller J, Grigorieva J, Tsypin M, Hunsucker SW, Caprioli R, Duncan MW, Hirsch FR, Bunn PA Jr, Carbone DP (2007) Mass spectrometry to classify non-small-cell lung cancer patients for clinical outcome after treatment with epidermal growth factor receptor tyrosine kinase inhibitors: a multicohort cross-institutional study. J Natl Cancer Inst 99(11):838–846. doi:10.1093/jnci/djk195
Terracciano R, Preiano M, Palladino GP, Carpagnano GE, Barbaro MP, Pelaia G, Savino R, Maselli R (2011) Peptidome profiling of induced sputum by mesoporous silica beads and MALDI-TOF MS for non-invasive biomarker discovery of chronic inflammatory lung diseases. Proteomics 11(16):3402–3414. doi:10.1002/pmic.201000828
Teunissen CE, Koel-Simmelink MJ, Pham TV, Knol JC, Khalil M, Trentini A, Killestein J, Nielsen J, Vrenken H, Popescu V, Dijkstra CD, Jimenez CR (2011) Identification of biomarkers for diagnosis and progression of MS by MALDI-TOF mass spectrometry. Mult Scler 17(7):838–850. doi:10.1177/1352458511399614
Timms JF, Arslan-Low E, Gentry-Maharaj A, Luo Z, T'Jampens D, Podust VN, Ford J, Fung ET, Gammerman A, Jacobs I, Menon U (2007) Preanalytic influence of sample handling on SELDI-TOF serum protein profiles. Clin Chem 53(4):645–656
Timms JF, Cramer R, Camuzeaux S, Tiss A, Smith C, Burford B, Nouretdinov I, Devetyarov D, Gentry-Maharaj A, Ford J, Luo Z, Gammerman A, Menon U, Jacobs I (2010) Peptides

generated ex vivo from serum proteins by tumor-specific exopeptidases are not useful biomarkers in ovarian cancer. Clin Chem 56(2):262–271

Timms JF, Menon U, Devetyarov D, Tiss A, Camuzeaux S, McCurrie K, Nouretdinov I, Burford B, Smith C, Gentry-Maharaj A, Hallett R, Ford J, Luo Z, Vovk V, Gammerman A, Cramer R, Jacobs I (2011) Early detection of ovarian cancer in samples pre-diagnosis using CA125 and MALDI-MS peaks. Cancer Genomics Proteomics 8(6):289–305

Tiss A, Smith C, Camuzeaux S, Kabir M, Gayther S, Menon U, Waterfield M, Timms JF, Jacobs I, Cramer R (2007) Serum peptide profiling using MALDI mass spectrometry: avoiding the pitfalls of coated magnetic beads using well-established ZipTip technology. Proteomics 7(Suppl 1):77–89

Tiss A, Smith C, Menon U, Jacobs I, Timms JF, Cramer R (2010) A well-characterised peak identification list of MALDI MS profile peaks for human blood serum. Proteomics 10(18):3388–3392

Vafadar-Isfahani B, Laversin SA, Ahmad M, Ball G, Coveney C, Lemetre C, Kathleen Miles A, van Schalkwyk G, Rees R, Matharoo-Ball B (2010) Serum biomarkers which correlate with failure to respond to immunotherapy and tumor progression in a murine colorectal cancer model. Proteomics Clin Appl 4(8-9):682–696. doi:10.1002/prca.200900218

Villanueva J, Lawlor K, Toledo-Crow R, Tempst P (2006a) Automated serum peptide profiling. Nat Protoc 1(2):880–891

Villanueva J, Martorella AJ, Lawlor K, Philip J, Fleisher M, Robbins RJ, Tempst P (2006b) Serum peptidome patterns that distinguish metastatic thyroid carcinoma from cancer-free controls are unbiased by gender and age. Mol Cell Proteomics 5(10):1840–1852

Villanueva J, Shaffer DR, Philip J, Chaparro CA, Erdjument-Bromage H, Olshen AB, Fleisher M, Lilja H, Brogi E, Boyd J, Sanchez-Carbayo M, Holland EC, Cordon-Cardo C, Scher HI, Tempst P (2006c) Differential exoprotease activities confer tumor-specific serum peptidome patterns. J Clin Invest 116(1):271–284

Xiao D, Meng FL, He LH, Gu YX, Zhang JZ (2011) Analysis of the urinary peptidome associated with Helicobacter pylori infection. World J Gastroenterol 17(5):618–624. doi:10.3748/wjg.v17.i5.618

MALDI Biotyping for Microorganism Identification in Clinical Microbiology

Arthur B. Pranada, Gerold Schwarz, and Markus Kostrzewa

Abstract In 1996, independent publications demonstrated the application of MALDI-TOF MS for microorganism identification using whole-cell profile mass spectra. As these were characteristic for distinct bacterial and fungal species, they could be used as a molecular fingerprint. Subsequently, sample preparation, data evaluation with bioinformatics and further aspects of the technology were further improved. Ease-of-use, rapidity and accuracy of this technology finally led to the implementation of MALDI-TOF MS into routine analytical work, particularly in clinical microbiology.

Today, microorganism identification by MALDI-TOF MS is performed in many clinical microbiology laboratories and is increasingly replacing the conventional methods utilized for decades. In this chapter, the impact of this technology on clinical diagnostics is described. Furthermore, exemplary protocols for microorganism identification are presented. Specialized protocols even allow identification of more demanding microorganisms like mycobacteria and filamentous fungi as well as identification from liquid culture media inoculated with patient specimen. Some analytical systems allow the extension of reference databases with own entries and an insight into evaluation algorithms, interpretation of results and creation of own references is given.

Further, MALDI-TOF MS has proven its utility in identification of food-borne and veterinary microorganisms. These topics as well as regulatory aspects together with the necessary steps for qualification and validation are also covered for the interested reader.

A.B. Pranada (✉)
Department of Medical Microbiology, MVZ Dr. Eberhard & Partner Dortmund (ÜBAG), Balkenstr. 17-19, 44137 Dortmund, Germany
e-mail: apranada@labmed.de

G. Schwarz
Elisabeth-Segelken-Str. 7, 28357 Bremen, Germany
e-mail: gerold-schwarz@web.de

M. Kostrzewa
Bruker Daltonik GmbH, Fahrenheitstr. 4, 28359 Bremen, Germany
e-mail: Markus.Kostrzewa@bruker.com

R. Cramer (ed.), *Advances in MALDI and Laser-Induced Soft Ionization Mass Spectrometry*, DOI 10.1007/978-3-319-04819-2_11

1 Introduction

Early in the development of mass spectrometry (MS) technologies their applicability to microorganism analysis, in particular the identification of microbes, was already discussed (Anhalt and Fenselau 1975; Sinha et al. 1985; Heller et al. 1987). The real breakthrough was achieved with the appearance of matrix-assisted laser desorption/ionization (MALDI)-time-of-flight (TOF) MS, a technology which combines soft ionization of large biomolecules with short analysis time. In 1996, three independent scientific publications showed ground-breaking results in the application of MALDI-TOF MS for microorganism identification using whole-cell profile mass spectra (Claydon et al. 1996; Holland et al. 1996; Krishnamurthy and Ross 1996).

The idea was to use the microbial MALDI-TOF profile mass spectra of whole cells as a molecular fingerprint, being characteristic for bacteria and fungi.

During the first years, several scientists worked on the improvement of sample preparation (Welham et al. 1998; Gantt et al. 1999; Smole et al. 2002; Vaidyanathan et al. 2002), reproducibility (Wang et al. 1998; Demirev et al. 1999; Saenz et al. 1999; Williams et al. 2003), bioinformatic data evaluation (Arnold and Reilly 1998; Jarman et al. 1999; Jarman et al. 2000; Pineda et al. 2000), and applicability of the technique to different microbial groups (Haag et al. 1998; Nilsson 1999; Amiri-Eliasi and Fenselau 2001; Conway et al. 2001; Bernardo et al. 2002; Hettick et al. 2004; Krader and Emerson 2004; Mandrell et al. 2005). Ryzhov and Fenselau (2001) found that most of the peaks which can be observed in typical bacterial profile spectra are derived from ribosomal and other high-abundant housekeeping proteins, which explains the robustness and reproducibility of approaches analysing profiles in the mass range of 2–20 kDa.

A first commercial system, launched by the company Micromass (Manchester, UK), analyzed fingerprints in the lower mass region (approximately up to *m*/*z* 4000) but did not succeed in routine laboratories. Therefore, MALDI-TOF MS profiling for microorganisms stayed a scientific topic in the area of mass spectrometry for quite some time but did not enter the area of routine (clinical) microbiology.

This situation started to change in 2007 with the introduction of a new generation of commercially available systems in clinical microbiology, equipped with easy-to-use software solutions and broadly applicable reference databases, supported with standard operating procedures (SOPs) for sample preparation. All these systems analyzed biomolecules in the mass range of approximately *m*/*z* 2000–20,000. Mellmann and co-workers compared the performance of the MALDI Biotyper system (Bruker Daltonik GmbH, Bremen, Germany) for the identification of Gram-negative non-fermenting bacteria with sequence analysis of 16S ribosomal DNA which is currently the 'gold standard' for identification of bacterial species (Mellmann et al. 2008). They found that for this bacterial group, which is very difficult to characterize by biochemical tests, 85.9% of strains were correctly identified. This performance was superior to the established biochemical routine analysis systems. Subsequently, the same group performed the first large multi-centre study for this technology (Mellmann et al. 2009). In this international study, eight laboratories analyzed 60 blind-coded non-fermenting bacteria samples each and achieved an inter-laboratory reproducibility of 98.75%. Only six of the 480 samples were misidentified due to sample interchanges (four samples) or contamination (one sample) or could not be identified because of insufficient signal intensity (one sample). The authors concluded that this level of reproducibility is usually only achievable with DNA sequence-based methods and that MALDI-TOF MS for microorganism identification is therefore a highly reproducible technology, a prerequisite for any routine utilization.

2 Applications

2.1 MALDI Biotyping in Medical Microbiology: An Overview

Medical microbiology plays an important role in the diagnosis and treatment of infectious diseases today. In case of an infection, patient specimens are collected and examined with laboratory methods to identify the infection-causing microorganisms like bacteria, fungi and viruses.

While viruses are usually identified using molecular methods, bacteria and fungi are generally identified from cultures grown on solid or in liquid media. Fast growing microorganisms form small colonies on solid media after 24–48 h. Shape and color of individual colonies as well as their microscopic features can be used for a first categorization of the microorganism. Gram staining is generally used for a further quick characterization step. Final identification is normally performed using biochemical reactions and tests that indicate the metabolic characteristics of the species of interest during growth for about another 6 h for fast growing bacteria (like enterobacteria, enterococci or staphylococci) in automated systems and up to 48 h for slower growing organisms or with fewer characteristic metabolic reactions.

MALDI biotyping, a modern, fast and accurate method suitable for routine analysis allows the identification of microorganisms in a few minutes from a colony. This has led to 'a fundamental shift in routine practice of clinical microbiology' as described by Clark et al. (Clark et al. 2013), while other authors call it an 'ongoing revolution' (Seng et al. 2009). The latter publication and work from Eigner and colleagues (Eigner et al. 2009) were the first reports of studies investigating the broad utilization of MALDI-TOF MS biotyping in a clinical microbiology laboratory. These studies revealed excellent results for the identification of microorganisms occurring in clinical routine. Subsequently, a number of studies have proven the superiority of the technology in comparison to traditional biochemical test systems (Marklein et al. 2009; Nagy et al. 2009; Bader et al. 2011; Bizzini et al. 2010; van Veen et al. 2010; Bille et al. 2012; Dhiman et al. 2011).

A particular strength of MALDI-TOF MS (with an appropriate database) is the potential to identify not only the frequently occurring but also rare and difficult-to-identify organisms which normally have to be characterized by DNA sequencing if identification is at all possible at species level (Bizzini et al. 2011). Thereby, species name can now be assigned for some bacteria, which were previously classified by many laboratories only to the genus level or an even more general systematic group, e.g. 'Gram-positive, non-sporulating rods' or 'Gram-negative non-fermenting rods'. Thus, MALDI-TOF MS profiling improves the quality of diagnostic microorganism identification through both greater accuracy and higher taxonomic resolution.

For example, in the group of yeasts, the *Candida parapsilosis* complex and the *Candida haemulonis* complex can now be resolved as distinct species (Quiles-Melero et al. 2012; Cendejas-Bueno et al. 2012). For bacteria, important examples are the corynebacteria (Alatoom et al. 2012; Vila et al. 2012) and *Haemophilus influenzae/haemolyticus* (Zhu et al. 2013; Bruin et al. 2014). For some groups of microorganisms, which usually had to be referred to reference laboratories, identification can now be performed by routine laboratories. This applies in particular to mycobacteria (Lotz et al. 2010; Saleeb et al. 2011; Balada-Llasat et al. 2013; Buchan et al. 2014) and filamentous fungi (Cassagne et al. 2011; Iriart et al. 2012; Lau et al. 2013; Schulthess et al. 2014), where MALDI-TOF MS profiling can also be employed to get a first-line identifier.

2.2 The Impact of MALDI-TOF MS on the Clinical Microbiology Laboratory

The methods and procedures described in this chapter are suitable for the identification of microorganisms in the microbiology laboratory. With the introduction of this technology, it has also been possible to optimize established routine workflows that use conventional methods. Formerly, microbial identification and antimicrobial susceptibility testing (AST) were performed in parallel after an initial 'triage' by colony morphology to decide which test panels could be the most appropriate. Today, in many cases where morphological assessment is difficult, AST can be started based on the precise identification of the microorganism by MALDI biotyping, and corresponding AST panels can be applied. Rapid identification also allows earlier therapy adjustment. This is particularly true for slow growing organisms or when differentiation by conventional biochemical methods is difficult.

Another important benefit of MALDI-TOF MS biotyping is its precision and high discriminatory power. Reviews of laboratory statistics of the MVZ (Medical Center) Dr. Eberhard & Partners (Dortmund, Germany)—one of the major independent laboratories in Germany—show a significant enhancement in identification of microorganisms after the introduction of MALDI-TOF MS. Sited in Dortmund this laboratory is located in the heart of the Ruhr valley which is one of the largest urban agglomerations in Europe with about five million inhabitants. The Department of Microbiology of the MVZ Dr. Eberhard & Partners serves as a diagnostic partner for several hospitals and many doctors in private practices as well as for other laboratories. In the year 2009, before the introduction of a MALDI-

TOF system at the Department of Microbiology, a total of 299 different species from 115 genera were found. By the end of 2009 first tests with a MALDI biotyping system had started, and the new technology was introduced into routine analysis a few months later. In 2010, the first year with MALDI-TOF MS as an additional routine identification method, the reports showed a total of 438 different species (+46%) from 146 different genera (+27%). Especially in the group of anaerobic bacteria and non-fermenters, which are usually difficult to identify by conventional methods, a broad variety of species can now be differentiated. Also, the commonly consolidated 'coryneform bacteria' can now be broken down into distinct species.

Arguably, this new accuracy in identification will also lead to an increase of knowledge about individual microorganisms and their infection potential. The possibility of extending the database with your own entries offers the potential for easy detection of uncommon microorganisms. To date, nucleic acid sequencing is the 'gold standard' in microorganism identification and remains the last option in selected cases where other identification methods are insufficient. With the subsequent creation of an own MSP ('main spectrum', i.e. reference database entry), uncommon microbes can easily be re-identified without the need for laborious and costly biomolecular sequencing. Even for yet unnamed species, recognition of recurrence of these microorganisms is now possible. Therefore, in addition to the manufacturers' databases major laboratories often also employ an increasing number of their own reference spectra for routine diagnostics. As the manufacturers' reference databases have grown over the years a broad range of identifications is now possible. A further extension of these reference libraries can be expected. Statistics from MVZ Dr. Eberhard & Partners, where the standard reference libraries are extended by about 200 self-acquired MSPs, show that in the year 2014 more than 700 different species from 190 genera were recorded. Compared to species identification before the implementation of MALDI-TOF MS, this is an increase of about 150%. In addition, MSPs of 21 different yet unnamed bacterial species detected these microorganisms in about 130 cases. This clearly stresses the great potential of MALDI-TOF MS in routine diagnostics of microbiology laboratories and its future prospect for a better knowledge about microorganisms in infections.

2.3 *Regulatory Aspects*

While in the beginning only research-use-only (RUO) systems were available for the identification of microorganisms by MALDI-TOF MS, this has changed in the following years. The pioneers who have introduced the technology into the diagnostic field validated the instrumentation, software and databases in thorough studies with reference strains and clinical isolates.

Aspects like the performance for specific microorganism groups as well as the robustness of the method and repeatability from day to day and user to user have to be proven and documented. If the system is labeled and distributed as RUO by the manufacturer and not intended for diagnostic purpose, the user takes full responsibility for the functionality and results of the system. On the other hand, users have the flexibility to define and control their own diagnostic device. For the improvement of results of a self-validated RUO system, several specialists have reported extensions of the manufacturer's database, modifications of preparation techniques or adaptations of acceptance criteria (Christensen et al. 2012; Khot et al. 2012; Theel et al. 2012; Ford and Burnham 2013; McElvania TeKippe et al. 2013; Murugaiyan et al. 2014).

In 2009, the first IVD-CE labeled system became available in Europe, for diagnostic identification of bacteria and yeast in clinical microbiology laboratories, the IVD MALDI Biotyper (Bruker Daltonik GmbH). Later, with the VITEK® MS IVD (bioMérieux, Marcy l'Etoile, France) a further system became available. With such a system, the manufacturer takes over the validation for clinical usage and later responsibility for the function of the system—provided that the system is utilized in the area of intended use and according to the instructions of the manufacturer.

In the USA, a clearance of such a device for in vitro diagnostic usage by the US Food and Drug Administration (FDA) is necessary and was obtained for both the MALDI Biotyper—Clinical Application (Bruker Daltonics, Billerica, MA, USA) and the VITEK® MS IVD (bioMérieux) in 2013. In particular, the FDA process involved detailed and expensive studies including investigations of robustness and repeatability, media compatibility and influence of all possible environmental conditions to the analyses. Furthermore, several thousands of samples had to be analyzed in clinical studies, and the results were compared to DNA sequencing as the current 'gold standard'. All studies resulted in the proof of excellent performance for the investigated MALDI-TOF MS systems which may therefore become a new 'gold standard' for microbial identification in the near future.

In Germany, laboratories offering diagnostic services to others need an accreditation according to DIN EN ISO 15189. Since 2009 accreditation of new methods using MALDI-TOF MS for microorganism differentiation has become possible, with sufficient familiarization as well as specific quality assurance being mandatory.

2.4 MALDI-TOF MS Profiling of Veterinary and Food-Borne Organisms

In parallel to the field of medical microbiology, MALDI-TOF MS has also conquered the field of microbial identification and characterization in veterinary and food microbiology. The reasons for success are the same as before: speed, accuracy and cost-effectiveness. As regulatory hurdles are lower in the veterinary field, in particular for university veterinary institutes, which are used to perform their own validation of methods, MALDI-TOF MS could even spread faster in veterinary microbiology, e.g. in the USA.

Besides studies to investigate the broad applicability of MALDI-TOF MS in these areas (Boehme et al. 2013; Wragg et al. 2014), studies for microorganisms from particular diseases and certain microorganisms, which are difficult to identify by traditional methods, have been performed (Alispahic et al. 2010; Taniguchi et al. 2014; Frey and Kuhnert 2015). It has to be mentioned that less sophisticated biochemical ready-to-use kits are available in veterinary microbiology because of the smaller commercial interest. Further, as not only pathogens but also hosts are very variable the diversity of microorganisms is significantly higher than in human microbiology. Both are very good reasons to use a universal technology like MALDI-TOF MS profiling.

Barreiro and colleagues found MALDI-TOF MS profiling very suitable for pathogens isolated from milk of cows with mastitis (Barreiro et al. 2010). Excellent identification of group D Streptococci could be demonstrated (Werner et al. 2012). Grosse-Herrenthey and co-workers investigated the performance of MALDI-TOF MS for the identification and differentiation of clostridia, a group of anaerobic, Gram-positive sporulating bacteria which comprises a number of pathogens with veterinary but also medical relevance (Grosse-Herrenthey et al. 2008). They analyzed a total of 64 clostridial strains of 31 different species and found an excellent differentiation of these. It was even possible to identify species, which are normally difficult to differentiate by traditional methods, such as *C. chauvoei* and *C. septicum*.

The group of Campylobacter species and the related genera Helicobacter and Arcobacter were also found to be well identified and differentiated by MALDI-TOF MS profiling. Alispahic and colleagues analyzed 144 clinical isolates of these genera using whole spectral profiles (Alispahic et al. 2010). It was found that correct identification could even be obtained from bacteria stored at room temperature or at 4 °C up to 9 days prior to being tested. Other groups also found MALDI-TOF MS an excellent tool for the identification of these species, which partially are also important human pathogens (Bessède et al. 2011; Martiny et al. 2011; Taniguchi et al. 2014). For *Campylobacter jejuni*, effective distinction of subgroups has recently been reported. It was possible to group specific *C. jejuni* subgroups of phylogenetically related isolates in distinct clusters by principle component analysis (PCA) and hierarchical clustering (Zautner et al. 2013).

Another study demonstrated that MALDI-TOF MS profiling represents a fast and reliable method for the identification and differentiation of Gallibacterium species, with applications in clinical diagnostics (Alispahic et al. 2011). Even the very difficult-to-identify pathogenic algae from the genus Prototheca, occurring in mastitis and sometimes infecting milkers, could be reliably identified (von Bergen et al. 2009; Murugaiyan et al. 2012). This is one of the examples where MALDI-TOF MS profiling can be of particular diagnostic power as the algae might initially be misidentified as yeast which could lead to an ineffective treatment with antimycotic drugs.

Another case for the importance of accurate technology can be made in the field of food-borne pathogens. The power of identifying accurately Campylobacter species has already been described. However, the most well-known bacteria causing food-borne diseases are Salmonella species. Reliable identification of the genus

Salmonella has been described by several researchers (Dieckmann et al. 2008; He et al. 2010) but a serovar differentiation which would be highly desirable for Salmonella species is currently questionable. There have been reports that the highly pathogenic serovar *Salmonella typhi* might be differentiated from the other residual members of the genus (Kuhns et al. 2012; Martiny et al. 2012; Schaumann et al. 2013), but this still has to be validated. A method described for the rapid detection of Salmonella species in stool samples of diseased people, which comprises the enrichment of the bacteria in a liquid selective broth (Sparbier et al. 2012), may also be adapted to food products but will need further development and validation work.

Identification of further food-borne pathogens as described for Listeria species (Barbuddhe et al. 2008; Hsueh et al. 2014), Bacillus species (Farfour et al. 2012) and Yersinia species (Ayyadurai et al. 2010; He et al. 2010; Lasch et al. 2010; Stephan et al. 2011) may be mainly applied for confirmation of clinical diagnostics and/or further developed towards fast detection tests as shown in a study with spiked samples for Listeria (Jadhav et al. 2014).

2.5 *Instrument Qualification and Method Validation* (for the Interested Reader: Can Be Skipped by the MALDI Novice)

2.5.1 Qualification as an Analytical Instrument: DQ-IQ-OQ-PQ

Instrument qualification of automated, computerized systems is a critical step and has to follow accepted official 'good manufacturing practices' (GMP) guidelines. All initial activities are usually summarized as design qualification (DQ) steps. This includes system documentation of hardware and software, intended use considerations/statements, user and/or functional requirement declarations and vendor assessments in coordination with a dedicated risk management defined by GMP guidelines for analytical procedures (e.g. USP, EP, ICH, PDA Tech Report33; cf. http://www.gmp-navigator.com/nav_guidelines.html; http://www.ich.org/home.html; http://www.usp.org/; http://www.pda.org/).

Next, installation qualification (IQ) documents as well as operational qualification (OQ) procedures are fully documented references to ensure that the system is installed and operational according to dedicated specifications to guarantee the intended use. It typically includes a system suitability test in a final performance qualification step (PQ) with several different freshly grown microorganisms.

In addition, a set of operational and administrative SOPs, preventive maintenance plans/activities and complete method validation scenarios are a crucial part of initial considerations. Once these steps are initiated any changes or amendments to software, hardware, methods, SOPs, etc. are not allowed or have to follow a stringent risk management-based approach usually handled by change control under well-defined, mandatory quality management (QM) rules and followed by a final approval through the responsible qualified person (QP). These QM rules are based on the aforementioned established GMP guidelines and conventions and are meant to show (objectively) whether any change in the complete process will influence the outcome/intended use of the system negatively. A small change in the process may require a complete new validation procedure.

2.5.2 Validation

The in-house validation part requires a setup of independent experiments generally starting with an equivalence study showing the sensitivity and specificity of the new method compared to an established standard method. Discrepancies between both methods have to be resolved by utilizing established DNA sequencing techniques (the current 'gold standard') or additional methods if required. Several strains of each relevant group (e.g. Gram-positive and Gram-negative bacteria, spore-formers and non-spore-formers, fungal species if relevant and additional relevant groups if existing) have to be freshly prepared and identified. Further, a system performance check or system suitability check needs to be run before every experiment to document correct system operation and demonstrate a fit-for-purpose state.

A system suitability test shall include at least one well characterized and officially certified strain. In addition, a negative test with a strain delivering a true detectable MALDI-TOF MS profile but resulting in 'no identification' after data processing (because it is not contained in the respective database) has to be successfully employed throughout all experiments.

To show method robustness, small but deliberate changes during the analytical process have to be demonstrated. These can be shown by experiments at different temperatures and humidity, using different pipettes if exact volumes are required, different time frames between sample handling and processing steps and/or utilizing different lots of matrix and solvents.

For the precision of the method, the degree of agreement of multiple runs of the same sample or different suspensions of the same bacterial material (usually in the same lab within a short time) will be recorded. If different people on different days (ideally in different labs) using different preparations of the same bacterial sample are included, the intra- and inter-assay repeatability (intermediate precision) can be shown and documented.

3 Materials and Protocols

The protocols usually applied for routine microbial identification by MALDI-TOF MS profiling are quite similar among the different commercially available systems. In the following, protocols are described for the MALDI Biotyper (Bruker Daltonics), one of the leading systems with about 1500 installed instruments at the end of 2014. Generally, these protocols can also be applied or transferred to other MALDI-TOF mass spectrometers.

3.1 Chemicals for Microorganism Identification by MALDI-TOF MS Biotyping

- HPLC water, ethanol, acetonitrile (ACN), trifluoroacetic acid and formic acid
- Matrix: α-Cyano-4-hydroxycinnamic acid (CHCA)
- Standard solvent composition: ACN:water:trifluoroacetic acid (50%:47.5%:2.5% (v:v:v))

All chemicals used should be of highest purity (intended for use in mass spectrometry or HPLC).

The matrix solution is typically prepared by adding 250 μL of the standard solvent to 2.5 mg of CHCA and vortexing the mixture until all matrix crystals are completely dissolved. The manufacturers usually offer small tubes with pre-portioned amounts of matrix to which just the adequate volume of solvent has to be added. The standard solvent is also available as a ready-to-use solution (cat. no. 19182; Sigma-Aldrich).

The matrix solution can be stored in the dark at room temperature for up to 2 weeks. As the organic solvents are volatile and ACN is strongly hygroscopic, it is recommended to review storage conditions. Storing small volumes in big tubes or containers as well as working for long time with opened containers for reagents can result in a shift in concentrations that render MALDI-TOF MS profiling measurements impossible.

3.2 Spectrum Acquisition

The MALDI-TOF mass spectrometer is typically operated in linear mode. The spectrum is acquired in an *m*/*z* range of 2000–20,000 targeting ribosomal proteins, which are highly conserved and in general independent on culture conditions. It is recommended to perform regular calibrations with a standard, covering the *m*/*z* range of interest. A total of up to 240 laser shots from different positions of the sample is sufficient to obtain a representative spectrum with many characteristic peaks. Latest software versions allow an early termination of the spectral acquisition as soon as enough peaks with a satisfactory resolution or amplitude have been acquired. Statistics from routine diagnostics in medical microbiology show that about 25% of the measurements are finished after 40–120 shots, while 50% of the measurements need less than 200 shots (Data source: MVZ Dr. Eberhard & Partners). This saves time and extends the laser life time as well as intervals between servicing for source cleaning.

3.3 Direct Transfer (DT) Procedure

Besides the speed of identification, the ease-of-use in daily routine analysis has been one of the main reasons for the success of MALDI-TOF MS in medical microbiology. This is especially true for the direct transfer (DT) procedure, which can be applied to most cases of routine analysis in medical microbiology. From the colony to be identified some biological material is directly transferred to the steel target and smeared as a very thin film on the sample spot area. Best results can be obtained from cultures that are still growing. So in case of fast growing bacteria, fresh overnight cultures are suitable and identifications from slow growing organisms can be performed from cultures over several days. As proteins can quickly degrade from organisms stored in a refrigerator where they enter the lag phase, spectrum quality for such organisms can also quickly degrade. However, measurements from colonies stored at room temperature and kept in a metabolic active state show acceptable spectra even after days. For smearing the colony material onto the steel target, inoculation loops, pipette tips or wooden toothpicks are used in most laboratories. Sometimes, it needs some training to establish the right technique and amount of biological material that has to be applied to the sample spot. Usually, very thin films that can sometimes hardly be seen by the naked eye provide better results. Some users also prefer to employ ground steel targets instead of polished steel targets.[1] The sample spot is then directly overlaid with 1 μL of matrix solution, which is prepared according to the description in Sect. 3.1. After drying at room temperature, spectral acquisition (see Sect. 3.2) can be started. Figure 1 shows the steps of the DT method for a freshly grown colony picked from a blood agar plate.[2]

[1] The slightly rougher structure of ground steel targets can facilitate the smearing to the right thickness.

[2] Most culture media are blood-based. The haemoglobin in these media can produce artificial peaks disturbing the spectral acquisition or deteriorating the identification result. Thus, accidental transfers of parts of the culture media to the target should be avoided.

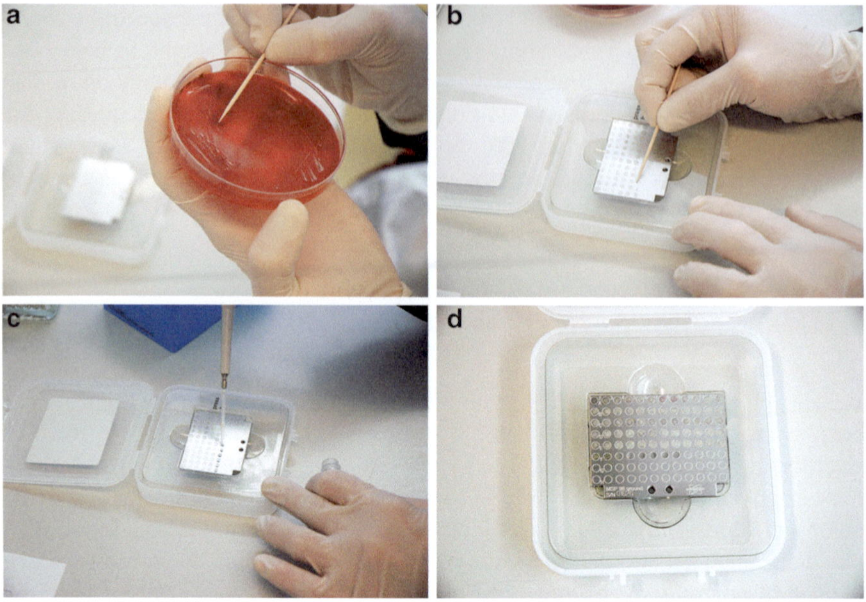

Fig. 1 Direct transfer (DT) procedure for microorganism identification by MALDI-TOF MS biotyping: (**a**) pick small amount of biomass from a bacterial colony of a freshly (overnight) grown culture, (**b**) transfer the biomass onto a sample spot area on a steel MALDI target plate and smear it out, producing a very thin film, (**c**) overlay with matrix, and (**d**) dry at room temperature before MALDI-TOF MS measurement

3.4 Extended Direct Transfer (EDT) Procedure

The DT procedure provides good-quality spectra for the majority of microorganisms in routine work. For some analytes like yeasts that have a more rigid cell wall, the extended direct transfer (EDT) procedure might be more advantageous. After smearing the colony material onto the MALDI target, the thin analyte film is overlaid with 1 μL of 70% formic acid. Then, after drying, 1 μL of the matrix solution is added and dried at room temperature. The formic acid can help to penetrate the cell walls and liberate the proteins of interest. Before applying the EDT method, it is advisable to check whether optimization of the standard DT procedure (cf. Sect. 3.3) could also improve results. In many cases, the amount of microorganism biomass transferred to the MALDI target is the more critical parameter.

3.5 Ethanol/Formic Acid Extraction (EFEx)

The best spectrum quality can be expected with the ethanol/formic acid extraction (EFEx) method. Compared to the DT and EDT procedure, it is more laborious and can be performed when the simpler methods give no satisfactory spectra. This elaborate extraction method is also appropriate for applications like strain sub-typing and the acquisition of reference spectra for extending the database when maximum spectrum quality is critical.

EFEx is performed in an Eppendorf® or similar tube with a volume of 1.5 mL by pipetting 300 μL of water into the tube and suspending a small amount of biological material (e.g. one single colony from a freshly grown culture of up to 5–10 mg) into it. Repeated pipetting (dispensing/aspirating) and/or vortexing facilitates the mixing process. Then, 900 μL of ethanol (absolute) are added and the solution is thoroughly mixed. The tube is then centrifuged at maximum speed for 2 min. The supernatant can be decanted and after another centrifugation step the remaining ethanol should be carefully removed without touching the pellet. Subsequently, the pellet is dried for some minutes at room temperature, followed by the addition of 1–80 μL of 70% formic acid. The exact amount depends on the amount of biological material entered into the extraction process: About 1–5 μL are sufficient for a small colony, while for bigger colonies 5–15 μL should be used. If a full 1-μL inoculation loop with colony material is used 10–40 μL are appropriate and for a 10-μL loop about 30–80 μL of 70% formic acid should be used and mixed well by pipetting (dispensing/aspirating) and/or vortexing. Then, pure ACN is added in the same amount as the formic acid and mixed. After centrifugation for 2 min, one should obtain a small pellet with the residual material and a supernatant with the extracted proteins. Then, 1 μL of the supernatant can be pipetted onto one sample spot of the steel target. After drying at room temperature, 1 μL of the matrix solution can be added and again dried at room temperature. To prevent chemical reactions like oxidation and methylation of the samples that lead to peak shifts in the spectrum, the supernatant with the extracted proteins should be used immediately and overlaid with the matrix solution as soon as possible (in general within 10 min).

Typical MALDI MS profile spectra using the EFEx procedure for the preparation of a variety of microorganisms from different phylogenetic branches are depicted in Fig. 2.

3.6 Extraction from Blood Cultures

3.6.1 Sepsityper Protocol

Blood cultures play an important role in medical microbiology. They allow the detection of pathogens in blood stream infections. Although these infections are severe and can lead to serious complications, there is usually only a small number

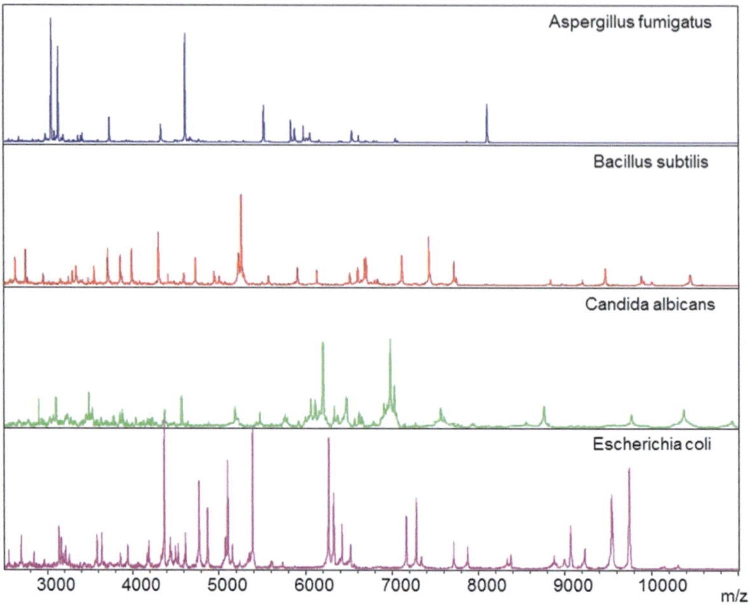

Fig. 2 MALDI-TOF MS profiles acquired using the ethanol/formic acid extraction (EFEx) method for four microorganisms belonging to different phylogenetic branches: the mold *Aspergillus fumigatus*, the Gram-positive bacterium *Bacillus subtilis*, the yeast *Candida albicans*, and the Gram-negative bacterium *Escherichia coli*

of 1–10 microorganisms per millilitre that can be detected. Therefore, liquid media in bottles are used as enrichment culture. A volume of 10–15 mL of blood is collected from the patient, inoculated into a bottle and incubated in an automated system that regularly checks for growth. As soon as the system detects growth (usually after several hours to days depending on the microorganism concentration and speed of growth), the liquid medium with the enriched microorganisms can be used for inoculation of usual solid media for further culture for at least another 24–48 h and subsequent identification and susceptibility testing. As blood stream infections can show a quite severe clinical picture an early initiation of antimicrobial therapy is essential. This is the reason why for an initial analysis direct microscopy with Gram staining is performed as soon as the blood culture has turned positive in order to estimate the group of organisms and a potentially effective therapy. MALDI-TOF MS now allows the identification directly from a positive blood culture bottle without subculture and identification results can be available about 12–24 h earlier.

There are several protocols published for the extraction of proteins for microorganism identification directly from positive blood cultures (La Scola and Raoult 2009; Christner et al. 2010; Moussaoui et al. 2010; Stevenson et al. 2010; Ferreira et al. 2011; Schubert et al. 2011). The main objective of all procedures is to enrich the cultured microorganisms and to separate them from blood cells and other blood components. Especially, the haemoglobin of the erythrocytes can perturb the MALDI-TOF measurement and need to be excluded.

For the MALDI Biotyper system, the manufacturer offers a ready-to-use kit (MBT Sepsityper IVD Kit) which has been investigated in more than 20 published studies (see e.g. Kok et al. 2011; Schubert et al. 2011; Yan et al. 2011; Buchan et al. 2012; Juiz et al. 2012; Klein et al. 2012; Loonen et al. 2012; Schieffer et al. 2014) and is currently the only IVD-CE kit for this purpose. The kit comprises a lysis buffer, a washing buffer and 1.5-mL plastic reaction/microcentrifugation tubes. As the procedure is based on the EFEx method, the same reagents as for EFEx should also be used.

The extraction is started by transferring about 1 mL of the blood culture fluid to the 1.5-mL plastic reaction tube. In the next step, 200 μL of the lysis buffer are added and mixed by vortexing for about 10 s.[3] The tube is then centrifugated at 13,000 rpm for about 2 min and the supernatant is removed. The remaining pellet is suspended in 1 mL of the washing buffer and mixed by pipetting several times up and down (i.e. dispensing/aspirating). After another centrifugation step for 1 min at 13,000 rpm, the supernatant is again removed. The EFEx procedure can now be performed using the resulting pellet by suspending it in 300 μL of HPLC-grade water, adding 900 μL of ethanol and mixing the solution. After finishing these extraction steps, 1 μL of the extract can be transferred to a spot on the target, dried and then overlaid with 1 μL of matrix solution (cf. Sect. 3.5). The MALDI-TOF MS profiling measurement can then be performed with standard methods.[4]

As haemoglobin will produce large artificial peaks in MALDI-TOF MS measurements, which render the MS spectra useless, several additional washing steps after lysis with the supplied washing buffer can help reducing such problems.

3.6.2 Alternative Protocols

There have been other methods published for the extraction of proteins from blood culture. For instance, saponin (Ferroni et al. 2010) or ammonium chloride (Prod'hom et al. 2010) can be used for lysis, while the separation of microorganisms from blood cells can be performed with differential centrifugation and gel separator tubes (Stevenson et al. 2010; Moussaoui et al. 2010) or just with simple sedimentation (La Scola and Raoult 2009; Christner et al. 2010; Ferreira et al. 2011).

---

[3] For blood culture bottles that contain charcoal for adsorbing antimicrobial substances, filtering the blood culture/lysis buffer mixture in the first steps with a spin column like SigmaPrep (cat. no. SC1000-1KT, Sigma-Aldrich; centrifugation for 2 min at 2000 rpm) can remove the distracting particles. For blood cultures with resin particles this procedure is not needed.

[4] Sometimes it is advisable to accept slightly lower score values. In the rare cases when more than one pathogenic microorganism is suspected in the blood culture or when Gram staining already shows multiple morphologies in microscopy, mixture algorithms can be applied for identification. However, in general, blood cultures only contain one microorganism.

3.7 *Limitations of Standard Extraction Methods*

The described procedures for microorganism profiling, especially the effortless DT method (cf. Sect. 3.3), have disburdened the routine microorganism identification process in medical microbiology. Furthermore, new workflows have been made possible as subsequent steps can now be based on precise identification results. Nevertheless, there are microorganisms that are resistant to standard extraction procedures and give low-quality spectra with no or only a low number of peaks.

As some of these pathogen groups can play an important role in infectious diseases adapted protocols have been developed or are still under review.

3.7.1 Filamentous Fungi

One of the above mentioned groups of microorganisms are the filamentous fungi. Usually, culture on solid media is performed with subsequent microscopy of the grown fungi. Identification is then based on the micromorphological structures observed in the different growth states of the fungus. Since these different growth states and also the sporulation of the fungi provide problems for MS spectrum acquisition, special culture conditions and extraction procedures can produce analytically more valuable spectra.

Different protocols have been published with several starting from mycel grown on solid media (Cassagne et al. 2011; Iriart et al. 2012; Lau et al. 2013; Ranque et al. 2014), one using a short incubation of mycel in liquid medium to unify the fungal material before extraction (Schulthess et al. 2014).

Figure 3 shows the identification of a fungus by conventional microscopy (a) and mass spectral analysis (b).

3.7.2 Mycobacteria

For mycobacteria, special extraction methods are currently under review. These bacteria have only a low number of ribosomes, and standard extraction methods cannot usually overcome their rigid cell walls. Generally, the currently adapted methods apply silica or zirkonia microbeads to disrupt cell aggregates and make the cell surface more accessible to the extraction solution (Lotz et al. 2010; Saleeb et al. 2011; El Khéchine et al. 2011; Balada-Llasat et al. 2013; Buchan et al. 2014). Reported identification rates are in the range of 80–90% of strains occurring in clinical routine work. The results show some limitations in distinguishing very closely related species, similar to the limitations of 16S rDNA sequencing but in some cases somewhat superior. As an example, the *Mycobacterium tuberculosis* complex members cannot be differentiated from each other but there are reports that *Mycobacterium abscessus* and *Mycobacterium massiliense*, both sometimes

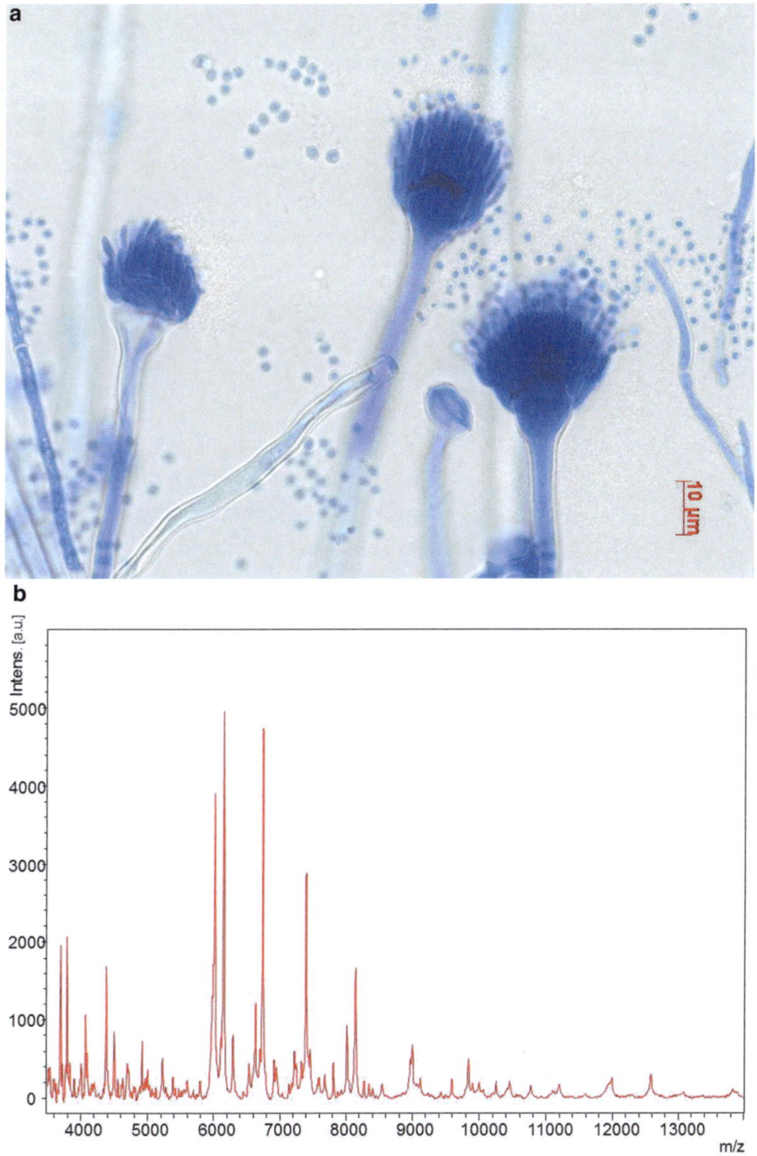

Fig. 3 Identification of the filamentous fungus *Aspergillus terreus*: (*top*) by microscopy using the characteristic size and shape of cells and micromorphological structures (e.g. from conidia (spores), which are formed on long cells sitting on round vesicles); (*bottom*) by MALDI-TOF MS profile analysis, allowing identification 2 days before characteristic morphological structures are observed by microscopy

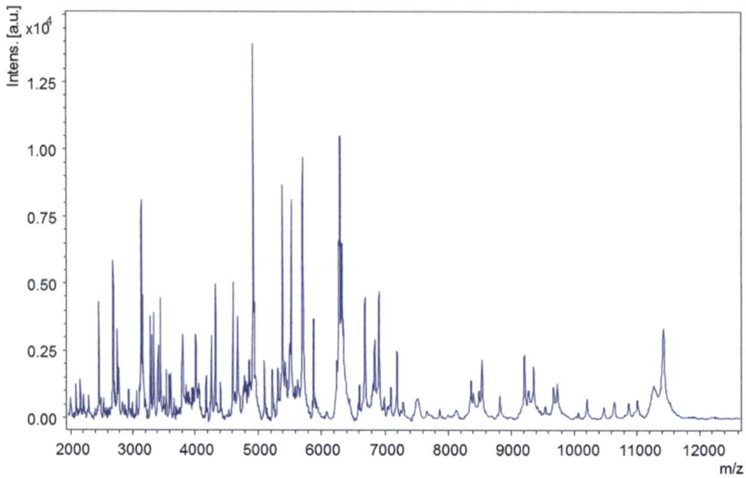

Fig. 4 Profile mass spectrum of the mycobacterium *Mycobacterium fortuitum* after bead-supported extraction of proteins. The microorganism was cultivated on solid Löwenstein–Jensen medium

regarded as subspecies *M. abscessus*, can be differentiated (Tseng et al. 2013; Fangous et al. 2014).

A further field of improvement is the identification of mycobacteria directly from liquid media which have been directly inoculated with patient sample material (generally sputum or aspirates from the patient's respiratory tracts). This might be the most significant contribution of MALDI-TOF MS towards mycobacteria identification as it will further accelerate microbiological routine analysis. However, problems with low cell count in the samples and interference from the patient's sample matrix may have to be overcome to achieve this.

A MALDI mass spectrum collected based on the microbead preparation method published by Buchan and co-workers (Buchan et al. 2014) is shown in Fig. 4.

3.8 Interpretation of Identification Results

Identification through the dedicated software of the different manufacturers is often based on proprietary algorithms. These algorithms differ significantly from each other. The calculation and interpretation of values for identification purposes is defined by the corresponding manufacturer.

With the MALDI Biotyper system a log(score) value is calculated to indicate reliability of the identification result. For the match of a spectrum with each database reference entry (MSP), a score is initially calculated by multiplication of three independent values. These values are (a) the proportion of peaks from the newly acquired spectrum, which match with peaks from the reference peak lists of the database entries (MSPs), (b) the proportion of the MSP peaks, which match with peaks from the unknown spectrum, and (c) the overall correlation of peak height of both the unknown spectrum and the MSPs. The result is a list of scores between 0 and 1000. The most similar MSP has the highest score value. By taking the logarithm the respective log(scores) are calculated and then ranked by value. The highest log(score) value (maximum of 3) is used for identification, while the following matches can be used for a reliability/quality check.

Initially, the manufacturer defined thresholds for identifications at the genus ($\geq$1.7) and species ($\geq$2.0) level. These were also the acceptance criteria applied in the respective IVD-CE study by the manufacturer in 2008/2009. Subsequent studies of users have shown that the thresholds might be too stringent and sensitivity (proportion of identifications) of the system can be increased without significantly losing specificity, i.e. inducing a higher rate of misidentifications (Theel et al. 2011; Alatoom et al. 2012; Fedorko et al. 2012). Szabados and co-workers even proposed the utilization of species-specific thresholds (Szabados et al. 2012). As the database has evolved and expanded over the years, the manufacturer removed the genus-specific limit and redefined the log(score) values of 1.7 and 2.0 as thresholds for low-confidence and high-confidence identification at the species level, respectively. Similarly, the threshold for identification of microorganisms directly isolated from positive blood cultures was adjusted.

In diagnostics, it is sometimes advisable to know which score level can be reached for different species under routine conditions. This can help with the interpretation of the results. For example, enterobacteria usually yield high log(score) levels up to 2.3 or more. In cases where identifications show lower log(score) levels than usual the identification result should be reviewed as the lower score can be a sign of impure or mixed cultures or other problems in spectra acquisition that can lead to a lower discriminatory power. Statistics of scores from prior identifications for particular species can be utilized to find additional species-specific criteria for acceptance of the identification results.

3.9 Databases

Acquired spectra are typically compared to a database of reference spectra (cf. Sect. 3.8). Current database versions have grown over time to several thousand references and are suitable for routine use in medical microbiology.

Nevertheless, due to the broad range of microorganisms that can occur in routine clinical microbiology analysis there are cases when good-quality spectra show no matches against the database. The software then shows implausible results with scores indicating low reliability of identification. Contrary to conventional biochemical identification systems, the software of MALDI-TOF MS-based systems in principle allows the extension of the database. As the databases of different manufacturers are generated differently, based on different mathematical algorithms and identification principles, there are also differences between the systems regarding their possibility to extend the database with own reference spectra.

The Vitek MS system (bioMérieux) is a pure IVD system, which allows no expansion of the database by the user. This is not only due to the restrictions imposed by the principles of an *in vitro* diagnostic system but also to a limitation of the algorithm used for the generation of the database. In short, the advanced spectra classifier algorithm uses a network of biomarkers to characterize a reference spectrum of a given species. For this, a spectrum is divided into thousands of bins. For each bin, the value for identification of the species as well as its value for identification of all other species in the database is calculated. The result is not a database of reference spectra but a kind of a 'biomarker network'. This network has to be calculated anew for each database extension which is only done by the manufacturer for database updates.

However, the VITEK MS plus system has a second, RUO, database which is based on SARAMIS (Spectral ARchive And Microbial Identification System) formerly developed by the company ANAGNOSTEC in Germany. SARAMIS stores raw peak lists from measured strains as well as the so-called 'Super Spectra' in its database. 'Super Spectra' are the main tool for identification. They consist of curated peak lists from multiple measurements with a manually introduced weighting of characteristic peaks for genus and species. The user can also introduce own 'Super Spectra' to extend the database. This is performed based on own measurements and a manual weighting process of the characteristic peaks.

The MALDI Biotyper database is available in two IVD versions, which do not support any extensions by the user. The one called MBT-CA (MALDI Biotyper—Clinical Application) is dedicated to the US diagnostic market, which is regulated by the FDA (US Food and Drug Administration). The other system is the IVD-CE MALDI Biotyper, which is intended for the European market and other countries where European certifications are applied. In addition, a research-use-only system (MALDI Biotyper RUO) is available. It uses the same algorithms as both diagnostic systems and contains the same general database as the IVD-CE MALDI Biotyper. The MALDI Biotyper RUO database can be extended by the user with own reference spectra (MSP).

3.9.1 Database Extension and Creation of Own MSPs for the MALDI Biotyper System

The creation of own MSPs in the MALDI Biotyper software is quite simple and compatible with the usual workflow in microorganism identification. For this, a spectrum with the best possible quality should be acquired. Thus, the optimum protein extraction method should be applied for the given microorganism. In general, the EFEx method is used for bacteria. For compensation of a potentially high measurement variance, the spectrum is acquired multiple times. Therefore, the extract is transferred to eight sample spots on the target. These eight spots are measured at least three times resulting in 24 spectra of the pure microorganism. An accurate calibration of the MALDI-TOF MS instrument before measurement is mandatory.

After alignment of the 24 microorganism spectra in the flexAnalysis spectra evaluation software (Bruker Daltonik GmbH), they can be reviewed visually and selected in a manual process. Spectra that show conspicuous deviations from the other spectra can then be excluded. In general, it is advisable to use about 20 spectra for the calculation and creation of the MSP in the MALDI Biotyper software for best database entry creation. This ensures an adequate weighting of frequent and less stable peaks. The resulting MSP should be checked by running an identification of itself against the existing database entries. If there are no implausible results and the new strain can be clearly differentiated from other MSPs, it can be used in routine identification. In case of database updates by the manufacturer, own database entries should be reviewed and checked for conflicts with new entries.

References

Alatoom AA, Cazanave CJ, Cunningham SA, Ihde SM, Patel R (2012) Identification of non-diphtheriae corynebacterium by use of matrix-assisted laser desorption ionization-time of flight mass spectrometry. J Clin Microbiol 50:160–163. doi:10.1128/JCM.05889-11

Alispahic M, Hummel K, Jandreski-Cvetkovic D, Nöbauer K, Razzazi-Fazeli E, Hess M, Hess C (2010) Species-specific identification and differentiation of Arcobacter, Helicobacter and Campylobacter by full-spectral matrix-associated laser desorption/ionization time of flight mass spectrometry analysis. J Med Microbiol 59:295–301. doi:10.1099/jmm.0.016576-0

Alispahic M, Christensen H, Hess C, Razzazi-Fazeli E, Bisgaard M, Hess M (2011) Identification of Gallibacterium species by matrix-assisted laser desorption/ionization time-of-flight mass spectrometry evaluated by multilocus sequence analysis. Int J Med Microbiol 301:513–522. doi:10.1016/j.ijmm.2011.03.001

Amiri-Eliasi B, Fenselau C (2001) Characterization of protein biomarkers desorbed by MALDI from whole fungal cells. Anal Chem 73:5228–5231

Anhalt JP, Fenselau C (1975) Identification of bacteria using mass spectrometry. Anal Chem 47:219–225. doi:10.1021/ac60352a007

Arnold RJ, Reilly JP (1998) Fingerprint matching of E. coli strains with matrix-assisted laser desorption/ionization time-of-flight mass spectrometry of whole cells using a modified correlation approach. Rapid Commun Mass Spectrom 12:630–636. doi:10.1002/(SICI)1097-0231(19980529)12:10<630::AID-RCM206>3.0.CO;2-0

Ayyadurai S, Flaudrops C, Raoult D, Drancourt M (2010) Rapid identification and typing of Yersinia pestis and other Yersinia species by matrix-assisted laser desorption/ionization time-of-flight (MALDI-TOF) mass spectrometry. BMC Microbiol 10:285. doi:10.1186/1471-2180-10-285

Bader O, Weig M, Taverne-Ghadwal L, Lugert R, Gross U, Kuhns M (2011) Improved clinical laboratory identification of human pathogenic yeasts by matrix-assisted laser desorption ionization time-of-flight mass spectrometry. Clin Microbiol Infect 17:1359–1365. doi:10.1111/j.1469-0691.2010.03398.x

Balada-Llasat JM, Kamboj K, Pancholi P (2013) Identification of mycobacteria from solid and liquid media by matrix-assisted laser desorption ionization-time of flight mass spectrometry in the clinical laboratory. J Clin Microbiol 51:2875–2879. doi:10.1128/JCM.00819-13

Barbuddhe SB, Maier T, Schwarz G, Kostrzewa M, Hof H, Domann E, Chakraborty T, Hain T (2008) Rapid identification and typing of Listeria species by matrix-assisted laser desorption ionization-time of flight mass spectrometry. Appl Environ Microbiol 74:5402–5407. doi:10.1128/AEM.02689-07

Barreiro JR, Ferreira CR, Sanvido GB, Kostrzewa M, Maier T, Wegemann B, Böttcher V, Eberlin MN, dos Santos MV (2010) Short communication: Identification of subclinical cow mastitis pathogens in milk by matrix-assisted laser desorption/ionization time-of-flight mass spectrometry. J Dairy Sci 93:5661–5667. doi:10.3168/jds.2010-3614

Bernardo K, Pakulat N, Macht M, Krut O, Seifert H, Fleer S, Hünger F, Krönke M (2002) Identification and discrimination of Staphylococcus aureus strains using matrix-assisted laser desorption/ionization-time of flight mass spectrometry. Proteomics 2:747–753. doi:10.1002/1615-9861(200206)2:6<747::AID-PROT747>3.0.CO;2-V

Bessède E, Solecki O, Sifré E, Labadi L, Mégraud F (2011) Identification of Campylobacter species and related organisms by matrix assisted laser desorption ionization-time of flight (MALDI-TOF) mass spectrometry. Clin Microbiol Infect 17:1735–1739. doi:10.1111/j.1469-0691.2011.03468.x

Bille E, Dauphin B, Leto J, Bougnoux M-E, Beretti J-L, Lotz A, Suarez S, Meyer J, Join-Lambert O, Descamps P, Grall N, Mory F, Dubreuil L, Berche P, Nassif X, Ferroni A (2012) MALDI-TOF MS Andromas strategy for the routine identification of bacteria, mycobacteria, yeasts, Aspergillus spp. and positive blood cultures. Clin Microbiol Infect 18:1117–1125. doi:10.1111/j.1469-0691.2011.03688.x

Bizzini A, Durussel C, Bille J, Greub G, Prod'hom G (2010) Performance of matrix-assisted laser desorption ionization-time of flight mass spectrometry for identification of bacterial strains routinely isolated in a clinical microbiology laboratory. J Clin Microbiol 48:1549–1554. doi:10.1128/JCM.01794-09

Bizzini A, Jaton K, Romo D, Bille J, Prod'hom G, Greub G (2011) Matrix-assisted laser desorption ionization-time of flight mass spectrometry as an alternative to 16S rRNA gene sequencing for identification of difficult-to-identify bacterial strains. J Clin Microbiol 49:693–696. doi:10.1128/JCM.01463-10

Boehme K, Fernández-No IC, Pazos M, Gallardo JM, Barros-Velázquez J, Cañas B, Calo-Mata P (2013) Identification and classification of seafood-borne pathogenic and spoilage bacteria: 16S rRNA sequencing versus MALDI-TOF MS fingerprinting. Electrophoresis 34:877–887. doi:10.1002/elps.201200532

Bruin JP, Kostrzewa M, van der Ende A, Badoux P, Jansen R, Boers SA, Diederen BMW (2014) Identification of Haemophilus influenzae and Haemophilus haemolyticus by matrix-assisted laser desorption ionization-time of flight mass spectrometry. Eur J Clin Microbiol Infect Dis 33:279–284. doi:10.1007/s10096-013-1958-x

Buchan BW, Riebe KM, Ledeboer NA (2012) Comparison of the MALDI Biotyper system using Sepsityper specimen processing to routine microbiological methods for identification of bacteria from positive blood culture bottles. J Clin Microbiol 50:346–352. doi:10.1128/JCM.05021-11

Buchan BW, Riebe KM, Timke M, Kostrzewa M, Ledeboer NA (2014) Comparison of MALDI-TOF MS with HPLC and nucleic acid sequencing for the identification of Mycobacterium species in cultures using solid medium and broth. Am J Clin Pathol 141:25–34. doi:10.1309/AJCPBPUBUDEW2OAG

Cassagne C, Ranque S, Normand A-C, Fourquet P, Thiebault S, Planard C, Hendrickx M, Piarroux R (2011) Mould routine identification in the clinical laboratory by matrix-assisted laser desorption ionization time-of-flight mass spectrometry. PLoS One 6, e28425. doi:10.1371/journal.pone.0028425

Cendejas-Bueno E, Kolecka A, Alastruey-Izquierdo A, Theelen B, Groenewald M, Kostrzewa M, Cuenca-Estrella M, Gómez-López A, Boekhout T (2012) Reclassification of the Candida haemulonii complex as Candida haemulonii (C. haemulonii group I), C. duobushaemulonii sp.

nov. (C. haemulonii group II), and C. haemulonii var. vulnera var. nov.: three multiresistant human pathogenic yeasts. J Clin Microbiol 50:3641–3651. doi:10.1128/JCM.02248-12

Christensen JJ, Dargis R, Hammer M, Justesen US, Nielsen XC, Kemp M, Danish MALDI-TOF MS Study Group (2012) Matrix-assisted laser desorption ionization-time of flight mass spectrometry analysis of Gram-positive, catalase-negative cocci not belonging to the Streptococcus or Enterococcus genus and benefits of database extension. J Clin Microbiol 50:1787–1791. doi:10.1128/JCM.06339-11

Christner M, Rohde H, Wolters M, Sobottka I, Wegscheider K, Aepfelbacher M (2010) Rapid identification of bacteria from positive blood culture bottles by use of matrix-assisted laser desorption-ionization time of flight mass spectrometry fingerprinting. J Clin Microbiol 48:1584–1591. doi:10.1128/JCM.01831-09

Clark AE, Kaleta EJ, Arora A, Wolk DM (2013) Matrix-assisted laser desorption ionization-time of flight mass spectrometry: a fundamental shift in the routine practice of clinical microbiology. Clin Microbiol Rev 26:547–603. doi:10.1128/CMR.00072-12

Claydon MA, Davey SN, Edwards-Jones V, Gordon DB (1996) The rapid identification of intact microorganisms using mass spectrometry. Nat Biotechnol 14:1584–1586. doi:10.1038/nbt1196-1584

Conway GC, Smole SC, Sarracino DA, Arbeit RD, Leopold PE (2001) Phyloproteomics: species identification of Enterobacteriaceae using matrix-assisted laser desorption/ionization time-of-flight mass spectrometry. J Mol Microbiol Biotechnol 3:103–112

Demirev PA, Ho YP, Ryzhov V, Fenselau C (1999) Microorganism identification by mass spectrometry and protein database searches. Anal Chem 71:2732–2738

Dhiman N, Hall L, Wohlfiel SL, Buckwalter SP, Wengenack NL (2011) Performance and cost analysis of matrix-assisted laser desorption ionization-time of flight mass spectrometry for routine identification of yeast. J Clin Microbiol 49:1614–1616. doi:10.1128/JCM.02381-10

Dieckmann R, Helmuth R, Erhard M, Malorny B (2008) Rapid classification and identification of salmonellae at the species and subspecies levels by whole-cell matrix-assisted laser desorption ionization-time of flight mass spectrometry. Appl Environ Microbiol 74:7767–7778. doi:10.1128/AEM.01402-08

Eigner U, Holfelder M, Oberdorfer K, Betz-Wild U, Bertsch D, Fahr A-M (2009) Performance of a matrix-assisted laser desorption ionization-time-of-flight mass spectrometry system for the identification of bacterial isolates in the clinical routine laboratory. Clin Lab 55:289–296

El Khéchine A, Couderc C, Flaudrops C, Raoult D, Drancourt M (2011) Matrix-assisted laser desorption/ionization time-of-flight mass spectrometry identification of mycobacteria in routine clinical practice. PLoS One 6, e24720. doi:10.1371/journal.pone.0024720

Fangous M-S, Mougari F, Gouriou S, Calvez E, Raskine L, Cambau E, Payan C, Héry-Arnaud G (2014) Classification algorithm for subspecies identification within the Mycobacterium abscessus species, based on matrix-assisted laser desorption ionization-time of flight mass spectrometry. J Clin Microbiol 52:3362–3369. doi:10.1128/JCM.00788-14

Farfour E, Leto J, Barritault M, Barberis C, Meyer J, Dauphin B, Le Guern A-S, Leflèche A, Badell E, Guiso N, Leclercq A, Le Monnier A, Lecuit M, Rodriguez-Nava V, Bergeron E, Raymond J, Vimont S, Bille E, Carbonnelle E, Guet-Revillet H, Lécuyer H, Beretti J-L, Vay C, Berche P, Ferroni A, Nassif X, Join-Lambert O (2012) Evaluation of the Andromas matrix-assisted laser desorption ionization-time of flight mass spectrometry system for identification of aerobically growing Gram-positive bacilli. J Clin Microbiol 50:2702–2707. doi:10.1128/JCM.00368-12

Fedorko DP, Drake SK, Stock F, Murray PR (2012) Identification of clinical isolates of anaerobic bacteria using matrix-assisted laser desorption ionization-time of flight mass spectrometry. Eur J Clin Microbiol Infect Dis 31:2257–2262. doi:10.1007/s10096-012-1563-4

Ferreira L, Sánchez-Juanes F, Porras-Guerra I, García-García MI, García-Sánchez JE, González-Buitrago JM, Muñoz-Bellido JL (2011) Microorganisms direct identification from blood culture by matrix-assisted laser desorption/ionization time-of-flight mass spectrometry. Clin Microbiol Infect 17:546–551. doi:10.1111/j.1469-0691.2010.03257.x

Ferroni A, Suarez S, Beretti J-L, Dauphin B, Bille E, Meyer J, Bougnoux M-E, Alanio A, Berche P, Nassif X (2010) Real-time identification of bacteria and Candida species in positive blood culture broths by matrix-assisted laser desorption ionization-time of flight mass spectrometry. J Clin Microbiol 48:1542–1548. doi:10.1128/JCM.02485-09

Ford BA, Burnham C-AD (2013) Optimization of routine identification of clinically relevant Gram-negative bacteria by use of matrix-assisted laser desorption ionization-time of flight mass spectrometry and the Bruker Biotyper. J Clin Microbiol 51:1412–1420. doi:10.1128/JCM.01803-12

Frey J, Kuhnert P (2015) Identification of animal Pasteurellaceae by MALDI-TOF mass spectrometry. Methods Mol Biol 1247:235–243. doi:10.1007/978-1-4939-2004-4_18

Gantt SL, Valentine NB, Saenz AJ, Kingsley MT, Wahl KL (1999) Use of an internal control for matrix-assisted laser desorption/ionization time-of-flight mass spectrometry analysis of bacteria. J Am Soc Mass Spectrom 10:1131–1137. doi:10.1016/S1044-0305(99)00086-0

Grosse-Herrenthey A, Maier T, Gessler F, Schaumann R, Böhnel H, Kostrzewa M, Krüger M (2008) Challenging the problem of clostridial identification with matrix-assisted laser desorption and ionization-time-of-flight mass spectrometry (MALDI-TOF MS). Anaerobe 14:242–249. doi:10.1016/j.anaerobe.2008.06.002

Haag AM, Taylor SN, Johnston KH, Cole RB (1998) Rapid identification and speciation of Haemophilus bacteria by matrix-assisted laser desorption/ionization time-of-flight mass spectrometry. J Mass Spectrom 33:750–756. doi:10.1002/(SICI)1096-9888(199808)33:8<750::AID-JMS680>3.0.CO;2-1

He Y, Li H, Lu X, Stratton CW, Tang Y-W (2010) Mass spectrometry biotyper system identifies enteric bacterial pathogens directly from colonies grown on selective stool culture media. J Clin Microbiol 48:3888–3892. doi:10.1128/JCM.01290-10

Heller DN, Cotter RJ, Fenselau C, Uy OM (1987) Profiling of bacteria by fast atom bombardment mass spectrometry. Anal Chem 59:2806–2809

Hettick JM, Kashon ML, Simpson JP, Siegel PD, Mazurek GH, Weissman DN (2004) Proteomic profiling of intact mycobacteria by matrix-assisted laser desorption/ionization time-of-flight mass spectrometry. Anal Chem 76:5769–5776. doi:10.1021/ac049410m

Holland RD, Wilkes JG, Rafii F, Sutherland JB, Persons CC, Voorhees KJ, Lay JO (1996) Rapid identification of intact whole bacteria based on spectral patterns using matrix-assisted laser desorption/ionization with time-of-flight mass spectrometry. Rapid Commun Mass Spectrom 10:1227–1232. doi:10.1002/(SICI)1097-0231(19960731)10:10<1227::AID-RCM659>3.0.CO;2-6

Hsueh P-R, Lee T-F, Du S-H, Teng S-H, Liao C-H, Sheng W-H, Teng L-J (2014) Bruker biotyper matrix-assisted laser desorption ionization-time of flight mass spectrometry system for identification of Nocardia, Rhodococcus, Kocuria, Gordonia, Tsukamurella, and Listeria species. J Clin Microbiol 52:2371–2379. doi:10.1128/JCM.00456-14

Iriart X, Lavergne R-A, Fillaux J, Valentin A, Magnaval J-F, Berry A, Cassaing S (2012) Routine identification of medical fungi by the new Vitek MS matrix-assisted laser desorption ionization-time of flight system with a new time-effective strategy. J Clin Microbiol 50:2107–2110. doi:10.1128/JCM.06713-11

Jadhav S, Sevior D, Bhave M, Palombo EA (2014) Detection of Listeria monocytogenes from selective enrichment broth using MALDI-TOF Mass Spectrometry. J Proteomics 97:100–106. doi:10.1016/j.jprot.2013.09.014

Jarman KH, Daly DS, Petersen CE, Saenz AJ, Valentine NB, Wahl KL (1999) Extracting and visualizing matrix-assisted laser desorption/ionization time-of-flight mass spectral fingerprints. Rapid Commun Mass Spectrom 13:1586–1594. doi:10.1002/(SICI)1097-0231(19990815)13:15<1586::AID-RCM680>3.0.CO;2-2

Jarman KH, Cebula ST, Saenz AJ, Petersen CE, Valentine NB, Kingsley MT, Wahl KL (2000) An algorithm for automated bacterial identification using matrix-assisted laser desorption/ionization mass spectrometry. Anal Chem 72:1217–1223

Juiz PM, Almela M, Melción C, Campo I, Esteban C, Pitart C, Marco F, Vila J (2012) A comparative study of two different methods of sample preparation for positive blood cultures for the rapid identification of bacteria using MALDI-TOF MS. Eur J Clin Microbiol Infect Dis 31:1353–1358. doi:10.1007/s10096-011-1449-x

Khot PD, Couturier MR, Wilson A, Croft A, Fisher MA (2012) Optimization of matrix-assisted laser desorption ionization-time of flight mass spectrometry analysis for bacterial identification. J Clin Microbiol 50:3845–3852. doi:10.1128/JCM.00626-12

Klein S, Zimmermann S, Köhler C, Mischnik A, Alle W, Bode KA (2012) Integration of matrix-assisted laser desorption/ionization time-of-flight mass spectrometry in blood culture diagnostics: a fast and effective approach. J Med Microbiol 61:323–331. doi:10.1099/jmm.0.035550-0

Kok J, Thomas LC, Olma T, Chen SCA, Iredell JR (2011) Identification of bacteria in blood culture broths using matrix-assisted laser desorption-ionization Sepsityper™ and time of flight mass spectrometry. PLoS One 6, e23285. doi:10.1371/journal.pone.0023285

Krader P, Emerson D (2004) Identification of archaea and some extremophilic bacteria using matrix-assisted laser desorption/ionization time-of-flight (MALDI-TOF) mass spectrometry. Extremophiles 8:259–268. doi:10.1007/s00792-004-0382-7

Krishnamurthy T, Ross PL (1996) Rapid identification of bacteria by direct matrix-assisted laser desorption/ionization mass spectrometric analysis of whole cells. Rapid Commun Mass Spectrom 10:1992–1996. doi:10.1002/(SICI)1097-0231(199612)10:15<1992::AID-RCM789>3.0.CO;2-V

Kuhns M, Zautner AE, Rabsch W, Zimmermann O, Weig M, Bader O, Groß U (2012) Rapid discrimination of Salmonella enterica serovar Typhi from other serovars by MALDI-TOF mass spectrometry. PLoS One 7, e40004. doi:10.1371/journal.pone.0040004

La Scola B, Raoult D (2009) Direct identification of bacteria in positive blood culture bottles by matrix-assisted laser desorption ionisation time-of-flight mass spectrometry. PLoS One 4, e8041. doi:10.1371/journal.pone.0008041

Lasch P, Drevinek M, Nattermann H, Grunow R, Stämmler M, Dieckmann R, Schwecke T, Naumann D (2010) Characterization of Yersinia using MALDI-TOF mass spectrometry and chemometrics. Anal Chem 82:8464–8475. doi:10.1021/ac101036s

Lau AF, Drake SK, Calhoun LB, Henderson CM, Zelazny AM (2013) Development of a clinically comprehensive database and a simple procedure for identification of molds from solid media by matrix-assisted laser desorption ionization-time of flight mass spectrometry. J Clin Microbiol 51:828–834. doi:10.1128/JCM.02852-12

Loonen AJM, Jansz AR, Stalpers J, Wolffs PFG, van den Brule AJC (2012) An evaluation of three processing methods and the effect of reduced culture times for faster direct identification of pathogens from BacT/ALERT blood cultures by MALDI-TOF MS. Eur J Clin Microbiol Infect Dis 31:1575–1583. doi:10.1007/s10096-011-1480-y

Lotz A, Ferroni A, Beretti J-L, Dauphin B, Carbonnelle E, Guet-Revillet H, Veziris N, Heym B, Jarlier V, Gaillard J-L, Pierre-Audigier C, Frapy E, Berche P, Nassif X, Bille E (2010) Rapid identification of mycobacterial whole cells in solid and liquid culture media by matrix-assisted laser desorption ionization-time of flight mass spectrometry. J Clin Microbiol 48:4481–4486. doi:10.1128/JCM.01397-10

Mandrell RE, Harden LA, Bates A, Miller WG, Haddon WF, Fagerquist CK (2005) Speciation of Campylobacter coli, C. jejuni, C. helveticus, C. lari, C. sputorum, and C. upsaliensis by matrix-assisted laser desorption ionization-time of flight mass spectrometry. Appl Environ Microbiol 71:6292–6307. doi:10.1128/AEM.71.10.6292-6307.2005

Marklein G, Josten M, Klanke U, Müller E, Horré R, Maier T, Wenzel T, Kostrzewa M, Bierbaum G, Hoerauf A, Sahl H-G (2009) Matrix-assisted laser desorption ionization-time of flight mass spectrometry for fast and reliable identification of clinical yeast isolates. J Clin Microbiol 47:2912–2917. doi:10.1128/JCM.00389-09

Martiny D, Dediste A, Debruyne L, Vlaes L, Haddou NB, Vandamme P, Vandenberg O (2011) Accuracy of the API Campy system, the Vitek 2 Neisseria-Haemophilus card and matrix-assisted laser desorption ionization time-of-flight mass spectrometry for the identification of Campylobacter and related organisms. Clin Microbiol Infect 17:1001–1006. doi:10.1111/j.1469-0691.2010.03328.x

Martiny D, Busson L, Wybo I, El Haj RA, Dediste A, Vandenberg O (2012) Comparison of the Microflex LT and Vitek MS systems for routine identification of bacteria by matrix-assisted laser desorption ionization-time of flight mass spectrometry. J Clin Microbiol 50:1313–1325. doi:10.1128/JCM.05971-11

McElvania Tekippe E, Shuey S, Winkler DW, Butler MA, Burnham C-AD (2013) Optimizing identification of clinically relevant Gram-positive organisms by use of the Bruker Biotyper matrix-assisted laser desorption ionization-time of flight mass spectrometry system. J Clin Microbiol 51:1421–1427. doi:10.1128/JCM.02680-12

Mellmann A, Cloud J, Maier T, Keckevoet U, Ramminger I, Iwen P, Dunn J, Hall G, Wilson D, Lasala P, Kostrzewa M, Harmsen D (2008) Evaluation of matrix-assisted laser desorption ionization-time-of-flight mass spectrometry in comparison to 16S rRNA gene sequencing for species identification of nonfermenting bacteria. J Clin Microbiol 46:1946–1954. doi:10.1128/JCM.00157-08

Mellmann A, Bimet F, Bizet C, Borovskaya AD, Drake RR, Eigner U, Fahr AM, He Y, Ilina EN, Kostrzewa M, Maier T, Mancinelli L, Moussaoui W, Prévost G, Putignani L, Seachord CL, Tang YW, Harmsen D (2009) High interlaboratory reproducibility of matrix-assisted laser desorption ionization-time of flight mass spectrometry-based species identification of nonfermenting bacteria. J Clin Microbiol 47:3732–3734. doi:10.1128/JCM.00921-09

Moussaoui W, Jaulhac B, Hoffmann A-M, Ludes B, Kostrzewa M, Riegel P, Prévost G (2010) Matrix-assisted laser desorption ionization time-of-flight mass spectrometry identifies 90% of bacteria directly from blood culture vials. Clin Microbiol Infect 16:1631–1638. doi:10.1111/j.1469-0691.2010.03356.x

Murugaiyan J, Ahrholdt J, Kowbel V, Roesler U (2012) Establishment of a matrix-assisted laser desorption ionization time-of-flight mass spectrometry database for rapid identification of infectious achlorophyllous green micro-algae of the genus Prototheca. Clin Microbiol Infect 18:461–467. doi:10.1111/j.1469-0691.2011.03593.x

Murugaiyan J, Walther B, Stamm I, Abou-Elnaga Y, Brueggemann-Schwarze S, Vincze S, Wieler LH, Lübke-Becker A, Semmler T, Roesler U (2014) Species differentiation within the Staphylococcus intermedius group using a refined MALDI-TOF MS database. Clin Microbiol Infect 20:1007–1015. doi:10.1111/1469-0691.12662

Nagy E, Maier T, Urban E, Terhes G, Kostrzewa M, ESCMID Study Group on Antimicrobial Resistance in Anaerobic Bacteria (2009) Species identification of clinical isolates of Bacteroides by matrix-assisted laser-desorption/ionization time-of-flight mass spectrometry. Clin Microbiol Infect 15:796–802. doi:10.1111/j.1469-0691.2009.02788.x

Nilsson CL (1999) Fingerprinting of Helicobacter pylori strains by matrix-assisted laser desorption/ionization mass spectrometric analysis. Rapid Commun Mass Spectrom 13:1067–1071. doi:10.1002/(SICI)1097-0231(19990615)13:11<1067::AID-RCM612>3.0.CO;2-N

Pineda FJ, Lin JS, Fenselau C, Demirev PA (2000) Testing the significance of microorganism identification by mass spectrometry and proteome database search. Anal Chem 72:3739–3744

Prod'hom G, Bizzini A, Durussel C, Bille J, Greub G (2010) Matrix-assisted laser desorption ionization-time of flight mass spectrometry for direct bacterial identification from positive blood culture pellets. J Clin Microbiol 48:1481–1483. doi:10.1128/JCM.01780-09

Quiles-Melero I, García-Rodríguez J, Gómez-López A, Mingorance J (2012) Evaluation of matrix-assisted laser desorption/ionisation time-of-flight (MALDI-TOF) mass spectrometry for identification of Candida parapsilosis, C. orthopsilosis and C. metapsilosis. Eur J Clin Microbiol Infect Dis 31:67–71. doi:10.1007/s10096-011-1277-z

Ranque S, Normand A-C, Cassagne C, Murat J-B, Bourgeois N, Dalle F, Gari-Toussaint M, Fourquet P, Hendrickx M, Piarroux R (2014) MALDI-TOF mass spectrometry identification of filamentous fungi in the clinical laboratory. Mycoses 57:135–140. doi:10.1111/myc.12115

Ryzhov V, Fenselau C (2001) Characterization of the protein subset desorbed by MALDI from whole bacterial cells. Anal Chem 73:746–750

Saenz AJ, Petersen CE, Valentine NB, Gantt SL, Jarman KH, Kingsley MT, Wahl KL (1999) Reproducibility of matrix-assisted laser desorption/ionization time-of-flight mass spectrometry for replicate bacterial culture analysis. Rapid Commun Mass Spectrom 13:1580–1585. doi:10.1002/(SICI)1097-0231(19990815)13:15<1580::AID-RCM679>3.0.CO;2-V

Saleeb PG, Drake SK, Murray PR, Zelazny AM (2011) Identification of mycobacteria in solid-culture media by matrix-assisted laser desorption ionization-time of flight mass spectrometry. J Clin Microbiol 49:1790–1794. doi:10.1128/JCM.02135-10

Schaumann R, Knoop N, Genzel GH, Losensky K, Rosenkranz C, Stîngu CS, Schellenberger W, Rodloff AC, Eschrich K (2013) Discrimination of Enterobacteriaceae and non-fermenting gram negative bacilli by MALDI-TOF Mass Spectrometry. Open Microbiol J 7:118–122. doi:10.2174/1874285801307010118

Schieffer KM, Tan KE, Stamper PD, Somogyi A, Andrea SB, Wakefield T, Romagnoli M, Chapin KC, Wolk DM, Carroll KC (2014) Multicenter evaluation of the Sepsityper™ extraction kit and MALDI-TOF MS for direct identification of positive blood culture isolates using the BD BACTEC™ FX and VersaTREK(®) diagnostic blood culture systems. J Appl Microbiol 116:934–941. doi:10.1111/jam.12434

Schubert S, Weinert K, Wagner C, Gunzl B, Wieser A, Maier T, Kostrzewa M (2011) Novel, improved sample preparation for rapid, direct identification from positive blood cultures using matrix-assisted laser desorption/ionization time-of-flight (MALDI-TOF) mass spectrometry. J Mol Diagn 13:701–706. doi:10.1016/j.jmoldx.2011.07.004

Schulthess B, Ledermann R, Mouttet F, Zbinden A, Bloemberg GV, Böttger EC, Hombach M (2014) Use of the Bruker MALDI Biotyper for identification of molds in the clinical mycology laboratory. J Clin Microbiol 52:2797–2803. doi:10.1128/JCM.00049-14

Seng P, Drancourt M, Gouriet F, La Scola B, Fournier P-E, Rolain JM, Raoult D (2009) Ongoing revolution in bacteriology: routine identification of bacteria by matrix-assisted laser desorption ionization time-of-flight mass spectrometry. Clin Infect Dis 49:543–551. doi:10.1086/600885

Sinha MP, Platz RM, Friedlander SK, Vilker VL (1985) Characterization of bacteria by particle beam mass spectrometry. Appl Environ Microbiol 49:1366–1373

Smole SC, King LA, Leopold PE, Arbeit RD (2002) Sample preparation of Gram-positive bacteria for identification by matrix assisted laser desorption/ionization time-of-flight. J Microbiol Methods 48:107–115

Sparbier K, Weller U, Boogen C, Kostrzewa M (2012) Rapid detection of Salmonella sp. by means of a combination of selective enrichment broth and MALDI-TOF MS. Eur J Clin Microbiol Infect Dis 31:767–773. doi: 10.1007/s10096-011-1373-0

Stephan R, Cernela N, Ziegler D, Pflüger V, Tonolla M, Ravasi D, Fredriksson-Ahomaa M, Hächler H (2011) Rapid species specific identification and subtyping of Yersinia enterocolitica by MALDI-TOF mass spectrometry. J Microbiol Methods 87:150–153. doi:10.1016/j.mimet.2011.08.016

Stevenson LG, Drake SK, Murray PR (2010) Rapid identification of bacteria in positive blood culture broths by matrix-assisted laser desorption ionization-time of flight mass spectrometry. J Clin Microbiol 48:444–447. doi:10.1128/JCM.01541-09

Szabados F, Tix H, Anders A, Kaase M, Gatermann SG, Geis G (2012) Evaluation of species-specific score cutoff values of routinely isolated clinically relevant bacteria using a direct smear preparation for matrix-assisted laser desorption/ionization time-of-flight mass spectrometry-based bacterial identification. Eur J Clin Microbiol Infect Dis 31:1109–1119. doi:10.1007/s10096-011-1415-7

Taniguchi T, Sekiya A, Higa M, Saeki Y, Umeki K, Okayama A, Hayashi T, Misawa N (2014) Rapid identification and subtyping of Helicobacter cinaedi strains by intact-cell mass spectrometry profiling with the use of matrix-assisted laser desorption ionization-time of flight mass spectrometry. J Clin Microbiol 52:95–102. doi:10.1128/JCM.01798-13

Theel ES, Hall L, Mandrekar J, Wengenack NL (2011) Dermatophyte identification using matrix-assisted laser desorption ionization-time of flight mass spectrometry. J Clin Microbiol 49:4067–4071. doi:10.1128/JCM.01280-11

Theel ES, Schmitt BH, Hall L, Cunningham SA, Walchak RC, Patel R, Wengenack NL (2012) Formic acid-based direct, on-plate testing of yeast and Corynebacterium species by Bruker Biotyper matrix-assisted laser desorption ionization-time of flight mass spectrometry. J Clin Microbiol 50:3093–3095. doi:10.1128/JCM.01045-12

Tseng S-P, Teng S-H, Lee P-S, Wang C-F, Yu J-S, Lu P-L (2013) Rapid identification of M. abscessus and M. massiliense by MALDI-TOF mass spectrometry with a comparison to sequencing methods and antimicrobial susceptibility patterns. Future Microbiol 8:1381–1389. doi:10.2217/fmb.13.115

Vaidyanathan S, Winder CL, Wade SC, Kell DB, Goodacre R (2002) Sample preparation in matrix-assisted laser desorption/ionization mass spectrometry of whole bacterial cells and the detection of high mass (>20 kDa) proteins. Rapid Commun Mass Spectrom 16:1276–1286. doi:10.1002/rcm.713

Van Veen SQ, Claas ECJ, Kuijper EJ (2010) High-throughput identification of bacteria and yeast by matrix-assisted laser desorption ionization-time of flight mass spectrometry in conventional medical microbiology laboratories. J Clin Microbiol 48:900–907. doi:10.1128/JCM.02071-09

Vila J, Juiz P, Salas C, Almela M, de la Fuente CG, Zboromyrska Y, Navas J, Bosch J, Agüero J, de la Bellacasa JP, Martínez-Martínez L (2012) Identification of clinically relevant Corynebacterium spp., Arcanobacterium haemolyticum, and Rhodococcus equi by matrix-assisted laser desorption ionization-time of flight mass spectrometry. J Clin Microbiol 50:1745–1747. doi:10.1128/JCM.05821-11

Von Bergen M, Eidner A, Schmidt F, Murugaiyan J, Wirth H, Binder H, Maier T, Roesler U (2009) Identification of harmless and pathogenic algae of the genus Prototheca by MALDI-MS. Proteomics Clin Appl 3:774–784. doi:10.1002/prca.200780138

Wang Z, Russon L, Li L, Roser DC, Long SR (1998) Investigation of spectral reproducibility in direct analysis of bacteria proteins by matrix-assisted laser desorption/ionization time-of-flight mass spectrometry. Rapid Commun Mass Spectrom 12:456–464. doi:10.1002/(SICI)1097-0231(19980430)12:8<456::AID-RCM177>3.0.CO;2-U

Welham KJ, Domin MA, Scannell DE, Cohen E, Ashton DS (1998) The characterization of micro-organisms by matrix-assisted laser desorption/ionization time-of-flight mass spectrometry. Rapid Commun Mass Spectrom 12:176–180. doi:10.1002/(SICI)1097-0231(19980227)12:4<176::AID-RCM132>3.0.CO;2-T

Werner G, Fleige C, Fessler AT, Timke M, Kostrzewa M, Zischka M, Peters T, Kaspar H, Schwarz S (2012) Improved identification including MALDI-TOF mass spectrometry analysis of group D streptococci from bovine mastitis and subsequent molecular characterization of corresponding Enterococcus faecalis and Enterococcus faecium isolates. Vet Microbiol 160:162–169. doi:10.1016/j.vetmic.2012.05.019

Williams TL, Andrzejewski D, Lay JO, Musser SM (2003) Experimental factors affecting the quality and reproducibility of MALDI TOF mass spectra obtained from whole bacteria cells. J Am Soc Mass Spectrom 14:342–351. doi:10.1016/S1044-0305(03)00065-5

Wragg P, Randall L, Whatmore AM (2014) Comparison of Biolog GEN III MicroStation semi-automated bacterial identification system with matrix-assisted laser desorption ionization-time of flight mass spectrometry and 16S ribosomal RNA gene sequencing for the identification of bacteria of veterinary interest. Journal of Microbiological Methods 105:16–21. doi: 10.1016/j.mimet.2014.07.003

Yan Y, He Y, Maier T, Quinn C, Shi G, Li H, Stratton CW, Kostrzewa M, Tang Y-W (2011) Improved identification of yeast species directly from positive blood culture media by combining Sepsityper specimen processing and Microflex analysis with the matrix-assisted laser desorption ionization Biotyper system. J Clin Microbiol 49:2528–2532. doi:10.1128/JCM.00339-11

Zautner AE, Masanta WO, Tareen AM, Weig M, Lugert R, Groß U, Bader O (2013) Discrimination of multilocus sequence typing-based Campylobacter jejuni subgroups by MALDI-TOF mass spectrometry. BMC Microbiol 13:247. doi:10.1186/1471-2180-13-247

Zhu B, Xiao D, Zhang H, Zhang Y, Gao Y, Xu L, Lv J, Wang Y, Zhang J, Shao Z (2013) MALDI-TOF MS distinctly differentiates nontypable Haemophilus influenzae from Haemophilus haemolyticus. PLoS One 8, e56139. doi:10.1371/journal.pone.0056139

Future Applications of MALDI-TOF MS in Microbiology

Markus Kostrzewa and Arthur B. Pranada

Abstract The appearance of MALDI-TOF MS in microbiological laboratories has rapidly induced new ideas of its potential for further applications to improve microbe characterization. Its introduction into routine analysis is well justified on the basis of accuracy, speed, and cost-effectiveness. Further, laboratory staff without any prior knowledge of mass spectrometry can be easily trained and convinced of the ease of using MALDI-TOF MS. Two applications in microbiology are most promising and shall be highlighted here.

One is the determination of microbial resistance/susceptibility. Traditional methods to detect susceptibility or resistance against antimicrobial drugs are time consuming. Frequently overnight incubation is required, leading to significantly delayed adapted therapy. Alternatively, for some resistance phenomena fast DNA-based methods are used which are often expensive and laborious. Further, there are limitations for molecular tests in cases of a multiplicity of different resistance genes or newly emerging resistance variants. Typical examples are emerging resistance types against β-lactam anti-infectives. Functional assays using MALDI-TOF MS for detecting the cleavage of the β-lactam ring can detect resistance in less than 2 h. Further assays using MALDI-TOF MS are under thorough evaluation. MALDI-TOF MS may also be used for typing microbial strains in the near future. Spectra acquired from bacteria and fungi do not only contain species-specific information but can also be used to differentiate subgroups within a species. In some cases, the identification of resistant or particular aggressive (virulent) strains is possible. In addition, sub-differentiation can be valuable for tracking strains in the course of infection control or epidemiological studies.

M. Kostrzewa (✉)
Bruker Daltonik GmbH, Fahrenheitstr. 4, 28359 Bremen, Germany
e-mail: Markus.Kostrzewa@bruker.com

A.B. Pranada
Department of Medical Microbiology, MVZ Dr. Eberhard & Partner Dortmund (ÜBAG), Balkenstr. 17-19, 44137 Dortmund, Germany
e-mail: apranada@labmed.de

R. Cramer (ed.), *Advances in MALDI and Laser-Induced Soft Ionization Mass Spectrometry*, DOI 10.1007/978-3-319-04819-2_12

1 Introduction

With the introduction of matrix-assisted laser desorption/ionization-time-of-flight (MALDI-TOF) mass spectrometry (MS) for fast microbe identification in the clinical microbiology laboratory the workflow in most labs has changed fundamentally. In particular, the information about the identity of a bacterium or yeast is now available at least about one day earlier than the information about antibiotic susceptibility or resistance, creating a significant time gap between both. While the fast identification is of interest and helpful for therapeutic decision making with physicians becoming quickly used to this new service, the successful application of antibiotic therapy is often still only possible after conventional, time-consuming antimicrobial susceptibility testing (AST). To date, phenotypic susceptibility testing with conventional methods is performed by culturing the microorganism together with the antimicrobial substance for a predefined time. The level of inhibition of growth of the microorganism caused by the antimicrobial substance allows a categorization into clinically 'susceptible' (high likelihood of therapeutic success with this antimicrobial against the microorganism), 'intermediate' (uncertain effect, high dosage level may be needed to be effective) or 'resistant' (therapeutic success is unlikely). Clinical breakpoints for those categories in combination with different antimicrobials and microorganisms are determined by organizations like EUCAST (European Committee on Antimicrobial Susceptibility Testing) or CLSI (Clinical and Laboratory Standards Institute) located in the USA. Incubation time for conventional AST is usually about 18–24 h, although with automated systems incubation time can sometimes be shorter. The demand for accelerating this part of diagnostics is obvious and has led to increased activities in developing adequate technologies.

In the race between technologies to fill this gap—a competition which still has not been concluded—one obvious option is to use MALDI-TOF MS. The necessary instrumentation has already been established in many laboratories and MALDI-TOF MS has been proven to be a fast, robust and cost-effective technology. Very different MALDI-TOF MS approaches have already been undertaken to enter the field of antibiotic resistance testing. Although thorough investigations and further studies still have to prove their performance in the clinical microbiological laboratory, it appears that some of them will make it into routine analysis.

2 Applications

2.1 Genotypic Methods: The Search for Antibiotic Resistance Markers in the MALDI–TOF MS Profile

Already in the very early days of MALDI-TOF MS profiling of microorganisms, a major goal was the differentiation of the two groups of *Staphylococcus aureus* that differ in their susceptibility or resistance against methicillin and related antibiotics,

MSSA and MRSA (**M**ethicillin **S**ensitive **S**taphylococcus **a**ureus and **M**ethicillin **R**esistant **S**taphylococcus **a**ureus). Staphylococci are facultative pathogenic bacteria which can colonize mucosa and skin. Under certain conditions, like small lesions of the skin, they can cause invasive infections. Among the staphylococci the species *S. aureus* is the most significant. With several virulence factors, this species has the potential to be responsible for many diseases. Furthermore, *S. aureus* has shown growing resistance against several important antibiotics over the years (Lowy 2003). While it was initially susceptible to penicillin, which was introduced in the 1940s, resistant isolates producing penicillinase, a penicillin-hydrolyzing enzyme, were already observed by the end of that decade (Barber and Rozwadowska-Dowzenko 1948). A similar history could be seen with the introduction of the penicillinase-stable anti-infective substance methicillin in 1959: 2 years later, in 1961, the first MRSA were found (Jevons 1961). The genetic key element for this was a gene called *mecA* encoding for an altered additional penicillin-binding protein PBP2a that has a low affinity to all β-lactam antibiotics (Enright et al. 2002). This renders the large and widely used group of β-lactam antibiotics, which are often applied in initial and empirical therapy, useless and is one of the reasons why rapid differentiation between MSSA and MRSA today is important for clinical practice.

The idea of using MALDI-TOF MS profiling is to differentiate both groups by characteristics in their protein profiles. Several reports indicated the possibility of this by a simple MS profile comparison (Edwards-Jones et al. 2000; Du et al. 2002) but other authors argued against it (Bernardo et al. 2002). In this context, Szabados et al. (2012) could demonstrate that two strains having the same genetic background did not show any difference in their protein mass fingerprint. For their experiments they used one strain with and one without the SCCmec cassette harboring *mecA*, resembling isogenic MRSA and MSSA strains, respectively. These results clearly argued against any characteristic difference between mass fingerprints of MRSA and MSSA strains. But new findings did change this view again. It was shown that a characterization of *Staphylococcus aureus* as MRSA with very high specificity, although limited sensitivity, can be done based on a single peak at *m/z* 2415 which is indicative for a small protein, psm, encoded on certain types of *mecA* cassettes (Josten et al. 2014). As only a subset of MRSA contains such a *mecA* cassette the sensitivity of this detection method is limited and dependent on the respective geographical region with its particular distribution of MRSA clonal groups. A further interesting aspect is that this peak can only be detected by simple 'smear' (direct transfer) preparation of the bacterial sample but not by tube extraction of its proteins, a fact which might have prevented its detection in earlier studies, e.g., by Szabados et al. (2012). The automated detection of the psm peak in *S. aureus* profiles is currently under development and, together with *Bacteroides fragilis* division II detection as described below, may be the first diagnostic resistance screening method based on simple MALDI-TOF MS profiling introduced to clinical microbiology practice. Such approaches are extremely valuable as the resistance information is virtually available at the same time as the identification, and respective clinical

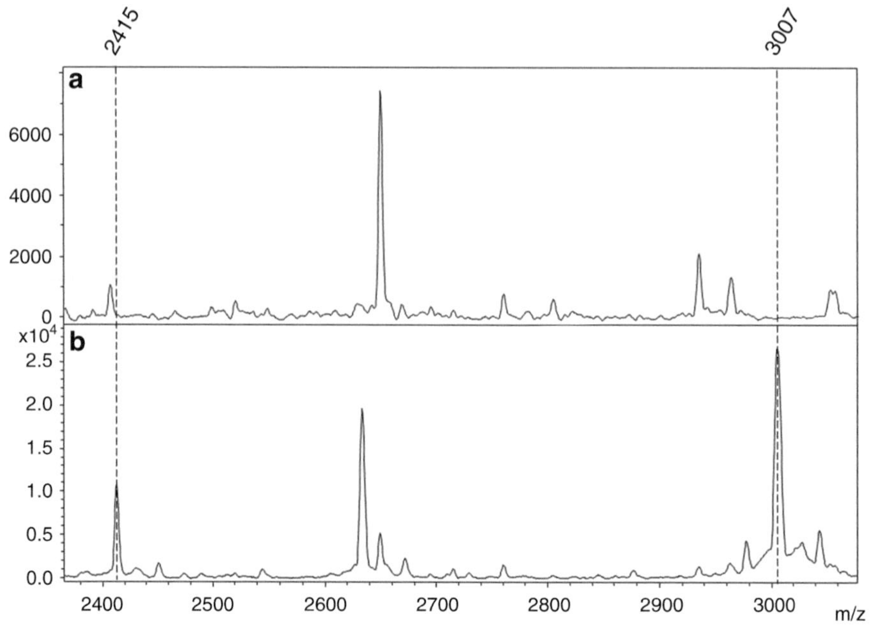

Fig. 1 Sections of MALDI profile mass spectra acquired for two *Staphylococcus aureus* strains. The methicillin-susceptible strain (MSSA) spectrum (*upper panel*) does not contain the peak for delta toxin at *m/z* 3007 nor the psm peptide peak at *m/z* 2415. The spectrum of the strain given in the *lower panel* exhibits both peaks which is typical for an *agr*-positive MRSA strain.

measures (e.g., adapted treatment or isolation of the patient) can be started at once. Examples of MALDI-TOF MS profiles with and without the psm indicator peak are given in Fig. 1.

Besides the detection of the psm-positive MRSA strains, to a limited extend the classification of a *Staphylococcus aureus* strain as MRSA or MSSA can be performed based on other typing information. Here, the applicability of MALDI-TOF MS profiling to determine mutations of *S. aureus* has been described by Josten et al. (2013).

In addition to the important MRSA case, there are two further examples for resistance prediction based on characteristic protein profile mass spectra. One has been published for the anaerobic gram-negative bacterium *Bacteroides fragilis*. For this species two molecular divisions are known, i.e., division I and II, respectively. The more rare division II carries a gene, *cfiA*, encoding for a very potent metallo-β-lactamase. This enzyme cleaves and thereby inactivates virtually all β-lactam antibiotics including carbapenems, a very powerful and today's last resort antibiotic. Two independent groups described the differentiation of both divisions by simple MALDI-TOF MS analysis (Nagy et al. 2011; Wybo et al. 2011).

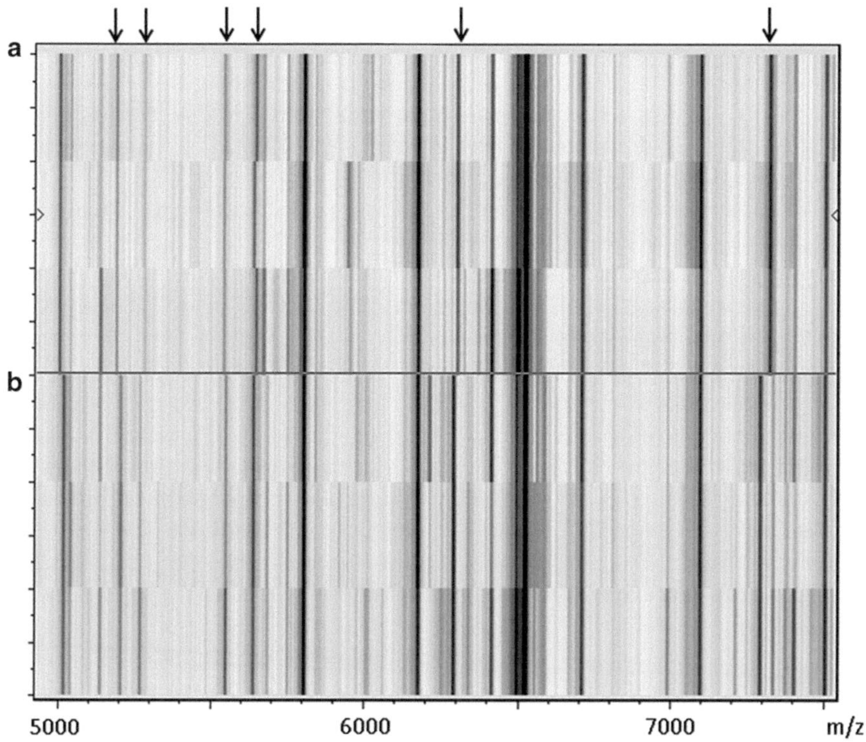

Fig. 2 Pseudo gel view of profile mass spectra of three strains each of *Bacteroides fragilis* division I (**a**) and division II (**b**). In the pseudo gel view, a *gray* scale is used for ion signal intensities. *Arrows* depict characteristic peak shifts between these two divisions

Division II *B. fragilis* profiles exhibit a number of characteristic peak shifts compared to division I which can be visually observed in the mass spectra, as depicted in Fig. 2, and also detected by different bioinformatics means, as described by the authors. This genotypic resistance prediction can be very useful in the future as the respective strains might not express the full β-lactamase activity in the beginning but could get activated under treatment. Implementation into routine analysis seems to be possible and eligible (Fenyvesi et al. 2014).

A further typing approach for resistance detection was described by Griffin and coworkers (2012). They used a commercial software package (ClinProTools, Bruker Daltonik GmbH, Bremen, Germany) to calculate a multivariate statistic model which should discriminate vancomycin-resistant *Enterococcus faecium* (vancomycin-resistant enterococci, VRE) from those that are vancomycin-susceptible. Their model worked very well in a first experimental internal validation (sensitivity of 92.4% and a specificity of 85.2%) and even better implementation into their routine analysis (sensitivity and specificity at 96.7% and 98.1%, respectively).

On the other hand, a German group has argued against the use of MALDI-TOF MS for vancomycin resistance detection by such a typing approach (Lasch et al. 2014). In their study, they did not identify any statistically relevant marker peaks linked to glycopeptide resistance determinants (vanA, vanB) in *E. faecium.* So it might be that the findings of Griffin and coworkers are based on a special epidemiological situation and not transferable to other parts of the world.

The beauty of peak-based profiling approaches as described above is their simplicity and speed. The information is easily available within the spectrum which has been acquired for identification and can be readily investigated by a second, dedicated algorithm adapted to finding the resistance marker of interest. Thereby, after introduction of such algorithms into commercially available systems, the idea of resistance detection at the same time as identification of a microbe can become true. However, this will only be possible for some particular species where a resistance phenomenon is linked to a specific peak pattern and thereby only a facet of MALDI-TOF MS-based resistance detection. A further drawback of such an approach, as for any genotypic method, is the fact that only a genetic predisposition for resistance can be quickly established. False positives, e.g., by inactive resistance genes, and false negatives, e.g., through non-detectable new variants, cannot be excluded. The real, active resistance status of a microorganism has to be determined by functional assays.

2.2 Functional Assays for Resistance Detection Using MALDI-TOF MS

2.2.1 Detection of β-Lactam Hydrolysis

The currently most near-to-routine MS-based resistance detection method is the cleavage analysis of β-lactam antibiotics (penicillin and its modern successors). This analysis will probably become the first widely used routine resistance detection by MALDI-TOF MS. The assay is based on the observation of the enzymatic activity of β-lactamases, destroying the antibiotic drugs by hydrolysis. β-lactamases are key to the most common resistance mechanism against the large group of β-lactam antibiotics. These enzymes are synthesized and excreted by resistant microorganisms. Point mutations in their genes can also cause resistance against newer classes of β-lactam antibiotics like the widely used broad-spectrum cephalosporins or the carbapenems, which have the broadest activity and should only be used in cases where other β-lactams are ineffective. Very often β-lactamases are encoded by genes located on plasmids. These are transferable DNA elements, which can easily transport the resistance from one bacterium to the other and thereby accelerate the spread of resistance. They are sometimes even linked with other genetic elements conferring resistance against other classes of antibiotic substances and thus causing resistance against multiple drugs.

Enzymatic cleavage by hydrolysis of the antibiotic's β-lactam ring in a first step leads to the addition of a water molecule to its chemical structure, resulting in an increase in molecular mass of about 18 Da. A further reaction, cleavage of a CO_2 from the molecule, frequently causes a mass change of −44 Da. Such mass changes, although tiny for classical technologies known from microbiology and molecular biology, can easily be detected by mass spectrometry. Thereby, MALDI-TOF MS enables a totally new way in microbiology to monitor this reaction.

Several variations of the assay have been described by different authors. The first publications have shown the applicability of the MALDI-TOF MS-based assay to carbapenemase activity detection (Hrabák et al. 2011; Burckhardt and Zimmermann 2011), focusing on the most powerful and threatening β-lactamases. Sparbier and coworkers described that MALDI-TOF MS can be used to investigate many different classes of β-lactams, making it broadly applicable in microbiology. Hrabák and colleagues (Hrabák et al. 2011) used 2,5-dihydroxybenzoic acid (DHB) as MALDI matrix, while Burckhardt and Zimmermann (2011) as well as Sparbier and coworkers (2012) preferred α-cyano-4-hydroxycinnamic acid (CHCA).

An exemplary protocol for the detection of β-lactamase activity is given in Sect. 3.2.

While a few authors have published on the utilization of this method for various bacterial species and β-lactam antibiotics (Hrabák et al. 2012; Kempf et al. 2012; Carvalhaes et al. 2012; Lee et al. 2013; Pavelkovich et al. 2014; Sauget et al. 2014a) others have worked on its extension towards a broader and more advanced applicability. One of these extensions is the usage of the MS-based assay for enzyme characterization. Here, the general principle of using an enzyme-specific inhibitor has already been shown in the publication of Sparbier and coworkers (2012) for several antibiotics using clavulanic acid as well as for piperacillin/tazobactam. Also, these authors showed the confirmation of carbapenemases of the Ambler class A (e.g., KPC, Klebsiella Pneumoniae Carbapenemase) using carbapenem antibiotics and the inhibitor APBA (amino-phenol-boronic acid). Further, scientists showed the applicability of this characterization assay with different antibiotic drug/inhibitor combinations and for different bacteria (Álvarez-Buylla et al. 2013; Hoyos-Mallecot et al. 2013; Johansson et al. 2014a, b). The possibility to detect and characterize β-lactamase activity at the molecular level might not only be interesting for clinical diagnostics but also for future drug development.

The second of the advanced applications, which might become the most important one in future, is the combination of the MALDI β-lactamase assay with the direct identification of bacteria from positively flagged blood cultures as it is described in the preceding Chapter 'MALDI Biotyping for Microorganism Identification in Clinical Microbiology'. The speed of resistance detection for very important groups of bacteria, in particular the detection of ESBL (Extended Spectrum β-Lactamase) and CRE (Carbapenem Resistant Enterobacteriaceae), directly after rapid identification can be a major step towards improved treatment of sepsis or bloodstream infections. A proof-of-concept experiment has already been shown by Sparbier et al. in 2012. Carvalhaes et al. (2014) detected 21 of 29 carbapenemase producers in 100 randomly selected positive blood culture bottles after 4 h of incubation time for the cleavage reaction. Jung and coworkers (2014b) investigated enterobacteria from 100 consecutive positive blood cultures. They were able to discriminate between bacteria susceptible or resistant for aminopenicillins with 100% sensitivity and specificity, also for resistance to third-generation cephalosporins in enterobacteriaceae that constitutively produced class C β-lactamases. Species expressing class A β-lactamases were characterized with a sensitivity and specificity for this group of 100% and 91.5%, respectively. For the anaerobe *Bacteroides fragilis*, the bacterial pellet recovered from spiked blood cultures was used in a totally MALDI-TOF MS-based sequential *'identification—resistance typing (for division I or II, as described above)—carbapenemase activity confirmation'* workflow in only 3 h (Johansson et al. 2014b). These results show the great potential of such a diagnostic test.

Finally, there is also some hope to apply a similar test to other types of resistance, namely for aminoglycosides where the main resistance mechanism is also based on an enzymatic modification of the antibiotic drug (Wright 2005). Aminoglycoside inactivation can be performed by different enzymatic reactions, i.e., modifications by acyltransferases, phosphotransferases, and nucleotidyltransferases. Future might also see MALDI-TOF MS as a diagnostic system here although respective assays still have to be established.

2.2.2 Other Antibiotic Resistance Detection Methods by Protein Profile Changes

The first publication about a resistance assay observing a protein profile change came from the area of fungal resistance testing. Here, a particular need exists as little fungal resistance testing is done in diagnostic laboratories at all and virtually no rapid test is available. Marinach and coworkers (2009) reported that the yeast *Candida albicans* grown for 24 h in a dilution series of an antimycotic drug, fluconazole, at a particular drug concentration changes its mass profile. The concentration at which the profile change occurred was in a maximum range of one dilution difference compared to the MIC (minimum inhibitory concentration) set for traditional tests by the CLSI (Clinical and Laboratory Standards Institute). The lowest concentration which caused the MALDI-TOF MS profile change was named MPCC (Minimal Profile Changing Concentration). An Italian group applied this assay with slight modifications to other Candida species and also Aspergillus species using the antifungal drug caspofungin (De Carolis et al. 2012). They used commercial software for data analysis (CCI, part of MALDI Biotyper, Bruker Daltonik GmbH) and thereby simplified the workflow significantly. Complete essential agreement was observed with the CLSI reference method. Essential agreement here was defined as agreement at a drug concentration within 2 twofold dilution steps. Categorical agreement with the breakpoints for susceptibility and resistance was observed for 94.1% of isolates tested. Although being an important first step, this assay still needed 15 h of incubation time. Later, the same group modified the test to simplify and accelerate it (Vella et al. 2013). First, they reduced the assay to a breakpoint assay, i.e., the number of incubation concentrations was reduced to three: no antimycotic, an intermediate drug concentration, and the maximum drug concentration, respectively. Resistance prediction was done by comparison of the profile at the breakpoint concentration with the profile at the two other concentrations. Resistant yeasts show identical profiles at the intermediate concentration and after incubation without any antimycotic drug. Second, they reduced the incubation time down to 3 h only. An excellent agreement (98.4%) with only one very major categorical error was achieved for 62 *Candida albicans* strains. These results show the high potential of the method but it still has to be reproduced by other routine laboratories. Also an MPCC assay has never been applied successfully to detect resistance in bacteria.

For bacteria, a novel method has recently been reported which is somewhat related to established metabolic labeling methods from MS-based quantitative proteomics such as SILAC (Hoedt et al. 2014). The new assay directly observes activity of protein synthesis in cells using media containing stable isotope-labeled nutrient compounds. In one publication, the protein synthesis was performed in a medium, which was entirely labeled with ^{13}C (Demirev et al. 2013). The experiment was performed in three vials, labeled medium with (1), and without (2) the antibiotic as well as in a control without antibiotic in unlabeled medium (3). A susceptible cell incubated in labeled medium is inhibited by the antibiotic drug and will not start protein synthesis. Thereby the cell will not synthesize proteins with an increased molecular mass, while a resistant microbe will quickly synthesize proteins with an increased molecular weight. In the protein profile, the increase of molecular masses can be observed. A drawback of this version of the assay are the entirely ^{13}C-labeled media which are not all commercially available today and are expensive to make. Sparbier and coworkers (Sparbier et al. 2013) published a somewhat different assay which they called MS-RESIST (Mass Spectrometry-based Resistance Testing with Stable Isotopes). They incubated the bacteria in a medium where only one amino acid was substituted by a labeled one ($^{13}C_6/^{15}N_2$-L-lysine) and showed the differentiation of MRSA and MSSA by visual as well as automated data analysis in 3 h only. Even 2 h is sufficient to detect *Klebsiella pneumoniae* carbapenem resistance (Kostrzewa et al. 2013). Further, the same group investigated the application of MS-RESIST to *Pseudomonas aeruginosa* strains and antibiotics of three different classes (meropenem, a carbapenem; tobramycin, an aminoglycoside; and ciprofloxacin, a fluoroquinolone) in a subsequent study (Jung et al. 2014a). For each antibiotic, 15 strains of the susceptible and 15 strains of the resistant range (according to EUCAST) were tested. Incubation times were 2.5–3 h only and all strains were correctly classified. These results point to a broad applicability of this novel assay although more tests by other labs and with different antibiotic/microorganism combinations have to be performed. Resistance prediction from positive blood cultures has not been demonstrated yet.

Just after the publication of the stable isotope-based assays another variant of microorganism resistance testing was reported (Lange et al. 2014). Here, growth in normal medium is monitored. Not protein synthesis but cell proliferation is observed by semi-quantification. In parallel, microorganisms are grown with and without an antibiotic. It is important to start with the same suspension to guarantee the same starting point (cell count). After a short growth phase of about 2 h, cells are harvested by centrifugation and proteins are extracted by a standard method for MALDI MS profiling. During this procedure, an equal amount of a standard protein is spiked into the bacterial extracts. Acquired MALDI MS profile spectra are then used for the differentiation of susceptible and resistant strains. Areas under the curves of the bacterial profiles compared to the peak of the standard protein can be used to semi-quantify the amount of bacterial protein. Due to the inhibition of growth by the antibiotic, susceptible strains will show significantly reduced profile intensities. In their study, with 94 Klebsiella strains and meropenem as antibiotic, Lange and coworkers could demonstrate a sensitivity of 97.3% and a specificity of 93.5% for resistance detection. A further benefit of this assay is that the profile spectrum without antibiotic can be used in parallel for identification confirmation and as growth control. Interestingly, this assay was even successful in a small proof-of-concept study with ten susceptible and eight resistant Klebsiella strains spiked and grown in blood culture bottles. Seventeen of these strains were correctly classified, one was invalidated by the lack of growth in the control. Incubation time for this test was 1 h only. Although these are very early results, which have to be confirmed in further studies, such an assay could essentially accelerate bacterial resistance detection from blood cultures and thereby improve treatment of sepsis, one of the most severe microbial infections with high morbidity and mortality rates.

The latest assays described here are similar to conventional phenotypic AST methods. They monitor growth or inhibition of growth by antimicrobials and so may have some advantages in the future compared to methods only detecting resistance mechanisms. Since an early initiation of an appropriate treatment can reduce mortality and morbidity in certain infections, the knowledge about susceptibility, i.e., efficacy of a drug, is sometimes preferred over the information about resistance towards anti-infectives or their mechanisms. Further studies are needed to clarify further if these assays can reliably detect susceptibility with such short incubation times.

2.3 *Epidemiological and Wider Typing Analyses by MALDI-TOF MS*

Besides resistance detection, the differentiation of microorganisms below the species level with MALDI-TOF MS is one of the objectives for the novel users in clinical as well as in food and veterinary microbiology. This is again driven by the promise of shorter analysis time and cost reduction. The aims range from detection of particular infectious or virulent strains or subgroups to the tracking of individual strains and epidemiological surveillance. Generally, there is variation in the MALDI-TOF MS profiles below the species level but the profile variance is very different from species to species, and sometimes the discriminatory power is insufficient.

2.3.1 Examples for Microbial Typing Using MALDI-TOF MS

Because of its importance as a pathogen in clinical microbiology and appearance in outbreaks, *S. aureus* was early in the focus of investigations on MALDI-TOF MS typing. The possibility to discriminate strains of *S. aureus* by MALDI-TOF MS profile spectra was already reported by Bernardo and coworkers in 2002 (Bernardo et al. 2002). Most of this work with *S. aureus* was focused on MRSA/MSSA discrimination as already described earlier (see Sect. 2.1). Therefore, it took some time until wider, less-specific typing was investigated in-depth. Investigation of the first outbreak of a community-associated non-multiresistant and PVL-positive MRSA strain in a neonatal intensive care unit in Australia revealed surprisingly good results with MALDI-TOF MS profiling. PVL (Pantone-Valentine Leucocidine) is a strong cytotoxin with the potential to cause severe skin and soft tissue infections or even necrotizing infections and therefore has a high impact on morbidity or even mortality. Correlation analysis of MALDI-TOF MS data led to a similar inter-strain relatedness as SNP-plus-binary gene typing (Schlebusch et al. 2010). In a more fundamental study with 25 representative MRSA isolates belonging to the 5 major hospital-acquired (HA-)MRSA clonal complexes (CC5, CC8, CC22, CC30, CC45) and 60 independent clinical MRSA isolates, Wolters and coworkers (2011) demonstrated reproducible spectral differences based on 13 characteristic *m*/*z* values leading to 15 reproducible specific mass profiles. This allowed them to discriminate clonal complexes in a robust way. The authors concluded that MALDI-TOF MS has the potential to become a valuable first-line infection control tool for inexpensive and rapid typing of MRSA. Another study showed that the US300 strain, an endemic community-acquired (CA-) MRSA strain in the USA (also harboring the gene for PVL), could be separated from other *S. aureus* strains by a multivariate statistical model with three peaks (*m*/*z* 5932, 6423 and 6592). Josten and colleagues (2013) showed that peak shifts differentiating the main *S. aureus* clonal complexes CC5, CC22, CC8, CC45, CC30, and CC1 correlate to point mutations in the respective genes. A retrospective study of an MRSA outbreak showed that it was possible to differentiate unrelated MSSA, MRSA, and borderline resistant *S. aureus* (BORSA) strains isolated from healthcare workers. The authors concluded that MALDI-TOF MS allows the detection of the epidemic lineages of *S. aureus* during species identification.

For typing of gram-positive bacteria, a big challenge are vancomycin-resistant enterococci (VRE). In addition to the detection of VRE as described above (see Sect. 2.1), Griffin and coworkers reported the utilization of MALDI-TOF MS to determine the relatedness of isolates contributing to an outbreak (Griffin et al. 2012). The capability of MALDI-TOF MS in combination with multivariate statistical data evaluation to differentiate vanA-positive *Enterococcus faecium* (VPEF) from vanA-negative *E. faecium* (VNEF) strains was also recently demonstrated (Nakano et al. 2014). Using a genetic algorithm, the separation of >90% of the two groups was achieved for 61 VPEF and 71 VNEF isolates. The authors concluded that this would enable the use of the technique in routine analysis. In contrast, Lasch and coworkers argued against the suitability of MALDI-TOF MS for *E. faecium* as well as *S. aureus* typing because they did not find suitable, robust markers for this purpose (Lasch et al. 2014).

In the case of *Streptococcus agalactiae*, highly specific marker peaks have been detected for sequence types ST-1 and ST-17 (Lartigue et al. 2011). The strains of these STs are major causes of meningitis and late-onset disease in neonates. Therefore, it could be of high impact to introduce a MALDI-TOF MS-based screening method for those sequence types in parallel to MALDI MS identification.

Propionibacterium acnes is a commensal anaerobic, gram-positive microorganism which is of increasing importance as a potential pathogen. Phylogroup-specific peaks and peak shifts were identified for types IA, IB, IC, II as well as III and successfully validated in comparison to multilocus sequence typing (MLST) with an independent set of 48 isolates (Nagy et al. 2013). The clinical impact of this typing has yet to be proven.

For *Escherichia coli* several groups have shown the successful application of MALDI-TOF MS typing. Karger and coworkers demonstrated the successful typing of Shiga toxin-producing *E. coli* (STEC) isolates of serotypes O165:H25, O26:H11/H32, and O156:H25 (Karger et al. 2010). *E. coli* O104:H4 from a large outbreak in northern Germany in 2011 could be reliably typed using MALDI-TOF MS profiling (Christner et al. 2014). Using libraries with entries specific for different *E. coli* pathotypes, most of the pathotypes could be identified correctly in a study with 136 isolates (Clark et al. 2013). Sauget and colleagues studied *E.coli* isolates representing the phylogroups A, B1, B2, and D (Sauget et al. 2014b). Phylogroup B2, which represents the vast majority of *E. coli* strains causing septicemia, urinary tract infections, and other extra-intestinal infections, could be differentiated from groups A, B1, and D by a single peak shift. By this marker peak they accurately classified 89% of the 656 isolates investigated. Using ClinProTools (Bruker Daltonik GmbH) to calculate a multivariate statistical model, they were able to differentiate strains belonging to the phylogroup B from the other phylogroups with an accuracy of 92%. Therefore, MALDI-TOF MS has the potential to detect the most virulent strains and to monitor the epidemiology of extra-intestinal pathogenic *E. coli*.

For the gram-negative pathogen *Yersinia enterocolitica* which is known as an important cause of food-borne gastro-intestinal infections, a specific identification of the non-pathogenic biotype 1A and pathogenic biotypes 2 and 4 was reported (Stephan et al. 2011).

With MALDI-TOF MS profiles from 535 *Klebsiella pneumoniae* strains from hospitals in France and Algeria, a dendrogram revealed five distinct clusters associated with particular phenotypes from different clinical and geographical sources and from different seasons (Berrazeg et al. 2013). Bernaschi and coworkers compared clustering of *K. pneumonia*e by pulsed-field gel electrophoresis (PFGE), automated repetitive-sequence-based PCR (rep-PCR) genotyping, and MALDI-TOF MS profile analysis (Bernaschi et al. 2013). All typing methods agreed on the generation of three different clusters of *K. pneumoniae* isogenetic/related multidrug resistant (MDR) strains. The authors suggested MALDI-TOF MS proteome profiling as a fast and valuable preliminary screening tool being able to support microbiologists during nosocomial outbreak surveys.

Investigation of *Acinetobacter baumanii*, a further important pathogenic bacterium involved in nosocomial outbreaks, MALDI-TOF MS has been evaluated in comparison with the rep-PCR-based Diversilab system (bioMérieux, Marcy l'Etoile, France) and a good correlation of results was found (Mencacci et al. 2012).

In the field of highly pathogenic bacteria, the successful differentiation of *Francisella tularensis* subspecies has been shown (Seibold et al. 2010). The subspecies *F. tularensis ssp. tularensis* (or type A), which is predominantly found in North America, is the most virulent of the four known subspecies and is associated with lethal pulmonary infections. MALDI-TOF MS was also reported to be able to differentiate the three biotypes of *Yersinia pestis* (Ayyadurai et al. 2010).

For fungi, much less experience exists with MALDI-TOF MS typing approaches than for bacteria. Pulcrano and colleagues genotyped 19 strains of *Candida parapsilosis* isolated from blood cultures of neonates (Pulcrano et al. 2012). Electrophoretic and MALDI-TOF mass spectrometric profile results were compared in order to identify similarities among the isolates and to study microevolutionary changes. Both methods were found to be rapid and effective.

In the area of food microbiology, MALDI protein profile cluster analysis of 33 *Saccharomyces cerevisiae* strains was compared to a DNA-based method (delta-PCR). It was shown that MALDI-TOF MS provides valuable information about the relationship between yeast strains (Usbeck et al. 2014).

In summary, there is growing evvidence that MALDI-TOF MS profiling-based typing has the potential to become a routine microbiological tool. In particular, serving as a fast and inexpensive pre-screening method, it might help to avoid expensive and laborious DNA-based tests and enable in some cases the typing in parallel to identification of the species. Still some work has to be done until typing by MALDI-TOF MS will be suitable for routine analysis as variations in the acquired spectra might also be due to slightly different culture conditions. A comparison of strains might also be difficult if they have been measured in different mass spectrometers or even in different runs of the same instrument as no standardization has been proposed so far. Furthermore, there are currently no cut-offs or thresholds to determine similarity of strains during typing. While it is simple to see that mass spectra of two clonal strains are (almost) identical it is more difficult to separate two similar mass spectra from each other. At the moment, there is no guideline that exactly defines how much difference is needed in two mass spectra to decide that their corresponding strains are epidemiologically not related to each other. This is also a difficult question for the currently used typing methods that are usually based on genetic characteristics. These established methods and the experience in their use make them a valuable tool in the analysis of outbreaks in healthcare facilities. Databases with genotypes or sequence types are common, and for other techniques there are proposed criteria for interpretation of typing patterns like the criteria of Tenover and coworkers for pulsed-field gel electrophoresis (PFGE) DNA restriction patterns (Tenover et al. 1995). In this regard, MALDI-TOF MS-based typing is still at the beginning but may show its full potential when robust mathematical algorithms and defined criteria for interpretation of mass spectra similarity become available.

3 Materials and Protocols

3.1 MALDI MS Profiling and Detection of MRSA Strains

A protocol for the detection of the psm profile peak is given below:

- Prepare a direct bacterial biomass transfer ('smear') of the respective strain onto a spot of a MALDI sample target as described in Chapter 'MALDI Biotyping for Microorganism Identification in Clinical Microbiology'.

> • **NOTE**: A well-calibrated MALDI-TOF MS instrument with regard to peptide analysis has to be used to be able to acquire quality spectra (intensity, resolution).
>
> Sample extraction with the ethanol/formic acid method will remove the indicator peptide and therefore create a false negative result!

- Load the spectrum into suitable software for mass spectra analysis.
- Calibrate the spectrum using a conserved, generally present peak in the spectrum.
- Interrogate the spectrum for a peak at *m/z* 2415. A mass deviation of a maximum up to 1 Da can be accepted.
- Control: a strain expressing the psm peptide should also show the peak for delta toxin at *m/z* 3007 (most strains) or *m/z* 3037 (CC1 strains)!

3.2 Functional Assay of β-lactamase Activity Using MALDI-TOF MS

A typical simple protocol for the detection of β-lactamase activity is given below:

- Start with fresh, isolated overnight culture of the test strain. Use a positive and negative control strain appropriate for the respective antibiotic.
- Add 50 μL of antibiotic solution with the appropriate buffer to a PCR microtube. A 1.5 mL-Eppendorf tube or a well of a PCR microplate can also be used. The appropriate concentration depends on the antibiotic and has to be determined before the experiment. The range is generally between 100 and 500 μg/mL.
- Pick three to five individual bacterial colonies with a 1 μL-inoculation loop.
- Suspend the bacteria in the respective antibiotic solution tube (corresponding to $3–6 \times 10^8$ cells/mL) and vortex for 5 s to mix the cell suspension.
- Incubate the cell suspension with the bacterial test strains at 37 °C. Treat positive and negative control strains the same way.

- **CAUTION**: Make sure to use plastic material which is compatible with MALDI-TOF MS!

- Centrifuge the reaction tubes for 2 min at 13,000 rpm at room temperature in a standard laboratory microcentrifuge. PCR microtubes/microplates should be centrifuged for 4 min at 4,000 rpm.
- For each sample, pipette 1 μL of the cell-free supernatant onto a polished steel MALDI target plate.
- Dry at room temperature.
- Overlay spots with 1 μL of CHCA matrix solution (10 mg/mL in 50% acetonitrile) and dry at room temperature.
- The MALDI-TOF measurement should be performed with a well-calibrated instrument in an m/z range of approximately 300–1000. An internal calibrant (e.g., matrix peak) improves the mass accuracy.

- **NOTE**: Depending on the type of carbapenemase (e.g., KPC, NDM, VIM, OXA-48), its activity and the buffer used (e.g., carbonate, citrate, pure water), the incubation time can vary between 0.5 and 4 h. Agitation (e.g., at 900 rpm) can accelerate the antibiotic's degradation.
- **NOTE**: The occurrence of different adducts has to be considered! The ions in the spectrum might occur as hydrogenated forms but also as sodium or potassium adducts, or as a combination of these.
- **NOTE**: Parallel incubation of the same bacteria/drug mixture with specific inhibitors for drug-degrading enzymes can be used to detect and characterize the β-lactamase type. If the specific inhibitor significantly reduces the hydrolysis activity, the sample is positive for the corresponding enzyme.
- **NOTE**: Identification confirmation should always be performed (preferably also by MALDI-TOF MS) using the same bacteria that were used to analyze β-lactamase activity.
- **NOTE**: Multiple measurements from multiple spots should be used and averaged per sample for calculating cleavage performance.

Figure 3 shows an example of spectra from a β-lactam cleavage assay. An alternative version of this assay has been published by Hooff and coworkers (2011). They introduced an extraction method to use only the enzyme but not whole cells for the β-lactam hydrolysis detection. This variant adds preparation steps extending the time to result significantly. Nevertheless, although this early protocol is lengthy and laborious, it might have advantages for special applications if it is simplified and shortened. For instance, the characterization of enzymes might be improved as effects of the cell wall and outer membrane on the result are eliminated, and thereby standardization is improved.

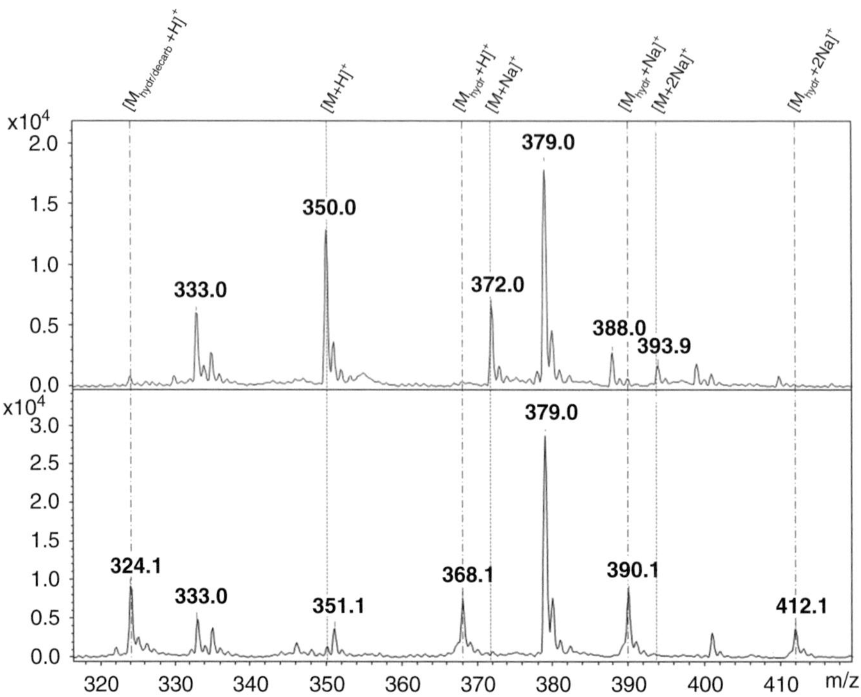

Fig. 3 Mass spectra of ampicillin after incubation with an ampicillin-susceptible (β-lactamase-negative) strain (*upper panel*) and an ESBL-producing strain which degrades ampicillin (*lower panel*). The susceptible strain does not cleave ampicillin. Therefore, the masses of intact ampicillin are observed as $[M+H]^+$, $[M+Na]^+$, and $[M+2Na]^+$ ions. After 2 h of incubation with the ESBL strain, the antibiotic is entirely cleaved and only peaks for the hydrolyzed forms $[M_{hydr}+H]^+$, $[M_{hydr}+Na]^+$, and $[M_{hydr}+2Na]^+$ and a mass peak indicating the hydrolyzed and decarboxylized ion $[M_{hydr/decarb}+H]^+$ are detected

References

Álvarez-Buylla A, Picazo JJ, Culebras E (2013) Optimized method for Acinetobacter species carbapenemase detection and identification by matrix-assisted laser desorption ionization-time of flight mass spectrometry. J Clin Microbiol 51:1589–1592. doi: 10.1128/JCM.00181-13

Ayyadurai S, Flaudrops C, Raoult D, Drancourt M (2010) Rapid identification and typing of Yersinia pestis and other Yersinia species by matrix-assisted laser desorption/ionization time-of-flight (MALDI-TOF) mass spectrometry. BMC Microbiol 10:285. doi: 10.1186/1471-2180-10-285

Barber M, Rozwadowska-Dowzenko M (1948) Infection by penicillin-resistant staphylococci. Lancet 2:641–644.

Bernardo K, Pakulat N, Macht M et al (2002) Identification and discrimination of Staphylococcus aureus strains using matrix-assisted laser desorption/ionization-time of flight mass spectrometry. Proteomics 2:747–753. doi:10.1002/1615-9861(200206)2:6<747::AID-PROT747>3.0.CO;2-V

Bernaschi P, Del Chierico F, Petrucca A et al (2013) Microbial tracking of multidrug-resistant Klebsiella pneumoniae isolates in a pediatric hospital setting. Int J Immunopathol Pharmacol 26:463–472

Berrazeg M, Diene SM, Drissi M et al (2013) Biotyping of multidrug-resistant Klebsiella pneumoniae clinical isolates from France and Algeria using MALDI-TOF MS. PLoS One 8, e61428. doi:10.1371/journal.pone.0061428

Burckhardt I, Zimmermann S (2011) Using MALDI-TOF mass spectrometry to detect Carbapenem resistance within one to two and a half hours. J Clin Microbiol 17(1):120–122. doi:10.1128/JCM.00287-11

Carvalhaes CG, Cayo R, Assis DM et al (2012) Detection of SPM-1-producing Pseudomonas aeruginosa and Class D-Lactamase-producing Acinetobacter baumannii isolates by use of liquid chromatography-mass spectrometry and matrix-assisted laser desorption ionization-time of flight mass spectrometry. J Clin Microbiol 51:287–290. doi:10.1128/JCM.02365-12

Carvalhaes CG, Cayô R, Visconde MF et al (2014) Detection of carbapenemase activity directly from blood culture vials using MALDI-TOF MS: a quick answer for the right decision. J Antimicrob Chemother 69:2132–2136. doi:10.1093/jac/dku094

Christner M, Trusch M, Rohde H et al (2014) Rapid MALDI-TOF mass spectrometry strain typing during a large outbreak of Shiga-Toxigenic Escherichia coli. PLoS One 9, e101924. doi:10.1371/journal.pone.0101924

Clark CG, Kruczkiewicz P, Guan C et al (2013) Evaluation of MALDI-TOF mass spectroscopy methods for determination of Escherichia coli pathotypes. J Microbiol Methods 94(3):180–191. doi:10.1016/j.mimet.2013.06.020

De Carolis E, Vella A, Florio AR et al (2012) Use of matrix-assisted laser desorption ionization-time of flight mass spectrometry (MALDI-TOF MS) for caspofungin susceptibility testing of Candida and Aspergillus species. J Clin Microbiol 50(7):2479–2483. doi:10.1128/JCM.00224-12

Demirev PA, Hagan NS, Antoine MD et al (2013) Establishing drug resistance in microorganisms by mass spectrometry. J Am Soc Mass Spectrom 24:1194–1201. doi:10.1007/s13361-013-0609-x

Du Z, Yang R, Guo Z et al (2002) Identification of Staphylococcus aureus and determination of its methicillin resistance by matrix-assisted laser desorption/ionization time-of-flight mass spectrometry. Anal Chem 74:5487–5491. doi:10.1021/ac020109k

Edwards-Jones V, Claydon MA, Evason DJ et al (2000) Rapid discrimination between methicillin-sensitive and methicillin-resistant Staphylococcus aureus by intact cell mass spectrometry. J Med Microbiol 49:295

Enright MC, Robinson DA, Randle G et al (2002) The evolutionary history of methicillin-resistant Staphylococcus aureus (MRSA). Proc Natl Acad Sci U S A 99:7687–7692. doi:10.1073/pnas.122108599

Fenyvesi VS, Urbán E, Bartha N et al (2014) Use of MALDI-TOF/MS for routine detection of cfiA gene-positive Bacteroides fragilis strains. Int J Antimicrob Agents 44:474–475. doi:10.1016/j.ijantimicag.2014.07.010

Griffin PM, Price GR, Schooneveldt JM et al (2012) The use of MALDI-TOF MS to identify vancomycin resistant enterococci and investigate the epidemiology of an outbreak. J Clin Microbiol 50(9):2918–2931. doi:10.1128/JCM.01000-12

Hoedt E, Zhang G, Neubert TA (2014) Stable isotope labeling by amino acids in cell culture (SILAC) for quantitative proteomics. Adv Exp Med Biol 806:93–106. doi:10.1007/978-3-319-06068-2_5

Hooff GP, van Kampen JJA, Meesters RJW et al (2011) Characterization of β-lactamase enzyme activity in bacterial lysates using MALDI-mass spectrometry. J Proteome Res 11(1):79–84. doi:10.1021/pr200858r

Hoyos-Mallecot Y, Cabrera-Alvargonzalez J, Miranda-Casas C et al (2013) MALDI-TOF MS, a useful instrument for differentiating metallo-β-lactamases in Enterobacteriaceae and Pseudomonas spp. Lett Appl Microbiol 58(4):325–329. doi:10.1111/lam.12203

Hrabák J, Walková R, Studentová V et al (2011) Carbapenemase activity detection by Matrix-Assisted Laser Desorption/Ionisation Time-of-Flight (MALDI-TOF) mass spectrometry. J Clin Microbiol 49(9):3222–3227. doi:10.1128/JCM.00984-11

Hrabák J, Studentová V, Walková R et al (2012) Detection of NDM-1, VIM-1, KPC, OXA-48, and OXA-162 carbapenemases by MALDI-TOF mass spectrometry. J Clin Microbiol 50(7):2441–2443. doi:10.1128/JCM.01002-12

Jevons MP (1961) 'Celbenin'-resistant Staphylococci. Br Med J 5219(1):124–125. doi:10.1136/bmj.1.5219.113

Johansson A, Nagy E, Sóki J (2014a) Detection of carbapenemase activities of Bacteroides fragilis strains with matrix-assisted laser desorption ionization–time of flight mass spectrometry (MALDI-TOF MS). Anaerobe 26:49–52

Johansson A, Nagy E, Sóki J (2014b) Instant screening and verification of carbapenemase activity in Bacteroides fragilis in positive blood culture, using matrix-assisted laser desorption ionization-time of flight mass spectrometry. J Med Microbiol 63:1105–1110. doi:10.1099/jmm.0.075465-0

Josten M, Reif M, Szekat C et al (2013) Analysis of the matrix-assisted laser desorption ionization-time of flight mass spectrum of Staphylococcus aureus identifies mutations that allow differentiation of the main clonal lineages. J Clin Microbiol 51:1809–1817. doi:10.1128/JCM.00518-13

Josten M, Dischinger J, Szekat C et al (2014) Identification of agr-positive methicillin-resistant Staphylococcus aureus harbouring the class A mec complex by MALDI-TOF mass spectrometry. Int J Med Microbiol 304(8):1018–1023. doi:10.1016/j.ijmm.2014.07.005

Jung JS, Eberl T, Sparbier K et al (2014a) Rapid detection of antibiotic resistance based on mass spectrometry and stable isotopes. Eur J Clin Microbiol Infect Dis 33:949–955. doi:10.1007/s10096-013-2031-5

Jung JS, Popp C, Sparbier K et al (2014b) Evaluation of matrix-assisted laser desorption ionization-time of flight mass spectrometry for rapid detection of -lactam resistance in Enterobacteriaceae derived from blood cultures. J Clin Microbiol 52:924–930. doi:10.1128/JCM.02691-13

Karger A, Ziller M, Bettin B et al (2010) Determination of serotypes of Shiga toxin-producing Escherichia coli isolates by intact cell matrix-assisted laser desorption ionization-time of flight mass spectrometry. Appl Environ Microbiol 77:896–905. doi:10.1128/AEM.01686-10

Kempf M, Bakour S, Flaudrops C et al (2012) Rapid detection of carbapenem resistance in Acinetobacter baumannii using matrix-assisted laser desorption ionization-time of flight mass spectrometry. PLoS One 7, e31676. doi:10.1371/journal.pone.0031676

Kostrzewa M, Sparbier K, Maier T, Schubert S (2013) MALDI-TOF MS: an upcoming tool for rapid detection of antibiotic resistance in microorganisms. Proteomics Clin Appl 7(11-12):767–778. doi:10.1002/prca.201300042

Lange C, Schubert S, Jung J et al (2014) Quantitative matrix-assisted laser desorption ionization-time of flight mass spectrometry for rapid resistance detection. J Clin Microbiol 52:4155–4162. doi:10.1128/JCM.01872-14

Lartigue M-F, Kostrzewa M, Salloum M et al (2011) Rapid detection of 'highly virulent' Group B Streptococcus ST-17 and emerging ST-1 clones by MALDI-TOF mass spectrometry. J Microbiol Methods 86:262–265. doi:10.1016/j.mimet.2011.05.017

Lasch P, Fleige C, Stämmler M et al (2014) Insufficient discriminatory power of MALDI-TOF mass spectrometry for typing of Enterococcus faecium and Staphylococcus aureus isolates. J Microbiol Methods 100:58–69. doi:10.1016/j.mimet.2014.02.015

Lee W, Chung H-S, Lee Y et al (2013) Comparison of matrix-assisted laser desorption ionization–time-of-flight mass spectrometry assay with conventional methods for detection of IMP-6, VIM-2, NDM-1, SIM-1, KPC-1, OXA-23, and OXA-51 carbapenemase-producing Acinetobacter spp., Pseudomonas aeruginosa, and Klebsiella pneumoniae. Diagn Microbiol Infect Dis 77(3):227–230. doi:10.1016/j.diagmicrobio.2013.07.005

Lowy FD (2003) Antimicrobial resistance: the example of Staphylococcus aureus. J Clin Invest 111:1265–1273. doi:10.1172/JCI18535

Marinach C, Alanio A, Palous M et al (2009) MALDI-TOF MS-based drug susceptibility testing of pathogens: the example of Candida albicans and fluconazole. Proteomics 9:4627–4631

Mencacci A, Monari C, Leli C et al (2012) Identification of nosocomial outbreaks of Acinetobacter baumannii using MALDI-TOF mass spectrometry. J Clin Microbiol 51(2):603–606. doi:10.1128/JCM.01811-12

Nagy E, Becker S, Soki J et al (2011) Differentiation of division I (cfiA-negative) and division II (cfiA-positive) Bacteroides fragilis strains by matrix-assisted laser desorption/ionization time-of-flight mass spectrometry. J Med Microbiol 60(Pt 11):1584–1590. doi:10.1099/jmm.0.031336-0

Nagy E, Urbán E, Becker S et al (2013) MALDI-TOF MS fingerprinting facilitates rapid discrimination of phylotypes I, II and III of Propionibacterium acnes. Anaerobe 20:20–26. doi:10.1016/j.anaerobe.2013.01.007

Nakano S, Matsumura Y, Kato K et al (2014) Differentiation of vanA-positive Enterococcus faecium from vanA-negative E. faecium by matrix-assisted laser desorption/ionisation time-of-flight mass spectrometry. Int J Antimicrob Agents 44:256–259. doi:10.1016/j.ijantimicag.2014.05.006

Pavelkovich A, Balode A, Edquist P et al (2014) Detection of carbapenemase-producing Enterobacteriaceae in the Baltic Countries and St. Petersburg area. Biomed Res Int 2014:1–7. doi:10.1155/2014/548960

Pulcrano G, Roscetto E, Iula VD et al (2012) MALDI-TOF mass spectrometry and microsatellite markers to evaluate Candida parapsilosis transmission in neonatal intensive care units. Eur J Clin Microbiol Infect Dis 31:2919–2928. doi:10.1007/s10096-012-1642-6

Sauget M, Cabrolier N, Manzoni M et al (2014a) Rapid, sensitive and specific detection of OXA-48-like-producing Enterobacteriaceae by matrix-assisted laser desorption/ionization time-of-flight mass spectrometry. J Microbiol Methods 105:88–91. doi:10.1016/j.mimet.2014.07.004

Sauget M, Nicolas-Chanoine M-H, Cabrolier N et al (2014b) Matrix-assisted laser desorption ionization-time of flight mass spectrometry assigns Escherichia coli to the phylogroups A, B1, B2 and D. Int J Med Microbiol 304(8):977–983. doi:10.1016/j.ijmm.2014.06.004

Schlebusch S, Price GR, Hinds S et al (2010) First outbreak of PVL-positive nonmultiresistant MRSA in a neonatal ICU in Australia: comparison of MALDI-TOF and SNP-plus-binary gene typing. Eur J Clin Microbiol Infect Dis 29:1311–1314. doi:10.1007/s10096-010-0995-y

Seibold E, Maier T, Kostrzewa M et al (2010) Identification of Francisella tularensis by whole-cell matrix-assisted laser desorption ionization-time of flight mass spectrometry: fast, reliable, robust, and cost-effective differentiation on species and subspecies levels. J Clin Microbiol 48:1061–1069. doi:10.1128/JCM.01953-09

Sparbier K, Schubert S, Weller U et al (2012) Matrix-assisted laser desorption ionization–time of flight mass spectrometry-based functional assay for rapid detection of resistance against β-Lactam antibiotics. J Clin Microbiol 50:927–937. doi:10.1128/JCM.05737-11

Sparbier K, Lange C, Jung J et al (2013) MALDI Biotyper based rapid resistance detection by stable isotope labeling. J Clin Microbiol 51(11):3741–3748. doi:10.1128/JCM.01536-13

Stephan R, Cernela N, Ziegler D et al (2011) Rapid species specific identification and subtyping of Yersinia enterocolitica by MALDI-TOF Mass spectrometry. J Microbiol Methods 87(2):150–153

Szabados F, Kaase M, Anders A, Gatermann SG (2012) Identical MALDI TOF MS-derived peak profiles in a pair of isogenic SCCmec-harboring and SCCmec-lacking strains of Staphylococcus aureus. J Infect 65(5):400–405. doi:10.1016/j.jinf.2012.06.010

Tenover FC, Arbeit RD, Goering RV et al (1995) Interpreting chromosomal DNA restriction patterns produced by pulsed-field gel electrophoresis: criteria for bacterial strain typing. J Clin Microbiol 33:2233–2239

Usbeck JC, Wilde C, Bertrand D et al (2014) Wine yeast typing by MALDI-TOF MS. Appl Microbiol Biotechnol 98:3737–3752. doi:10.1007/s00253-014-5586-x

Vella A, De Carolis E, Vaccaro L et al (2013) Rapid antifungal susceptibility testing by matrix-assisted laser desorption ionization-time of flight mass spectrometry analysis. J Clin Microbiol 51:2964–2969. doi:10.1128/JCM.00903-13

Wolters M, Rohde H, Maier T et al (2011) MALDI-TOF MS fingerprinting allows for discrimination of major methicillin-resistant Staphylococcus aureus lineages. Int J Med Microbiol 301:64–68. doi:10.1016/j.ijmm.2010.06.002

Wright GD (2005) Bacterial resistance to antibiotics: enzymatic degradation and modification. Adv Drug Deliv Rev 57:1451–1470. doi:10.1016/j.addr.2005.04.002

Wybo I, De Bel A, Soetens O et al (2011) Differentiation of cfiA-negative and cfiA-positive Bacteroides fragilis isolates by matrix-assisted laser desorption ionization-time of flight mass spectrometry. J Clin Microbiol 49:1961–1964. doi:10.1128/JCM.02321-10

Part V
MALDI Biotyping Beyond the Clinic

Whole/Intact Cell MALDI MS Biotyping in Mammalian Cell Analysis

Bogdan Munteanu and Carsten Hopf

Abstract Matrix-assisted laser desorption/ionization (MALDI) mass spectrometry (MS) biotyping of microorganisms has arguably emerged as the premier application of MALDI MS in recent years. It is now widely used in clinical microbiology as an approved diagnostic tool and has recently been applied beyond the clinic, e.g., in environmental microbiology. Not long ago the adaptation of MALDI MS biotyping methods to applications with non-preprocessed, whole mammalian cells has started. Important tasks such as standardization of cell culture, sample preparation, and data acquisition as well as development of suitable databases, classification algorithms, and software are still at an early stage. However, applications of mammalian cell biotyping by MALDI MS in the clinic and beyond have already become visible, in particular in the pharmaceutical and diagnostics industry. They range from general quality assurance of mammalian assay cell lines, to at-line quality control of bioprocesses during development or production, and to the development of MALDI MS-based assays. In this chapter, we describe general methods for whole-cell MALDI MS biotyping of mammalian cells and highlight their application in development of cell-based assays for discovery and profiling of histone deacetylase (HDAC) inhibitors.

1 Introduction

Since its first description in 1975 by Anhalt and Fenselau (Anhalt and Fenselau 1975), whole-cell (WC) MS has revolutionized the identification, characterization, and diagnosis of health threatening pathogens (Demirev and Fenselau 2008a; Welker 2011; La Scola 2011). It was the development of soft ionization techniques such as MALDI MS (Karas et al. 1987) in the mid-1980s that was critical for today's

B. Munteanu • C. Hopf (✉)
Institute of Medical Technology, University of Heidelberg and Mannheim University of Applied Sciences, Center for Applied Research in Biomedical Mass Spectrometry (ABIMAS), Paul-Wittsack-Str. 10, 68163 Mannheim, Germany

Instrumental Analytics and Bioanalytics, Mannheim University of Applied Sciences, Paul-Wittsack-Str. 10, 68163 Mannheim, Germany
e-mail: carsten.hopf@medtech.uni-heidelberg.de

R. Cramer (ed.), *Advances in MALDI and Laser-Induced Soft Ionization Mass Spectrometry*, DOI 10.1007/978-3-319-04819-2_13

increased relevance of WC MS in clinical applications and beyond. MALDI MS has enabled the analysis of intact, labile biomolecules such as proteins without extensive fragmentation. Before that, the analysis had been restricted to low molecular mass species such as lipids and ubiquinone, which were analyzed by plasma desorption (PD) MS. Pioneering work demonstrated that MALDI was a suitable ionization technique for MS to generate species-specific protein fingerprints from whole cells that could be used for diagnostic applications (Cain et al. 1994; Holland et al. 1996). The expression 'whole' or 'intact' cell refers to the fact that the target cells (e.g., bacterial or mammalian cells) are analyzed without cell fractionation or complete lysis prior to sample preparation for MALDI MS analysis (Holland et al. 1996). However, the direct exposure to organic solvents, acids, and MALDI matrices results in cell perforation, protein extraction and lysis (Ryzhov and Fenselau 2001).

In the following years, due to rigorous method standardization and substantial volume increase of bacterial reference spectra, WC MALDI MS evolved to become a fast, specific, and automated method for the characterization and identification of microorganisms in the clinical (Walker et al. 2002), the environmental (Koubek et al. 2012) and the biodefense fields (Demirev and Fenselau 2008b; Ryzhov et al. 2000), thus shifting classical diagnostic procedures towards MS-based systems. The main advantages of WC MALDI MS are its simplicity of use, speed, and the lack of expensive additives (e.g., antibodies, purification, or labeling reagents).

2 Applications

Because of the complexity of the eukaryotic proteome, direct analysis of mammalian cells by MALDI MS biotyping was deemed not feasible for a long time. In 2006, however, Zhang and colleagues described the discriminatory potential of mammalian WC MALDI MS analysis in a landmark study (Zhang et al. 2006). Since then mammalian WC MALDI MS biotyping is being steadily developed for possible clinical and industrial applications (reviewed in Munteanu and Hopf 2013). Published applications were and still are mainly restricted to the biotyping of purified blood cells (Munteanu et al. 2012; Ouedraogo et al. 2010) and single types of primary cells or continuous cell lines (Zhang et al. 2006; Dong et al. 2011; Marvin-Guy et al. 2008; Hanrieder et al. 2011). Consequently, the true discriminatory potential of WC MALDI MS biotyping is not yet known, as no systematic studies of defined cell mixtures or very closely related cell types have been published yet. Nevertheless, much progress has been made (Munteanu and Hopf 2013), and these major improvements in mammalian WC MALDI MS biotyping beyond clinical applications shall be discussed here. In particular, two recent industrial applications have been described for mammalian WC MALDI MS biotyping, namely (1) the monitoring of product expression and cell viability in bioprocesses and (2) a rapid label-free cellular assay, e.g., for histone deacetylase inhibitors (HDACi).

2.1 WC MALDI MS Biotyping for at-Line Analytics of Bioprocesses with Mammalian Production Cell Lines

WC MALDI MS biotyping studies using Chinese Hamster Ovarian (CHO) cells, a common cell line used in biotechnological processes (Kim et al. 2012; Rodrigues et al. 2014), have recently suggested that this technique could be a fast, cheap, and complementary tool for 'at-line' process analytics. In one example, WC MALDI MS biotyping has been used to monitor product expression as part of product control analytics (Feng et al. 2010; Feng et al. 2011). Most recently it has been suggested that modeling based on partial least square discriminant analysis (PLS-DA) may be able to predict a cell line's growth or recombinant protein productivity in large bioprocessing using data from early cell line construction at small culture scale (Povey et al. 2014). Other laboratories have utilized the method to distinguish between several cell states (healthy, apoptotic, necrotic) and to analyze cell culture viability (Dong et al. 2011; Schwamb et al. 2013). It has been suggested that changes in cellular protein patterns that can be assessed by WC MALDI MS biotyping enable the prediction of upcoming cell viability changes about a day earlier than standard cell culture monitoring would (Schwamb et al. 2013).

2.2 Reagent-Free Cell-Based Drug Discovery and Profiling Assays

Target engagement describes the process of drug target activation after drug binding and its demonstration represents a crucial enabling step during drug discovery. The development of new drugs takes up to 15 years and requires a capital investment of approximately one billion dollars. The failure rate is high and currently nine out of ten drugs tested in human clinical trials do not obtain market approval, often failing due to a lack of efficacy in phase II and phase III clinical trials (Paul et al. 2010). Therefore, analytical platforms are required to allow monitoring of target engagement and activation in cell culture and in-vivo studies during lead optimization and, ideally, in human samples during clinical development. Currently applied techniques such as chemoproteomics-based profiling (Bantscheff et al. 2011) or the recently introduced cellular (and ex-vivo) thermal shift assays (Martinez Molina et al. 2013; Savitski et al. 2014), are multi-step procedures and require expensive reagents as well as extensive sample preparation and data analysis. Reagent-free in-situ analysis of target-proximal pharmacodynamic signatures in cells and tumors remains challenging.

In a recently published study, we provided proof-of-concept evidence suggesting that drug target activation by HDACi could be quantified by WC MALDI MS biotyping using label-free analysis of pharmacodynamic biomarkers in cancer cell lines without any purification, extraction, or labeling procedures (Munteanu et al. 2014). We also demonstrated that the combination of WC MALDI MS biotyping

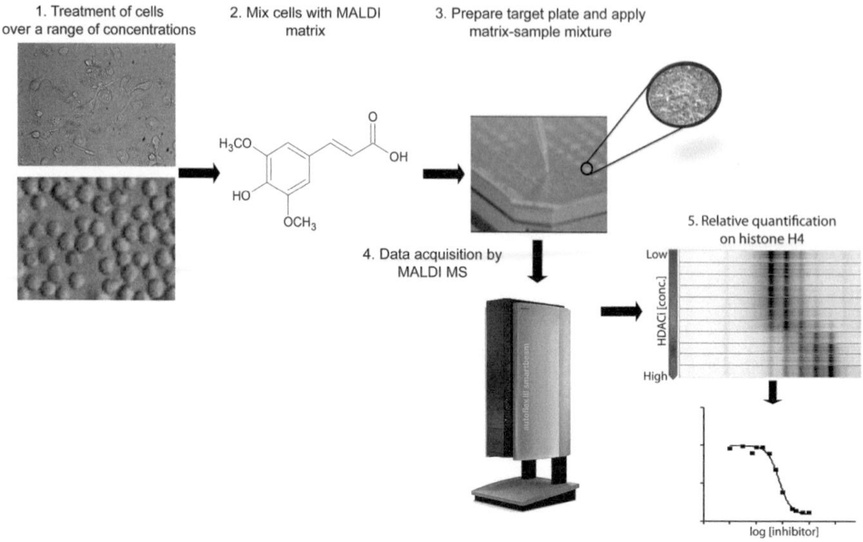

Fig. 1 Schematic representation of the whole-cell MALDI MS workflow for the evaluation of HDACi potency in whole cells

with MALDI MS imaging (MSI) could provide evidence of target activation in a gastric cancer mouse model in-vivo.

In the following part of this chapter we will describe in detail the general WC MALDI biotyping procedure as well as its adaptation for the analysis of drug target activation of HDACi (Fig. 1)—with a major focus on sample preparation and data analysis. In order to distinguish the WC MALDI MS technique for HDACi evaluation from the standard biotyping technique, differences in workflow, sample preparation, and data analysis will be indicated in each section of the chapter.

3 Materials and Protocols

3.1 Materials and Equipment

All listed solvents are of HPLC grade and all reagents are of the highest available purity from various suppliers.

3.1.1 Cell Culture

1. −80 °C Refrigerator
2. Cell culture laminar flow hood
3. Diverse cell culture plates or flasks for cell lines growing in suspension or adherently (e.g., from Greiner Bio-One, Solingen, Germany)

4. Small-molecule drugs (e.g., histone deacetylase inhibitors (HDACi) such as LBH-589) (e.g., from Selleck Chemicals, Houston, TX, USA, or LcLabs, Woburn, MA, USA)
5. DMSO (for cell culture)
6. Liquid nitrogen
7. Neubauer Improved (VWR Intern, Darmstadt, Germany) counting chamber (or similar device)
8. Phosphate-Buffered Saline (PBS)
9. Cell culture medium
10. 1.5- or 2-mL centrifuge tubes
11. Refrigerated laboratory microcentrifuge
12. Trypan blue solution (0.4% (w/v) in PBS) (Applichem, Darmstadt, Germany)
13. Trypsin/EDTA or alternative cell culture dissociation enzymes (e.g., Accutase)
14. Vacuum aspirator

3.1.2 MALDI Biotyping Sample Preparation

1. 0.65-mL reagent tubes
2. ddH_2O
3. Ethanol, methanol, acetic acid, acetonitrile
4. MTP 384 ground steel target plates (Bruker Daltonik, Bremen, Germany)
5. Sinapinic acid
6. Target cleaning solution (50% methanol/5% acetic acid; (v/v))
7. Trifluoroacetic acid
8. Ultrasonic bath

3.1.3 MS Data Acquisition and Analysis

1. MALDI-TOF mass spectrometer (e.g., Autoflex Speed; Bruker Daltonik)
2. ClinProTools 3.0 (Bruker Daltonik)
3. FlexAnalysis 3.4 (Bruker Daltonik)
4. FlexControl 3.4 (Bruker Daltonik)
5. GraphPad Prism 5.0 (GraphPad, La Jolla, CA, USA)
6. Protein Calibration Standard I (Bruker Daltonik)

3.2 Protocols

3.2.1 Cell Culture

This protocol section describes in detail the cell culture process for the generation of the required cell material for mammalian WC MALDI MS biotyping and also includes the required experimental steps necessary for quantitative evaluation of

HDACi. The latter can be performed both on adherent cells as well as on cell lines growing in suspension. The required cell culture conditions (e.g., medium composition) will not be discussed here, since each cell line has its individual requirements. Relevant information can be obtained from external sources (e.g., American Type Culture Collection, www.atcc.org).

1. *If treatment of cell cultures with drugs is desired*: Dissolve HDACi in DMSO to prepare a 100-mM stock solution. DMSO stocks can be stored at −20 °C or for longer time periods at −80 °C.
2. Seed suspension or adherently growing cell lines in corresponding cell culture dishes at a density of 5×10^5 cells per well in 6- or 24-well plates 12 h prior to treatment.[1]
3. *If treatment of cell cultures with drugs is desired*: Add inhibitor to a final dilution of 1:1000 (v/v) into the cell culture media. For concentration-dependence studies, the final inhibitor concentration should cover a wide range of concentrations, for example from 100 μM to 1 nM (as cell viability permits and drug potency requires). In addition, a control is prepared by adding 1 μL of DMSO into 1 mL of cell culture medium.
4. *If treatment of cell cultures with drugs is desired*: For qualitative evaluation incubate the cell culture for at least 2 h in a cell culture incubator (5% CO_2, 37 °C). However, quantitative statements (EC_{50} or IC_{50}) require an incubation time of 24 h or longer.

Harvest of Suspension Cells

1. Transfer all cells into a 1.5- or 2-mL tube.
2. Spin the cell suspension for 5 min at $130 \times g$ at 4 °C. Remove the supernatant using a pipette or a vacuum aspirator.
3. Resuspend the cell pellet in PBS (~ ½ of the volume in step 1).
4. Spin the cell suspension for 5 min at 130 × g at 4 °C. Remove the supernatant using a pipette or a vacuum aspirator.
5. Repeat steps 3 and 4 twice.
6. After the last washing step, resuspend the cell pellet in a small volume of PBS and determine cell viability and density by trypan blue staining in a Neubauer counting chamber or another device such as a CASY counter. Prepare cell aliquots at defined cell counts.
7. Collect the cells by centrifugation for 5 min at $300 \times g$ at 4 °C. Remove the supernatant carefully and completely without damaging the pellet.
8. Freeze cell pellet aliquots in liquid nitrogen and store at −80 °C.

[1] In order to avoid the attachment of suspension growing cell lines to the plate, suspension cell lines must be cultivated in dishes especially designed for suspension growing cell lines while adherent cell lines require cell culture dishes for adherent growing.

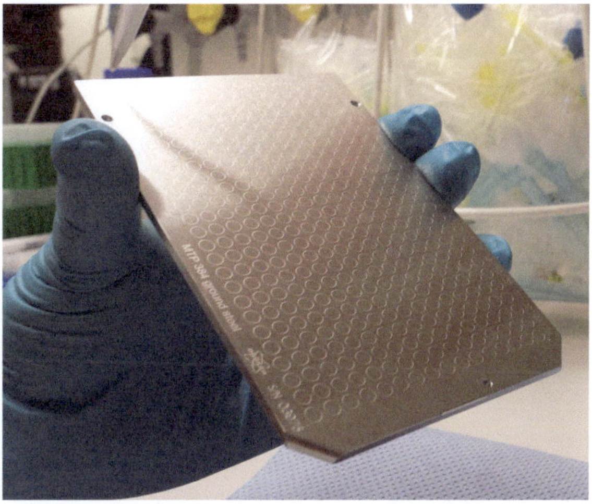

Fig. 2 Ground steel MALDI target, being held inclined for thin-layer sample preparation

Harvest of Adherent Cells

1. Remove the cell culture medium.
2. Wash the wells with PBS in order to remove cell culture medium remains.
3. Add 500 μL of trypsin/EDTA or Accutase solution and incubate at 37 °C. Monitor the detachment process under a light microscope.[2]
4. Transfer the detached cells into a fresh tube.
5. Perform steps 2–8 as described for the harvest of suspension cells (see above).

3.2.2 Sample Preparation for WC MALDI MS Biotyping

MALDI sample preparation has a critical impact on the reproducibility and quality of the MALDI MS biotyping measurement. The first step consists of the preparation of a sinapinic acid (SA) thin layer for a homogenous sample and sensitive MALDI MS (Fig. 2). An image of an optimal thin-layer sample preparation is presented in Fig. 3a, b, while 'worst' cases are illustrated in Fig. 3c, d. The second step consists of applying the cells-matrix mixture to the earlier prepared thin matrix layer on the MALDI target.

[2] In order to ensure the cell integrity during the harvest process, adherent growing cell lines should be detached by enzymatic detachment (Trypsin/Accutase) and not by scraping. Please consult standard cell culture protocols for details. Proteolysis due to residual enzymes is reduced by working at low temperatures and diluting the enzyme. Serum inhibition should be avoided in order to reduce contamination with serum proteins.

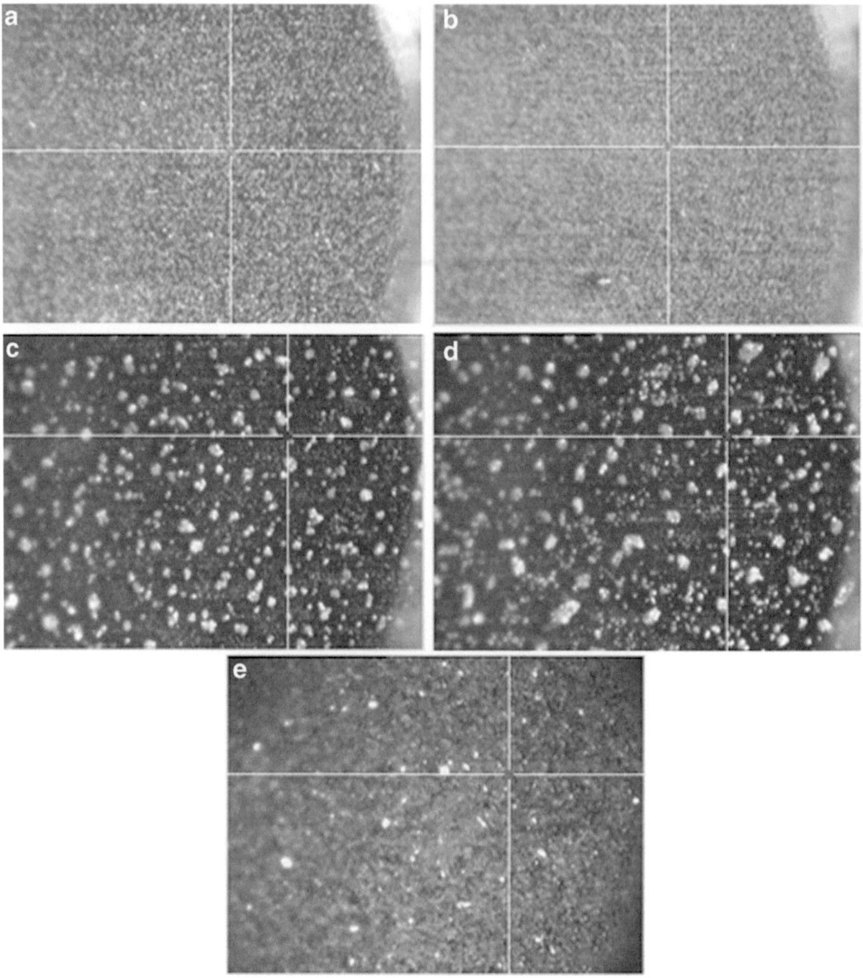

Fig. 3 Sinapinic acid thin-layer MALDI samples. (**a**, **b**) Examples of homogenous thin-layer crystallization (optimal); (**c**, **d**) Examples of 'worst-case' crystallization; (**e**) Representative WC MALDI sample after crystallization of the cells-matrix mixture on a thin matrix layer (see text for details)

Initial Thin-Layer Preparation for General WC MALDI MS Biotyping

1. Immerse a MALDI target plate in the cleaning solution (50% methanol/ 5% acetic acid; v/v) for 30 min in a sonic bath. Thereafter, wipe the plate with methanol using clean lint-free tissue paper. Finally, rinse with methanol and air-dry in a laminar-flow hood.
2. After the washing step, prepare 1 mL of saturated SA solution using absolute ethanol as a solvent and a sonic bath.

3. Centrifuge the matrix solution at maximum speed for 2 min.
4. Transfer the supernatant onto the entire previously cleaned MALDI target plate using a 1-mL pipette under a laminar-flow hood. In order to facilitate the cover process, the MALDI target plate should be held inclined during the matrix solution transfer onto the plate (Fig. 2). The matrix solution will quickly evaporate and small homogenous crystals will form, covering the entire plate (Fig. 3a, b).[3] Clean the edges of the target plate with absolute ethanol in order to remove excess matrix.[4]

Sample Preparation for General WC MALDI MS Biotyping

1. Prepare a 38 mg/mL SA solution in 60/40 (ACN/0.5% TFA; v/v). *For drug-treated cells:* Prepare a 38 mg/mL SA solution in 80/20 (ACN/12.5% TFA; v/v).[5]
2. Resuspend the frozen cell pellet in ddH_2O to achieve a cell density of 5000 cells/µL, place the cell suspension on ice.[6]
3. Transfer 15 µL of the cell suspension into a fresh 0.65-mL tube.
4. Add 15 µL of the MALDI matrix solution that was prepared in step 1 and mix by vortexing.
5. Saturate a 10 µL pipette tip with the cells-matrix mixture by repeated resuspension of the cells-matrix mixture and apply 1 µL each at eight different target positions onto the top of the thin matrix layer prepared earlier on the target plate.[7] Air-dry at room temperature (Fig. 3e). In addition, a calibration spot containing Protein Calibration Standard I is mixed in a ratio 1:1 (v/v) with the MALDI matrix solution that was prepared in step 1 and applied onto the target plate.

3.2.3 MS Data Acquisition and Analysis

MS instrument settings typically need to be adjusted according to the target analyte. In the example below, the histone protein molecular weights, ranging between 11 and 16 kDa, need to be considered. Protein analysis on mass spectrometers equipped with a MALDI ion source is mostly performed using a linear time-of-flight (TOF) mass analyzer, and instruments are available from different manufacturers. In this protocol, instrument settings will only be discussed in detail for the Bruker Autoflex

[3] Slow evaporation is a sign for increased water content and mostly results in an inhomogeneous thin layer affecting reproducibility.

[4] Matrix crystals at the contact surface between steel target and target frame might negatively affect the conductivity, resulting in reduced sensitivity and increased noise.

[5] The increased solvent acidity for the latter is required to facilitate quantitative extraction of histone core proteins from whole cells.

[6] Please note that hypotonic cell lysis is desirable at this step.

[7] Saturation of the cells-matrix mixture at the pipette tip reduces further non-specific binding to the tip surface and improves the technical reproducibility.

Speed mass spectrometer. The data acquisition parameters presented here represent guidelines and can vary between instruments.

Data Acquisition and Mass Spectrometer Optimization

1. Introduce the target plate into the mass spectrometer.
2. Wait until source vacuum reaches at least 3×10^{-7} mbar.[8]
3. Load a linear TOF positive ion mode acquisition method for the mass range of 4–20 kDa.
4. *In case of HDACi evaluation*: Adjust the acquisition method for optimal resolution of the histone H4 signal (FWHM < *m/z* 15 at *m/z* 11,306) using the FlexControl Software. The values for our set-up were:

(a) Ion source 1 voltage:	20.0 kV
(b) Ion source 2 voltage:	18.6 kV
(c) Lens voltage:	7.8 kV
(d) Pulsed ion extraction delay:	200–300 ns
(e) Matrix suppression mode:	Gating or deflection
(f) Real-time smoothing:	High
(g) Sample rate and digitizer setting:	1 GS/s
(h) Adjust linear detector voltage and laser intensity for maximal resolution[9]	

5. Calibrate the instrument with the optimized acquisition method.

   (a) Acquire a MALDI MS reference spectrum from the Protein Calibration Standard I spot, containing the following proteins listed with their average reference *m/z* values:

Insulin	$[M+H]^+$	5734.52
Cytochrome C	$[M+2H]^{2+}$	6181.05
Myoglobin	$[M+2H]^{2+}$	8476.66
Ubiquitin	$[M+H]^+$	12,360.97
Myoglobin	$[M+H]^+$	16,952.31

   (b) Calibrate the mass spectrometer by using the quadratic calibration method, which should provide mass accuracies ≪500 ppm for the linear acquisition mode.

6. Design an AutoXecute method.

---

[8] High vacuum generally improves the mass measurement resolution of high molecular species (>10 kDa). It may therefore be beneficial to wait for higher and more stable vacuum conditions.

[9] A guidance value regarding a suitable detector voltage can be obtained by performing a detector check on a CHCA thin layer using the Detector Check Tool within the FlexControl software. The detector check evaluates the detector sensitivity by measuring the noise, which reaches the detector and can be used as a guidance value.

The AutoXecute method is a feature within the FlexControl software for automated data acquisition and encompasses the following major steps:

(a) Load the optimized linear acquisition method
(b) Add the optimal laser intensity
(c) Add quality parameters for optimal data acquisition such as peak half-widths (e.g., <20 Da)
(d) Introduce the intended number of laser shots per sample (2000–4000 shots ensure optimal spot-to-spot reproducibility)
(e) Set the measurement to random walk movement in order to avoid 'hot spots' and to increase shot-to-shot reproducibility[10]
(f) Enable automated sample measurement stop and move to next sample after ten subsequently failed acquisitions as a measure against low-quality samples
(g) Select the desired sample spots and start measurement

Data Analysis

At the data analysis stage, the objectives are to (a) recalibrate the acquired data, (b) generate average spectra normalized by total ion count, (c) extract the intensity of the ion signal of interest (e.g., non-acetylated histone H4), and *in the case of drug-treated cells* (d) to calculate IC_{50} values for each tested HDACi.

1. Transfer the acquired MALDI MS spectra files to FlexAnalysis.
2. Perform baseline-subtraction and external calibration using the Protein Calibration Standard I ion signals.
3. Group the recalibrated data in separate file folders.
4. Load individual folders into ClinProTools.
5. Process data as follows:

(a) Resolution:	800
(b) Mass range:	*m/z* 4000–20,000
(c) Smoothing:	Savitzky-Golay with width 6 (*m/z*) and five cycles
(d) Null spectra exclusion:	Enable
(e) Noise spectra exclusion:	Enable
(f) Recalibration:	500 ppm maximal peak shift
(g) Peak picking *S/N* threshold:	>5 on total average spectra

6. During the standard WC MALDI MS evaluation process ClinProTools can be used for statistical evaluation. Groups of spectra can be analyzed for similarity or variety by principal component analysis (PCA) or unsupervised hierarchical cluster analysis (using the Euclidean distance method and the Ward linkage algorithms).

[10] 'Hot spots' describe sample spot areas, which are characterized by high analyte signal intensities compared to surrounding areas.

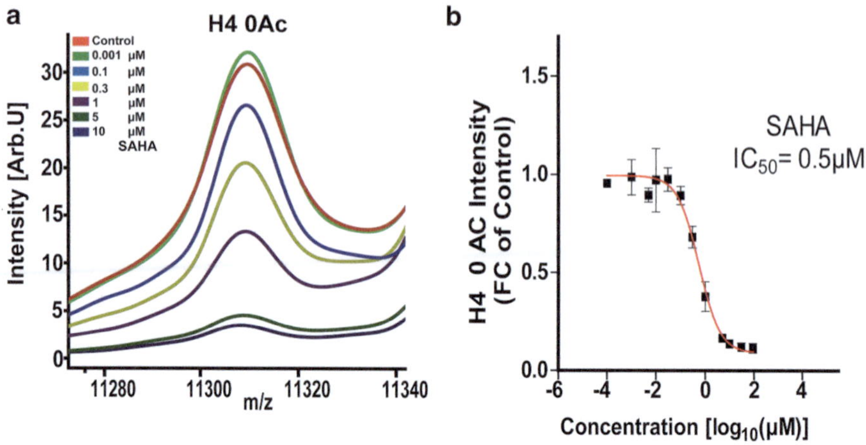

Fig. 4 Determination of HDAC inhibitor potencies under whole-cell conditions by whole-cell MALDI-TOF MS biotyping. (**a**) Average fingerprint spectra of non-acetylated histone H4 ion signals after 24 h treatment of K562 cells with vorinostat/suberanilohydroxamic acid (SAHA) over a range of concentrations (as an example shown for a concentration range of 10 μM–1 nM). (**b**) Concentration-response curve and IC_{50} value for SAHA determined from the same dataset displayed in (**a**) (with additional data points, i.e., the entire measured concentration range from 90 μM to 0.1 nM). Data are presented as Mean ± SD from at least three independent experiments. *FC* fold change

7. *In case of drug treatment*: Extract the intensity of the non-acetylated (0 Ac) histone H4 ion signal (*m/z* 11,306 ± 5) from the average peak list (Fig. 4a) for all tested compound concentrations.
8. Relative quantification is then performed by determining the intensity fold change (FC) of the peak corresponding to the non-acetylated histone H4 (0 Ac) ion of the compound-treated sample versus control.
9. IC_{50} values are then calculated by performing a nonlinear regression curve fit *(log(inhibitor concentration) vs intensity of response)* by Graphpad Prism software (GraphPad Software Inc., CA, USA) (Fig. 4b) to obtain a sigmoidal curve distribution.

References

Anhalt J, Fenselau C (1975) Identification of bacteria using mass spectrometry. Anal Chem 47(2):219–225. doi:10.1021/ac60352a007

Bantscheff M, Hopf C, Savitski MM, Dittmann A, Grandi P, Michon AM, Schlegl J, Abraham Y, Becher I, Bergamini G, Boesche M, Delling M, Dumpelfeld B, Eberhard D, Huthmacher C, Mathieson T, Poeckel D, Reader V, Strunk K, Sweetman G, Kruse U, Neubauer G, Ramsden NG, Drewes G (2011) Chemoproteomics profiling of HDAC inhibitors reveals selective targeting of HDAC complexes. Nat Biotechnol 29(3):255–265. doi:10.1038/nbt.1759

Cain T, Lubman DM, Weber WJ (1994) Differentiation of bacteria using protein profiles from matrix-assisted laser desorption/ionization time-of-flight mass spectrometry. Rapid Commun Mass Spectrom 8:1026–1030

Demirev PA, Fenselau C (2008a) Mass spectrometry for rapid characterization of microorganisms. Annu Rev Anal Chem 1:71–93. doi:10.1146/annurev.anchem.1.031207.112838

Demirev PA, Fenselau C (2008b) Mass spectrometry in biodefense. J Mass Spectrom 43(11): 1441–1457. doi:10.1002/jms.1474

Dong H, Shen W, Cheung MT, Liang Y, Cheung HY, Allmaier G, Kin-Chung Au O, Lam YW (2011) Rapid detection of apoptosis in mammalian cells by using intact cell MALDI mass spectrometry. Analyst 136(24):5181–5189. doi:10.1039/c1an15750g

Feng HT, Wong NS, Sim LC, Wati L, Ho Y, Lee MM (2010) Rapid characterization of high/low producer CHO cells using matrix-assisted laser desorption/ionization time-of-flight. Rapid Commun Mass Spectrom 24(9):1226–1230. doi:10.1002/rcm.4506

Feng HT, Sim LC, Wan C, Wong NS, Yang Y (2011) Rapid characterization of protein productivity and production stability of CHO cells by matrix-assisted laser desorption/ionization time-of-flight mass spectrometry. Rapid Commun Mass Spectrom 25(10):1407–1412. doi:10.1002/rcm.5011

Hanrieder J, Wicher G, Bergquist J, Andersson M, Fex-Svenningsen A (2011) MALDI mass spectrometry based molecular phenotyping of CNS glial cells for prediction in mammalian brain tissue. Anal Bioanal Chem 401(1):135–147. doi:10.1007/s00216-011-5043-y

Holland RD, Wilkes JG, Rafii F, Sutherland JB, Persons CC, Voorhees KJ, Lay JO Jr (1996) Rapid identification of intact whole bacteria based on spectral patterns using matrix-assisted laser desorption/ionization with time-of-flight mass spectrometry. Rapid Commun Mass Spectrom 10(10):1227–1232. doi:10.1002/(SICI)1097-0231(19960731)10:10<1227::AID-RCM659>3.0.CO;2-6

Karas M, Bachmann D, Bahr U, Hillenkamp F (1987) Matrix-assisted ultraviolet laser desorption of non-volatile compounds. Int J Mass Spectrom Ion Process 78(24):53–68. doi:10.1016/0168-1176(87)87041-6

Kim JY, Kim YG, Lee GM (2012) CHO cells in biotechnology for production of recombinant proteins: current state and further potential. Appl Microbiol Biotechnol 93(3):917–930. doi:10.1007/s00253-011-3758-5

Koubek J, Uhlik O, Jecna K, Junkova P, Vrkoslavova J, Lipov J, Kurzawova V, Macek T, Mackova M (2012) Whole-cell MALDI-TOF: rapid screening method in environmental microbiology. Int Biodeter Biodegr 69:82–86. doi:10.1016/j.ibiod.2011.12.007

La Scola B (2011) Intact cell MALDI-TOF mass spectrometry-based approaches for the diagnosis of bloodstream infections. Expert Rev Mol Diagn 11(3):287–298. doi:10.1586/erm.11.12

Martinez Molina D, Jafari R, Ignatushchenko M, Seki T, Larsson EA, Dan C, Sreekumar L, Cao Y, Nordlund P (2013) Monitoring drug target engagement in cells and tissues using the cellular thermal shift assay. Science 341(6141):84–87. doi:10.1126/science.1233606

Marvin-Guy LF, Duncan P, Wagniere S, Antille N, Porta N, Affolter M, Kussmann M (2008) Rapid identification of differentiation markers from whole epithelial cells by matrix-assisted laser desorption/ionisation time-of-flight mass spectrometry and statistical analysis. Rapid Commun Mass Spectrom 22(8):1099–1108. doi:10.1002/rcm.3479

Munteanu B, Hopf C (2013) Emergence of whole-cell MALDI-MS biotyping for high-throughput bioanalysis of mammalian cells? Bioanalysis 5(8):885–893. doi:10.4155/bio.13.47

Munteanu B, von Reitzenstein C, Hansch GM, Meyer B, Hopf C (2012) Sensitive, robust and automated protein analysis of cell differentiation and of primary human blood cells by intact cell MALDI mass spectrometry biotyping. Anal Bioanal Chem. doi:10.1007/s00216-012-6357-0

Munteanu B, Meyer B, von Reitzenstein C, Burgermeister E, Bog S, Pahl A, Ebert MP, Hopf C (2014) Label-free in situ monitoring of histone deacetylase drug target engagement by matrix-assisted laser desorption ionization-mass spectrometry biotyping and imaging. Anal Chem 86(10):4642–7. doi:10.1021/ac500038j

Ouedraogo R, Flaudrops C, Ben Amara A, Capo C, Raoult D, Mege JL (2010) Global analysis of circulating immune cells by matrix-assisted laser desorption ionization time-of-flight mass spectrometry. PLoS One 5(10), e13691. doi:10.1371/journal.pone.0013691

Paul SM, Mytelka DS, Dunwiddie CT, Persinger CC, Munos BH, Lindborg SR, Schacht AL (2010) How to improve R&D productivity: the pharmaceutical industry's grand challenge. Nat Rev Drug Discov 9(3):203–214. doi:10.1038/nrd3078

Povey JF, O'Malley CJ, Root T, Martin EB, Montague GA, Feary M, Trim C, Lang DA, Alldread R, Racher AJ, Smales CM (2014) Rapid high-throughput characterisation, classification and selection of recombinant mammalian cell line phenotypes using intact cell MALDI-ToF mass spectrometry fingerprinting and PLS-DA modelling. J Biotechnol 184:84–93. doi:10.1016/j.jbiotec.2014.04.028

Rodrigues AF, Carrondo MJ, Alves PM, Coroadinha AS (2014) Cellular targets for improved manufacturing of virus-based biopharmaceuticals in animal cells. Trends Biotechnol 32(12):602–607. doi:10.1016/j.tibtech.2014.09.010

Ryzhov V, Fenselau C (2001) Characterization of the protein subset desorbed by MALDI from whole bacterial cells. Anal Chem 73(4):746–750

Ryzhov V, Hathout Y, Fenselau C (2000) Rapid characterization of spores of Bacillus cereus group bacteria by matrix-assisted laser desorption-ionization time-of-flight mass spectrometry. Appl Environ Microbiol 66(9):3828–3834

Savitski MM, Reinhard FB, Franken H, Werner T, Savitski MF, Eberhard D, Martinez Molina D, Jafari R, Dovega RB, Klaeger S, Kuster B, Nordlund P, Bantscheff M, Drewes G (2014) Tracking cancer drugs in living cells by thermal profiling of the proteome. Science 346(6205):1255784. doi:10.1126/science.1255784

Schwamb S, Munteanu B, Meyer B, Hopf C, Hafner M, Wiedemann P (2013) Monitoring CHO cell cultures: cell stress and early apoptosis assessment by mass spectrometry. J Biotechnol 168(4):452–461. doi:10.1016/j.jbiotec.2013.10.014

Walker J, Fox AJ, Edwards-Jones V, Gordon DB (2002) Intact cell mass spectrometry (ICMS) used to type methicillin-resistant Staphylococcus aureus: media effects and inter-laboratory reproducibility. J Microbiol Methods 48(2–3):117–126, 10.1016/s0167-7012(01)00316-5

Welker M (2011) Proteomics for routine identification of microorganisms. Proteomics 11(15):3143–3153. doi:10.1002/pmic.201100049

Zhang X, Scalf M, Berggren TW, Westphall MS, Smith LM (2006) Identification of mammalian cell lines using MALDI-TOF and LC-ESI-MS/MS mass spectrometry. J Am Soc Mass Spectrom 17(4):490–499. doi:10.1016/j.jasms.2005.12.007

Food Authentication by MALDI MS: MALDI-TOF MS Analysis of Fish Species

Rosa Anna Siciliano, Diego d'Esposito, and Maria Fiorella Mazzeo

Abstract Consumer demand for healthy and well-sourced food has been growing in recent times. In particular, demand for fish is constantly increasing due to the awareness of beneficial effects of fishery products on human health. Furthermore, the opening of new markets and the use of a larger number of fish species are strong and timely reminders for the urgent need to guarantee safety, traceability, and authenticity of seafood. Recent European Union directives and regulations for quality control of food products have prompted the development of new methods for large-scale tests to ensure consumer protection. MALDI-TOF MS has provided a significant contribution to food science, proving to be a key tool in the analysis of several food matrices, including fish, especially in studies aimed to assess food quality, safety, and authenticity.

This chapter is focused on an innovative molecular profiling strategy based on MALDI-TOF MS analysis of sarcoplasmic protein extracts from fish muscle, successfully applied to fish authentication. The described method allows to rapidly discriminate different fish species, to verify commercial product authenticity and to detect fraudulent substitutions.

1 Introduction

Consumption of fish products that enhances the intake of long chain and shorter chain omega 3 fatty acids can positively affect human health, in particular contributing to the prevention of cancer and cardiovascular events (Hooper et al. 2006). Therefore, in the last decades, consumer demands for seafood has rapidly increased. To meet these demands, the market has enlarged fish species assortment including fish captured in Asian and African seas and using a growing number of species to produce transformed fish-based foods. Unfortunately, as a consequence, globalization and freer markets have favored a seafood mislabeling phenomenon that is

R.A. Siciliano (✉) • D. d'Esposito • M.F. Mazzeo
Centro di Spettrometria di Massa Proteomica e Biomolecolare, Istituto di Scienze dell'Alimentazione, CNR, via Roma 64, 83100 Avellino, Italy
e-mail: rsiciliano@isa.cnr.it; diegodesposito@isa.cnr.it; fmazzeo@isa.cnr.it

R. Cramer (ed.), *Advances in MALDI and Laser-Induced Soft Ionization Mass Spectrometry*, DOI 10.1007/978-3-319-04819-2_14

associated with potential health risks and, at the same time, illegal economic gains. These issues emphasize the need to determine fish authenticity and origin in order to guarantee proper quality and safety controls and to protect consumers. Reliable quality control methods are crucial to detect deceptive practices of seafood substitution that occurs when one species of fish, crustacean or shellfish is sold as another species (Arvanitoyannis et al. 2005b; Herrero 2008).

Conventional identification methods rely on the analysis of anatomical and morphological characteristics, such as the head, fins, skin, or bones, which are lost during processing, thus making any identification impossible. In agreement with European Union directives and regulations on fishery and aquaculture products, the species, geographical origin, and production method (wild or cultivated) must be provided in fish labeling (Council Regulation (EC) No.104/2000 and 2065/2001 of the European Parliament) to ensure market transparency. Furthermore, the European Food Safety Authority provided complete procedures for the traceability of food (including fishery and aquaculture products) and feed businesses to guarantee food safety at all stages (EC regulation no. 178/2002 of the European Parliament reviewed in Arvanitoyannis et al. 2005a).

Classical methodologies for fish authentication are based on the analysis of protein extracts by electrophoretic, chromatographic, and immunological methods. The isoelectric focusing (IEF) analysis of the sarcoplasmic proteins has been applied to fish and shrimp authentication (Etienne et al. 2000; Piñeiro et al. 2000; Rehbein et al. 2000; Ortea et al. 2010) and has been adopted by the Association of Official Analytical Chemistry as the validated method for species identification purposes (Helrich 1990). DNA-based procedures (mainly DNA sequencing of the cytochrome-b gene) are also routinely used for the authentication of fish species, as they present a number of advantages over protein-based methods, in particular for the analysis of highly processed samples (Rasmussen and Morrissey 2009; Carrera et al. 2013a). However, a fundamental drawback in DNA-based methods is the difficulty to standardize protocols and techniques. This is mandatory to rule out inconsistencies in results from different laboratories that could have regulatory or legal implications and to obtain a rigorous standard operating procedure (SOP) applicable across different countries (Griffiths et al. 2014).

The need for rapid screening of a large number of samples requires the development of high-tech approaches with minimal time consumption, low costs, and high reliability, which can successfully complement or substitute methods already in use.

In the last decades, the introduction of omics platforms has significantly contributed to research activities in food science, cumulating in the term 'foodomics' being coined in 2009 to indicate 'a new discipline that studies food and nutrition domains through the application of advanced omics technologies to improve consumer's well-being, health, and confidence' (Cifuentes 2009; Herrero et al. 2010). Analytical methodologies based on mass spectrometry (MS) play a central role in foodomics. In particular, proteomics has been used to investigate several aspects of food quality and safety, including, traceability, authenticity, absence of contaminating, and/or adulterating agents and impact of the processing/storage methods (Herrero et al. 2012).

2 Applications

Recently, 'molecular profiling' strategies based on matrix-assisted laser desorption/ionization time-of-flight (MALDI-TOF) MS have emerged as a general tool for the discovery of biomarkers that are potentially useful as indicators of authenticity for several food matrices (Cozzolino et al. 2001; Wang et al. 2009; Nunes-Miranda et al. 2012; Ciarmiello et al. 2014). A similar strategy has also recently been applied to the identification of shrimp at the species level (Salla and Murray 2013).

In this chapter, we describe an innovative molecular profiling approach based on MALDI-TOF MS analysis of sarcoplasmic protein extracts from fish muscle, developed by our group and successfully applied to the authentication of fish species (Mazzeo et al. 2008). The main strengths of this approach are the straightforward sample preparation protocol and the low demands on time and cost for the analysis (only a few minutes are necessary for sample preparation and mass spectra acquisition). It does not require any preliminary information on the sample under investigation or the identification of the biomarker-generating proteins. The methodology is highly accurate and sensitive, and due to the high unambiguity of mass spectrometric results, fish identification can be achieved with high confidence. Therefore, this strategy holds the potential to become a reliable first-line authenticity test for fish.

Our previously published data and the data presented here demonstrate that MALDI-TOF MS can be employed as a powerful tool in fish authentication. The presented methodology can be upgraded by exploiting the outstanding performance of modern MALDI-TOF instruments that provide high accuracy and resolution in molecular mass measurements, improving the definition of the biomarker pattern, and thus increasing confidence and reliability in the identification of unknown samples. Furthermore, these instruments assure faster analysis, perform completely automated data acquisition and processing and are specifically designed to be user-friendly, so that highly specialized operators would not be required to carry out mass spectrometric analyses.

As a matter of fact, these innovations and the development of specific software have paved the way to the introduction of MALDI-TOF MS-based methodologies for bacterial identification as routine tests in clinical microbiological laboratories (see Seng et al. 2009). Similarly, we can foresee that the creation of a specific database containing mass spectra and/or reference peak lists of a growing number of fish species, as well as ad hoc bioinformatics tools for database querying will prompt the application of this analytical approach as a routine method in fish and other food products authentication. In addition, as such database can be continuously updated by the operator, this method will be quite flexible and easily adaptable to specific analytical needs of the market. As proof of concept, a commercial MALDI mass spectral fingerprint matching software has been applied for the first time in food science to the discrimination at species level of 72 shrimp samples from the market (Salla and Murray 2013).

In conclusion, these studies strongly suggest that in the near future analytical strategies based on MALDI-TOF MS will play a key role in the assessment of food quality and safety and are going to represent robust tools suitable to be integrated in or substitute current official screening methods.

3 Materials and Protocols

The robustness of the method has been assessed by analyzing protein extracts from 40 different fish species, selected from widely consumed products of high commercial value or commonly involved in frauds. The presented data were obtained from fish species belonging to seven different Orders and twenty Families, representing one of the most comprehensive repertoires of species analyzed in fish authentication studies (Table 1), including species previously reported in a similar study (Mazzeo et al. 2008). The general workflow employed in our lab is outlined in Fig. 1.

Table 1 List of the analyzed fish species[a]

Order	Family	Genus	Species	Common name
Perciformes	Serranidae	*Dicentrarchus*	*Dicentrarchus labrax*	Seabass
		Epinephelus	*Epinephelus marginatus*	Dusky grouper
	Sparidae	*Sparus*	*Sparus auratus*	Seabream
		Pagellus	*Pagellus acarne*	Axillary seabream
			Pagellus erythrinus	Common pandora
		Diplodus	*Diplodus sargus*	White seabream
			Diplodus vulgaris	Common two-banded seabream
			Diplodus puntazzo	Sharpsnout seabream
			Diplodus annularis	Annular seabream
		Dentex	*Dentex dentex*	Common dentex
		Pagrus	*Pagrus pagrus*	Red porgy
	Centracantidae	*Spicara*	*Spicara maena*	Blotched picarel
	Mullidae	*Mullus*	*Mullus barbatus*	Red mullet
	Uranoscopidae	*Uranoscopus*	*Uranoscopus scaber*	Atlantic stargazer
	Percidae	*Perca*	*Perca fluviatilis*	European perch
	Triglidae	*Aspitriglia*	*Aspitrigla cuculus*	East Atlantic red gurnard
	Cichlidae	*Tilapiini*	*Tilapiine cichlids*	Tilapias
	Scombridae	*Auxis*	*Auxis thazard*	Frigate tuna
		Sarda	*Sarda sarda*	Atlantic bonito
		Scomber	*Scomber scombrus*	Atlantic mackerel
	Coryphaenidae	*Coryphaena*	*Coryphaena hippurus*	Common dolphinfish
	Carangidae	*Trachurus*	*Trachurus trachurus*	Atlantic horse mackerel

(continued)

Table 1 (continued)

Order	Family	Genus	Species	Common name
Gadiformes	Gadidae	*Gadus*	*Gadus morhua*	Atlantic cod
		Merluccius	*Merluccius capensis*	Shallow-water cape hake
			Merluccius merluccius	European hake
			Merluccius hubbsi	Argentine hake
			Merluccius paradoxus	Deep-water cape hake
		Trisopterus	*Trisopterus minutus minutus*	Poor cod
		Micromesistius	Micromesistius poutassou	Blue whiting
		Molva	*Molva elongata*	Mediterranean ling
		Phycis	*Phycis blennioides*	Greater forkbeard
Pleuronectiformes	Bothidae	*Arnoglassus*	*Arnoglossus laterna*	Scaldfish
	Pleuronectidae	*Reinhardtius*	*Reinhardtius hippoglossoides*	Greenland halibut
		Pleuronectes	*Pleuronectes platessa*	European plaice
	Soleidae	*Solea*	*Solea solea*	Common sole
Lophiiformes	Lophiidae	*Lophius*	*Lophius piscatorius*	Angler
Salmoniformes	Salmonidae	*Salmo*	*Salmo salar*	Atlantic salmon
Clupeiformes	Engraulidae	*Engraulis*	*Engraulis encrasicolus*	European anchovy
	Clupeidae	*Sardina*	*Sardina pilchardus*	European pilchard
Siluriformes	Pangasiidae	*Pangasius*	*Pangasius pangasius*	Striped catfish

[a]Adapted with permission from Mazzeo et al. (2008) J Agric Food Chem 56:11071-11076. Copyright (2014) American Chemical Society

3.1 Extraction of Sarcoplasmic Proteins from Muscle Tissue

The protocol can be applied to the analysis of fresh and frozen fish samples. One gram of white tissue muscle is taken from three individuals of each species without damaging any organ in order to avoid any contamination and stored at −20 °C. Sarcoplasmic protein extraction is carried out by vortexing 0.1–0.2 g fish muscle in 100–200 μL of 0.1% trifluoroacetic acid (TFA) for 1 min. Protein extracts are then centrifuged at 13,000 rpm for 5 min and the recovered supernatants are diluted 1:10 in 0.1% TFA and immediately analyzed by MALDI-TOF MS. For fish

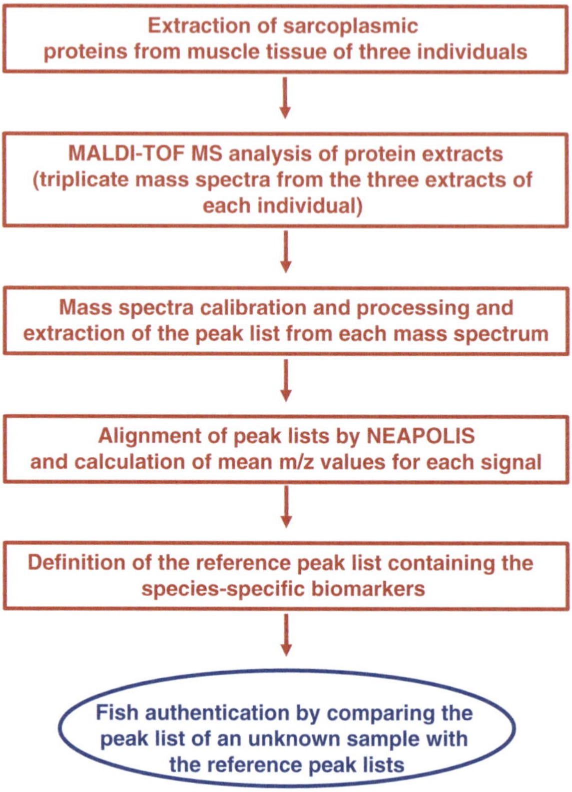

Fig. 1 Analytical scheme of the described authentication method

species particularly rich in fats, sample extracts are subjected to a defatting step with chloroform, i.e., using 0.1% TFA/chloroform (1:4; v/v) that improves mass spectra quality.

3.2 MALDI-TOF MS Analysis

MALDI-TOF MS analyses are carried out using as matrix solution a solution of α-cyano-4-hydroxycinnamic acid (CHCA) in 50% acetonitrile (ACN)/0.1% TFA (10 mg/mL) that contains 1 pmol/μL of cytochrome C as internal standard. One microliter of the analyte extract is mixed with 1 μL of matrix solution and deposited onto a MALDI target plate so that co-crystallization occurs under ambient conditions. In some cases, mass spectra quality is improved by adding 1 μL of matrix solution directly on the crystallized samples and analyses are carried out after subsequent recrystallization. In our experiments, mass spectra are acquired on a Voyager-DE PRO MALDI-TOF mass spectrometer (AB-SCIEX, Foster City, CA, USA), operating in linear, positive-ion mode with delayed extraction, using a pulsed nitrogen laser (337 nm; 3 ns). Parameters for data acquisition are the following: laser intensity set just above the ion generation threshold, low mass gate at 1990, delay time at 500 ns, accelerating voltages at 25,000 V, grid voltage and guide wire voltage at 95% and 0.1% of the accelerating voltage, respectively.

Mass spectra are typically acquired by accumulating spectra obtained from 100 laser shots in the *m/z* range of 2000–15,000. Internal calibration is performed using the doubly and singly charged ions of cytochrome C (*m/z* 6181.05 and 12,361.10, respectively). All *m/z* values are recorded as average values.

Reference molecular profiles are constructed from triplicate MALDI-TOF MS analysis of analyte extracts from three individuals for each species.

3.3 Data Analysis

Mass spectra are processed applying baseline subtraction and smoothing algorithms and transformed into a list containing the *m/z* values of signals present in the *m/z* range of 8000–15,000 with an ion signal intensity of >10%. In our lab, DataExplorer 5.1 software (AB-SCIEX) is used for these data processing steps. For each species, the peak lists obtained from the nine processed mass spectra are aligned along the *m/z* axis using the NEAPOLIS software (www.bioinformatics.org/bioinfo-af-cnr/NEAPOLIS) (Mangerini et al. 2011) which calculates the mean *m/z* value for each signal. The threshold mass tolerance value for the alignment is fixed to 500 ppm, so that, among the aligned signals, the minimum and maximum *m/z* values differ by <500 ppm and the standard deviations of mass measurement are <3 Da. The mean *m/z* values of signals present in all replicate mass spectra are included in the reference peak list for each analyzed species. A direct comparison of the reference peak lists shows that the pattern of signals included in individual reference peak lists is unique for each analyzed species and therefore unequivocally identifies that species. This pattern, containing as few as one to four signals, is considered the species-specific biomarker pattern and can be used to discriminate fish species unambiguously. Table 2 details the reference peak lists for the 40 fish species studied in our lab.

To identify an unknown sample, its peak list is obtained following the previously described method and compared to all the reference peak lists in the pre-recorded database. A positive match is obtained if the peak list of the unknown sample completely matches one database reference peak list (Table 2, Fig. 1).

3.4 General Remarks and Examples

MALDI-TOF MS analyses of sarcoplasmic protein extracts yield mass spectra characterized by a pattern of a few highly intense signals, mainly in the *m/z* range of about 11,000–12,000, that can be considered as species-specific biomarkers. These specific molecular profiles are suitable for fish authentication and allow the differentiation of

Table 2 Biomarker patterns useful for discriminating fish species, as obtained by MALDI-TOF MS analysis[a]

Scientific name	Common name	Biomarker pattern[b]		SD	
Dicentrarchus labrax	Seabass	8032.0	±	0.8	
		11,404.7	±	1.0	M*[c]
		11,495.9	±	1.1	
Epinephelus marginatus	Dusky grouper	11,606.8	±	0.2	M*
Sparus auratus	Seabream	11,370.5	±	1.1	
		11,442.3	±	0.9	M*
Pagellus acarne	Axillary seabream	11,407.4	±	0.7	
		11,563.3	±	0.8	M*
Pagellus erythrinus	Common pandora	11,429.9	±	0.7	
		11,588.2	±	2.3	
		11,606.4	±	0.9	M*
Diplodus sargus sargus	White seabream	11,301.5	±	1.4	
		11,456.7	±	1.3	M*
Diplodus vulgaris	Common two-banded seabream	11,343.4	±	1.0	
		11,457.5	±	0.8	M*
Diplodus puntazzo	Sharpsnout seabream	11,380.7	±	1.7	
		11,457.8	±	1.5	M*
Diplodus annularis	Annular seabream	11,270.7	±	1.2	
		11,461.5	±	1.5	M*
		11,488.2	±	1.2	
Dentex dentex	Common dentex	11,519.8	±	1.0	M*
		11,567.4	±	1.0	
Pagrus pagrus	Red porgy	11,489.8	±	2.2	
		11,582.2	±	1.4	M*
Spicara maena	Blotched picarel	11,376.2	±	2.5	
		11,507.9	±	1.2	M*
Mullus barbatus	Red mullet	11,383.5	±	2.1	M*
		11,544.9	±	2.0	
Uranoscopus scaber	Atlantic stargazer	11,731.1	±	1.0	M*
		12,078.9	±	1.0	
Perca fluviatilis	European perch	11,403.0	±	1.1	
		11,434.6	±	0.8	M*
Aspritrigla cuculus	East Atlantic red gurnard	11,519.7	±	0.6	
		11,639.4	±	0.6	M*
Tilapiine cichids	Tilapias	11,382.0	±	0.5	M*
		11,559.9	±	0.9	
Auxis thazard	Frigate tuna	11,377.1	±	0.8	M*
		11,419.5	±	0.7	M*
Sarda sarda	Atlantic bonito	11,459.3	±	0.4	
Scomber scombrus	Atlantic mackerel	11,456.2	±	0.5	M*
		11,468.1	±	0.6	M*
Coryphaena hippurus	Common dolphinfish	11,452.6	±	0.2	
		11,638.3	±	0.1	M*

(continued)

Table 2 (continued)

Scientific name	Common name	Biomarker pattern[b]		SD	
Trachurus trachurus	Atlantic horse mackerel	11,295.1	±	1.7	
		11,483.1	±	1.5	M*
Gadus morhua	Atlantic cod	11,366.6	±	0.5	M*
		11,464.0	±	0.6	
Merluccius capensis	Shallow-water cape hake	8432.7	±	0.9	
		11,338.4	±	0.5	M*
		11,361.4	±	0.6	
		11,387.9	±	0.8	
Merluccius merluccius	European hake	8432.3	±	0.5	
		11,338.6	±	0.4	M*
		11,388.2	±	0.3	
		11,361.0	±	0.8	
Merluccius hubbsi	Argentine hake	8437.5	±	1.2	
		11,339.3	±	0.3	M*
		11,362.7	±	0.9	
		11,387.4	±	1.1	
Merluccius paradoxus	Deep-water cape hake	8476.9	±	0.5	
		11,339.1	±	0.5	M*
		11,389.2	±	0.7	
Trisopterus minutus minutus	Poor cod	11,310.7	±	1.5	
		11,351.3	±	0.6	M*
Micromesistius poutassou	Blue whiting	11,350.8	±	0.7	
		11,448.3	±	0.4	M*
Molva elongata	Mediterranean ling	11,550.7	±	0.6	
Phycis blennoides	Greater forkbeard	11,447.4	±	0.8	
		11,553.3	±	0.6	M*
Arnoglossus lanterna	Scaldfish	11,478.7	±	1.8	
		11,548.2	±	2.0	M*
		11,783.0	±	1.3	
Reinhardtius hippooglossoides	Greenland halibut	11,433.8	±	1.2	
Pleuronectes platessa	European plaice	11,351.6	±	0.6	M*
		11,764.1	±	0.8	
Solea solea	Common sole	11,976.3	±	1.4	
Lophius piscatorius	Angler	11,522.6	±	0.6	M*
		11,588.9	±	0.7	
Salmo salar	Atlantic salmon	11,295.7	±	0.9	
		11,825.3	±	0.9	M*
Engraulis encrasicolus	European anchovy	11,537.2	±	0.3	
Sardina pilchardus	European pilchard	11,360.2	±	0.2	M*
		11,731.4	±	0.4	M*
Pangasius pangasius	Striped catfish	11,555.1	±	0.8	M*
		12,075.8	±	2.4	

[a]Adapted with permission from Mazzeo et al. (2008) J Agric Food Chem 56:11071-11076. Copyright (2014) American Chemical Society

[b]Mean *m/z* values of signals present in the replicate mass spectra are reported. Average *m/z* values are recorded in the mass spectra

[c]M* indicates the most intense signal in MALDI-TOF mass spectra

fish species, even phylogenetically closely related, as well as the authentication of commercial products. In a few cases, the identification is achieved by taking into account also signals present in a broader *m/z* range. The fish species analyzed up to now and their species-specific biomarker pattern are reported in Table 2. The potential of the method is well demonstrated in the following examples.

The mass spectrum of *Dicentrarchus labrax* (seabass) obtained from applying the above method exhibits three intense signals at *m/z* 8031.9, 11,403.7, and 11,495.3, whereas for *Sparus auratus* (seabream), two strong peaks were detected at *m/z* 11,441.2 and 11,370.0 (Fig. 2). Therefore, these two species that are widely consumed and of high commercial value are easily discriminated.

The method specificity assures the discrimination of very closely related species as shown by the analysis of four different species belonging to the *Diplodus* genus. Mass spectra of *Diplodus sargus sargus* (white seabream), *Diplodus vulgaris* (common two-banded seabream) and *Diplodus puntazzo* (sharpsnout seabream) share a major peak at *m/z* 11,457 while minor intense peaks can be detected at *m/z* 11,301.5

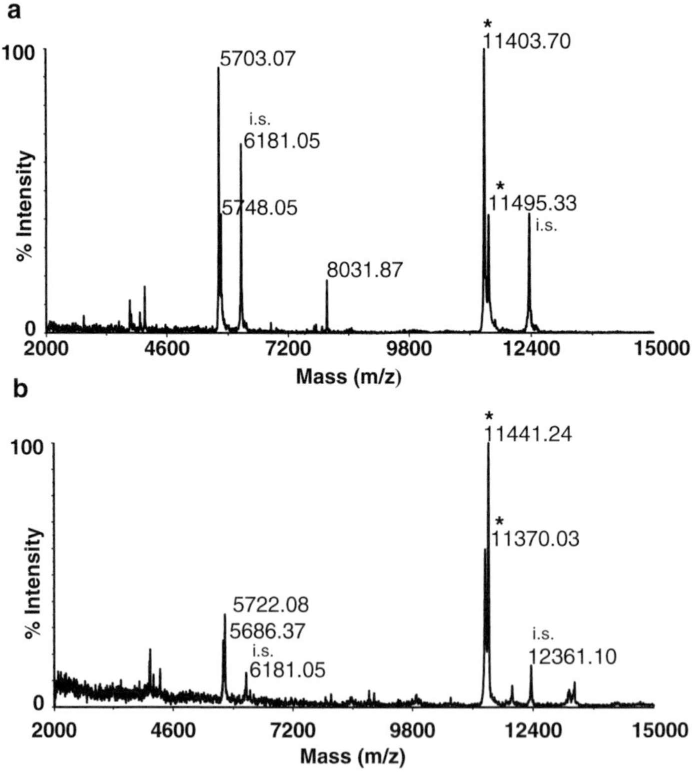

Fig. 2 MALDI-TOF mass spectra obtained from the analysis of *Dicentrarchus labrax* (seabass) (**a**) and *Sparus auratus* (seabream) (**b**). Signals selected as biomarkers are indicated with *asterisks*. Adapted with permission from Mazzeo et al. (2008) J Agric Food Chem 56:11071-11076. Copyright (2014) American Chemical Society

for *Diplodus sargus sargus*, *m*/*z* 11,343.4 for *Diplodus vulgaris*, and *m*/*z* 11,380.7 for *Diplodus puntazzo*. The molecular profile of *Diplodus annularis* is quite different showing two signals with almost the same intensities at *m*/*z* 11,461.5 and 11,488.2 and a minor one at *m*/*z* 11,270.7 Therefore a species-specific biomarker pattern can be defined for the *Diplodus* species (Table 2).

Similarly, two phylogenetically related species of the *Pagellus* genus can be discriminated by peaks at *m*/*z* 11,563.3 and 11,407.4 present in the mass spectrum of *Pagellus acarne* (axillary seabream) and absent in that of *Pagellus erythrinus* (common pandora), which is characterized instead by three signals at *m*/*z* 11,606.4, 11,588.2, and 11,429.9.

The proposed method also allows a rapid authentication of widely consumed species within the *Gadidae* family, such as *Merluccius* species, *Gadus morhua* (atlantic cod), *Trisopterus minutus minutus* (poor cod), *Phycis blennoides* (greater forkbeard), *Molva elongata* (mediterranean ling), and *Micromesistius poutassou* (blue whiting) (Table 2).

It is noteworthy that this method takes advantage of a fast extraction protocol and of some important technical features of MALDI-TOF MS, such as sensitivity of analysis and tolerance to contaminants. In fact, the obtained protein extracts could be directly analyzed without any prior purification and/or concentration step. Moreover, although the analyzed protein mixtures are quite complex, MALDI-TOF mass spectra show a few strong signals. Therefore, the biomarker pattern is quite simple and fish identification can be easily and rapidly achieved.

3.4.1 Detection of Frauds

The ability of this approach to verify the authenticity of fish and commercial products (such as fillets and fishsticks) is demonstrated by the following exemplary case studies.

A common fraudulent practice is the substitution of the high value species *Dentex dentex* (common dentex), generally present in the market as fillets, with the low-cost species *Pagrus pagrus* (red porgy). Figure 3 shows the mass spectral profiles of the two species, characterized by the presence of two intense signals at *m*/*z* 11,519.9 and 11,567.2 for *Dentex dentex* and at *m*/*z* 11,488.0 and 11,581.5 for *Pagrus pagrus*. These signals immediately and unambiguously allow to discriminate the two species.

Pangasius pangasius (pangas catfish) and *Tilapiine cichids* (tilapias) fillets are often mislabeled and sold as fillets of a vast number of fish products, including cod fillets. MALDI-TOF MS analysis of pangas catfish led to the identification of a very intense signal at *m*/*z* 11,555.1 and a minor one at *m*/*z* 12,075.8 as species-specific biomarkers, while the signals at *m*/*z* 11,382.0 and 11,559.9 were characteristic of tilapias. Therefore, the straightforward discrimination of these fish species from more valuable ones, such as species belonging to Gadiformes, could be achieved (see biomarker pattern in Table 2).

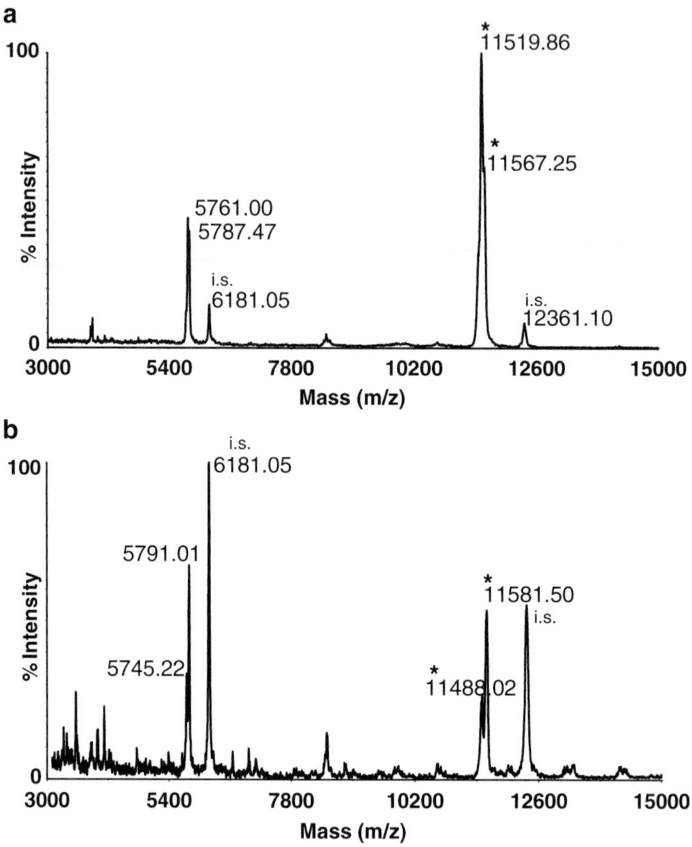

Fig. 3 MALDI-TOF mass spectra obtained from the analysis of *Dentex dentex* (common dentex) (**a**) and *Pagrus pagrus* (red porgy) (**b**). Signals selected as biomarkers are indicated with *asterisks*

Similarly, it was possible to differentiate among fillets of sole (*Solea solea*; biomarker at *m/z* 11,976.3), European plaice (*Pleuronectes platessa*; biomarkers at *m/z* 11,351.6 and 11,764.1), and Greenland halibut (*Reinhardtius hippoglossoides*; biomarker at *m/z* 11,433.8) based on the specific signal patterns.

Furthermore, the developed method permits to verify the correctness of what is declared on labels as demonstrated by the mass spectra of some commercial cod fishsticks, which were identical to those obtained from *Merluccius capensis*, and thus in agreement with the label (Mazzeo et al. 2008).

It is worth stressing that the presented method can be applied to the analysis of cooked products as mass spectra obtained from some heat-treated samples (*Sparus auratus*, *Merluccius merluccius*, *Molva elongata*, *Phycis blennoides*, *Micromesistius poutassou*, and *Solea solea*) show the same biomarker patterns as the untreated samples (Mazzeo et al. 2008). The method can also be applied to the analysis of processed products such as homogenized baby foods.

3.4.2 Parvalbumins as Species-Specific Biomarkers for Fish Authentication

Earlier analyses based on tandem mass spectrometric experiments determined the primary structure of a few protein isoforms (such as that from *Trisopterus minutus minutus*) present in the fish muscle extracts and demonstrated that biomarker signals in the *m/z* range of 11,000-12,000 originate from parvalbumins (Mazzeo et al. 2008). Parvalbumins are calcium-binding proteins with molecular weights in the 11–12 kDa range, relatively abundant in muscle tissues and known as the major allergy-eliciting proteins in fish.

These proteins can be regarded as the most suitable biomarkers for fish species authentication due to the interspecies variability of their sequences, their high concentration in the fish muscle, and solubility in aqueous buffers that make the extraction protocol extremely fast and easy. Moreover, parvalbumins exhibit a high ionization efficiency in MALDI-TOF MS analysis so that, regardless of the complexity of the analyzed sarcoplasmic extracts, the obtained mass spectra predominantly show signals originating from these proteins (Mazzeo et al. 2008; Carrera et al. 2013b). The structural stability of parvalbumins even under harsh conditions such as heat is paramount to utilize these biomarkers also for the authentication of fish species sold as thermally processed products (Elsayed and Bennich 1975; Kawai et al. 1992; Carrera et al. 2010).

The interspecies variability of parvalbumin sequences is fundamental for the discrimination of different fish species as assessed by earlier proteomic analyses performed on species belonging to the Merlucciidae family (Piñeiro et al. 2001). Proteomic studies integrating two-dimensional electrophoresis (2-DE) with MALDI-TOF MS peptide mass mapping for protein identification allowed the characterization of the 2-DE parvalbumin-specific pattern and the definition of a set of specific tryptic peptides suitable for the identification of nine hake species (Carrera et al. 2006). More recently, exploiting the improved performance of new instruments such as Fourier-transform ion-cyclotron resonance (FTICR) mass spectrometers and linear ion trap (LIT) mass spectrometers, innovative strategies for the extensive characterization of parvalbumins have been proposed. These studies led to the de novo sequencing of 25 isoforms from all commercial species of the Merlucciidae family and to the rapid and direct detection of the presence of fish allergens in all of the investigated food products (Carrera et al. 2010; Carrera et al. 2012).

These latest results provide the structural evidence and demonstrate the potential of parvalbumins as suitable biomarkers for fish authentication and their importance in MALDI-TOF molecular profiling strategies such as the one described in this chapter.

References

Arvanitoyannis IS, Choreftaki S, Tserkezou P (2005a) An update of EU legislation (Directives and Regulations) on food-related issues (safety, hygiene, packaging, technology, GMOs, additives, radiation, labelling): presentation and comments. Int J Food Sci Technol 40:1021–1112

Arvanitoyannis IS, Tsitsika EV, Panagiotaki P (2005b) Implementation of quality control methods (physicochemical, microbiological and sensory) in conjunction with multivariate analysis towards fish authenticity. Int J Food Sci Technol 40:237–263

Carrera M, Cañas B, Piñeiro C et al (2006) Identification of commercial hake and grenadier species by proteomic analysis of the parvalbumin fraction. Proteomics 6:5278–5287

Carrera M, Cañas B, Vázquez J, Gallardo JM (2010) Extensive de novo sequencing of new parvalbumin isoforms using a novel combination of bottom-up proteomics, accurate molecular mass measurement by FTICR-MS, and selected MS/MS ion monitoring. J Proteome Res 9:4393–4406

Carrera M, Cañas B, Gallardo JM (2012) Rapid direct detection of the major fish allergen, parvalbumin, by selected MS/MS ion monitoring mass spectrometry. J Proteomics 75:3211–3220

Carrera M, Cañas B, Gallardo JM (2013a) Fish Authentication. In: Toldrá F, Nollet LM (eds) Proteomics in foods: principles and applications. Springer, New York, pp 205–222

Carrera M, Cañas B, Gallardo JM (2013b) The sarcoplasmic fish proteome: pathways, metabolic networks and potential bioactive peptides for nutritional inferences. J Proteomics 78:211–220

Ciarmiello LF, Mazzeo MF, Minasi P et al (2014) Analysis of different European hazelnut (*Corylus avellana* L.) cultivars: authentication, phenotypic features, and phenolic profiles. J Agric Food Chem 62:6236–6246

Cifuentes A (2009) Food analysis and foodomics. J Chromatogr A 1216:7109–7110

Cozzolino R, Passalacqua S, Salemi S et al (2001) Identification of adulteration in milk by matrix-assisted laser desorption/ionization time-of-flight mass spectrometry. J Mass Spectrom 36:1031–1037

Elsayed S, Bennich H (1975) The primary structure of allergen M from cod. Scand J Immunol 4:203–208

Etienne M, Jérôme M, Fleurence J et al (2000) Identification of fish species after cooking by SDS – PAGE and urea IEF: a collaborative study. J Agric Food Chem 48:2653–2658

Griffiths AM, Sotelo CG, Mendes R et al (2014) Current methods for seafood authenticity testing in Europe: is there a need for harmonisation? Food Control 45:95–100

Helrich K (ed) (1990) Official method of analysis of the Association of Official Analytical Chemists (AOAC), vol 2, 15th edn. AOAC, Arlington

Herrero AM (2008) Raman spectroscopy: a promising technique for quality assessment of meat and fish: a review. Food Chem 107:1642–1651

Herrero M, García-Cañas V, Simó C, Cifuentes A (2010) Recent advances in the application of capillary electromigration methods for food analysis and foodomics. Electrophoresis 31:205–228

Herrero M, Simó C, García-Cañas V et al (2012) Foodomics: MS-based strategies in modern food science and nutrition. Mass Spectrom Rev 31:49–69

Hooper L, Thompson RL, Harrison RA et al (2006) Risks and benefits of omega 3 fats for mortality, cardiovascular disease, and cancer: systematic review. Br Med J 332:752–760

Kawai Y, Uematsu S, Shinano H (1992) Effect of heat-treatment on some physicochemical properties and emulsifying activity of carp sarcoplasmic protein. Nippon Suisan Gakkai 58: 1327–1331

Mangerini R, Romano P, Facchiano A et al (2011) The application of atmospheric pressure matrix-assisted laser desorption/ionization to the analysis of long-term cryopreserved serum peptidome. Anal Biochem 417:174–181

Mazzeo MF, Giulio BD, Guerriero G et al (2008) Fish authentication by MALDI-TOF mass spectrometry. J Agric Food Chem 56:11071–11076

Nunes-Miranda JD, Santos H, Reboiro-Jato M (2012) Direct matrix assisted laser desorption ionization mass spectrometry-based analysis of wine as a powerful tool for classification purposes. Talanta 91:72–76

Ortea I, Cañas B, Calo-Mata P et al (2010) Identification of commercial prawn and shrimp species of food interest by native isoelectric focusing. Food Chem 121:569–574

Piñeiro C, Barros-Velázquez J, Pérez-Martín RI, Gallardo JM (2000) Specific enzyme detection following isoelectric focusing as a complimentary tool for the differentiation of related Gadoid fish species. Food Chem 70:241–245

Piñeiro C, Vázquez J, Marina AI et al (2001) Characterization and partial sequencing of species-specific sarcoplasmic polypeptides from commercial hake species by mass spectrometry following two-dimensional electrophoresis. Electrophoresis 22:1545–1552
Rasmussen RS, Morrissey MT (2009) Application of DNA-based methods to identify fish and seafood substitution on the commercial market. Compr Rev Food Sci Food Safety 8:118–154
Rehbein H, Kündiger R, Pineiro C, Perez-Martin RI (2000) Fish muscle parvalbumins as marker proteins for native and urea isoelectric focusing. Electrophoresis 21:1458–1463
Salla V, Murray KK (2013) Matrix-assisted laser desorption ionization mass spectrometry for identification of shrimp. Anal Chim Acta 794:55–59
Seng P, Drancourt M, Gouriet F et al (2009) Ongoing revolution in bacteriology: routine identification of bacteria by matrix-assisted laser desorption ionization time-of-flight mass spectrometry. Clin Infect Dis 49:543–551
Wang J, Kliks MM, Qu W et al (2009) Rapid determination of the geographical origin of honey based on protein fingerprinting and barcoding using MALDI TOF MS. J Agric Food Chem 57:10081–10088

Index

R. Cramer (ed.), *Advances in MALDI and Laser-Induced Soft Ionization Mass Spectrometry*, DOI 10.1007/978-3-319-04819-2

MIX
Papier aus verantwortungsvollen Quellen
Paper from responsible sources
FSC® C105338

FSC
www.fsc.org

If you have any concerns about our products,
you can contact us on
ProductSafety@springernature.com

In case Publisher is established outside the EU,
the EU authorized representative is:
Springer Nature Customer Service Center GmbH
Europaplatz 3, 69115 Heidelberg, Germany

Printed by Libri Plureos GmbH
in Hamburg, Germany